Lecture Notes in Physics

Edited by J. Ehlers, München K. Hepp, Zürich
R. Kippenhahn, München H. A. Weidenmüller, Heidelberg
and J. Zittartz, Köln
Managing Editor: W. Beiglböck, Heidelberg

139

Differential Geometric Methods in Mathematical Physics

Proceedings of the International Conference
Held at the Technical University of Clausthal,
Germany, July 1978

Edited by H. D. Doebner

Springer-Verlag
Berlin Heidelberg New York 1981

Editor

Heinz-Dietrich Doebner
Institut für Theoretische Physik, Technische Universität Clausthal
D-3392 Clausthal-Zellerfeld

ISBN 3-540-10578-6 Springer-Verlag Berlin Heidelberg New York
ISBN 0-387-10578-6 Springer-Verlag New York Heidelberg Berlin

Printing and binding: Beltz Offsetdruck, Hemsbach/Bergstr.
2153/3140-543210

PREFACE

The present volume contains the proceedings of the International Conference on "Differential Geometric Methods in Mathematical Physics" held at the Technical University of Clausthal in July 1978. The conference continues the tradition of the Bonn conference series devoted to an exchange between physics and mathematics, particularly in the fields of geometry and topology applied to gravitation, particle physics and quantization methods. According to their tradition these conferences are not only an occasion to communicate physical and mathematical results and their interrelation, but also to report on mathematical structures and techniques which could help to understand and to unite experimental results and, using the momentum of successful application of a mathematical structure in physics, to develop and to extend it.

The conference, organized in cooperation with K. Bleuler, Bonn, and W.H. Greub, Toronto, was centered around the following topics:

- Quantization Methods and Special Quantum Systems

  geometric quantization, vectorfield quantization, quantization of stochastic phase spaces, dynamics of magnetic monopoles, spectrum generating groups

- Gauge Theories

  phase space of the classical Yang-Mills equation, nonlinear σ - models, gauging geometrodynamics, exceptional gauge groups

- Elliptic Operators, Spectral Theory and Applications

  the Atiyah-Singer theorem applied to quantum-field theory, spectral theory applied to phase transitions

- Geometric Methods and Global Analysis

  systems on non-Hausdorff spaces and on non-Euclidean spaces, Weyl geometry, Lorentz manifolds, manifolds of embeddings.

The contributions in this volume cover almost all the material presented in the conference; one paper is included through its abstract. The responsibility for the final preparation of the manuscripts for the printing was in the hands of the editor. I thank B. Angermann for his assistance and W. Weihrauch for typing the manuscripts.

The organizers wish to express their gratitude to the Volkswagen-stiftung and to the Technische Universität Clausthal for their most generous financial help. They are indebted to Mrs. Jutta Müller for the excellent and invaluable work as conference secretary, to all lecturers and participants, and to the members of the Clausthal Institute for Theoretical Physics whose effort made the conference what it was: lively and stimulating, i.e., successful.

H. Doebner

TABLE OF CONTENTS

1. QUANTIZATION METHODS AND SPECIAL QUANTUM SYSTEMS

2. GAUGE THEORIES

3. ELLIPTIC OPERATORS, SPECTRAL THEORY AND PHYSICAL APPLICATIONS

4. GEOMETRIC METHODS AND GLOBAL ANALYSIS

List of participants

S.T.Ali, Toronto, Canada
E.Aguirre, Madrid, Spain
B.Angermann, Clausthal, FRG
A.O.Barut, Boulder, USA
L.C.Biedenharn, Durham, USA
E.Binz, Mannheim, FRG
K.Bleuler, Bonn, FRG
P.Campbell, Lancaster, England
P.Cotta-Ramusino, Mailand, Italy
H.D.Doebner, Clausthal, FRG
R.Domiaty, Graz, Austria
K.Drühl, Starnberg, FRG
M.Forger, FU Berlin, FRG
P.L.Garcia-Perez, Salamanca, Spain
G.Gerlich, Braunschweig, FRG
W.Greub, Toronto, Canada
G.C.Hegerfeldt, Göttingen, FRG
K.-E.Hellwig, TU Berlin, FRG
J.D.Hennig, Köln, FRG
H.Heß, FU Berlin, FRG
Y.Ingvason, Göttingen, FRG
G.Karrer, Zürich, Switzerland
S.R.Komy, Riyad, Saudi Arabia
D.Krausser, TU Berlin, FRG
K.Just, Tuscon, USA
W.Lücke, Clausthal, FRG
E.W.Mielke, Kiel, FRG
F.B.Pasemann, Clausthal, FRG
H.R.Petry, Bonn, FRG
T.Rasetti, Turin, Italy
H.Römer, Cern, Switzerland
I.E.Segal, Cambridge, USA
H.J.Schmidt, Osnabrück, FRG
A.Schober, TU Berlin, FRG
J.Slawianowski, Warschau, Poland
J.Sniatycki, Calgary, Canada
J.Tarski, Clausthal, FRG
R.B.Teese, Austin, USA
R.Wilson, München, FRG
J.-E.Werth, Clausthal, FRG

and other participants from the Technical University of Clausthal.

On a geometric quantization scheme generalizing those of Kostant-Souriau and Czyz

Harald Hess
Freie Universität Berlin
FB 20, WE 4
Arnimallee 3
D-1000 Berlin 33

Abstract: A quantization method (strictly generalizing the Kostant-Souriau theory) is defined, which may be applied in some cases where both Kostant-Souriau prequantum bundles and metaplectic structures do not exist. It coincides with the Czyz theory for compact Kähler manifolds with locally constant scalar curvature. Quantization of dynamical variables is defined without use of intertwining operators, extending either the Kostant map or some ordering rule like that of Weyl or Born-Jordan.

0. Introduction:

The aim of this article is to present a new method for geometric quantization extending that of Kostant-Souriau in two respects. First, the Kostant-Souriau theory cannot be applied to classical phase spaces with non-vanishing second Stiefel-Whitney class, since in this case metaplectic structures and half-forms do not exist.

This problem arises for quantization of energy surfaces of the n-dimensional harmonic oscillator, where the reduced phase space is $P^{n-1}(\mathbb{C})$, for odd n $(n > 1)$. In case of the Schrödinger energy levels, even Kostant-Souriau prequantum bundles do not exist, which has been recognized by Czyz [7], [8] who invented another geometric quantization theory for compact Kähler manifolds without the mentioned disadvantages. As in the Kostant-Souriau theory, the quantizing Hilbert space there is also built from sections in a complex line bundle, but the latter is directly chosen to satisfy some basic postulates, while in the Kostant-Souriau theory it is the tensor product of the prequantum bundle and the bundle of the half-forms. In addition, the connection on the line bundle is an ordinary one in the Czyz theory, while in the Kostant-Souriau theory it is only a partial connection, which can be evaluated only along the respective polarization.

Secondly, the Kostant-Souriau theory does not yield self-adjoint operators for moderately general functions on phase space. In fact, the quantizing operators are not even formally self-adjoint when the function in question is (roughly) a polynomial in the momentum variables of order strictly greater than 2, see Kostant [18] for this statement.

To cure these defects, the basic philosophy of our new approach is to examine closely the relationship between conventional quantum theory and geometric quantization. The latter will be obtained from the former applied in the tangent spaces, being locally curved and globally twisted. In sophisticated terms, conventional quantum theory deals with symplectic vector spaces and irreducible Weyl systems (representations of the CCR in exponential form) thereon. The global twisting has to be performed with automorphisms of the given Weyl system. We denote the automorphism group by $Mp^c(2n,\mathbb{R})$. It has been studied extensively by A. Weyl [32] , and it is the precise symplectic analogue of the (extended) orthogonal spinor group known as $Spin^c(2n)$, cf. [2] .

How the global twisting has to be done will be coded in a principal $Mp^c(2n,\mathbb{R})$-bundle $\tilde{P}$ adapted to the given 2n-dimensional symplectic manifold (M,ω). To construct differential operators on complex line bundles arising from $\tilde{P}$, the latter should be equipped with an ordinary connection. Only one part of this connection will be fairly uniquely determined by (M,ω), the other one will be yielded by polarizations. Existence and classification of such $Mp^c(2n,\mathbb{R})$-bundles with the polarization-independent part of the connection is discussed in section 1, where it is also shown how to get these data when Kostant-Souriau prequantum bundles and metaplectic structures are given.

In section 2, it will be seen that two transverse polarizations determine a unique torsion-free symplectic connection. Together with the data of section 1, it allows to construct complex line bundles with connection, generalizing those of the Kostant-Souriau and Czyz theories. The construction is done in section 3 by a procedure of reducing the structure group of the principal $Mp^c(2n,\mathbb{R})$-bundle and subsequent building associated bundles. It is somewhat complicated, but very similar to the way of getting half-form bundles from metaplectic structures in the Kostant-Souriau theory. The complex line bundles yielded by this procedure satisfy the dogma of having $\left[-\frac{1}{2\pi}\omega\right] + \frac{1}{2}c_1(TM,\omega)$ as their first (real) Chern class, where

$c_1(TM, \omega)$ is a symplectic invariant of (M, ω).

The rôle of this dogma is to some extent clarified in section 4, where we indicate how to assign differential operators on the above line bundles to certain functions on M. Such a map will be obtained by generalizing ordering rules of conventional quantum theory, like those of Weyl or Born-Jordan, via replacing ordinary differentiations by the connections from above. One of the possible ordering rules gives a map similar to that considered by Kostant [18] . Note that we don't need any kind of intertwining operators as long as the functions on the phase space are of a special type.
Our approach is a strict extension only of the Kostant-Souriau theory. In contrast, it generalizes the Czyz theory just for some restricted class of Kähler manifolds, which contains all examples explicitly investigated by Czyz.

The results of section 4 and most of those in section 3 have not been part of the original conference talk. The material presented here will be treated in more detail in the author's doctor thesis [13] ; see also [9] for another view of a special case.

1. Prequantization

The notion of prequantization used here is not quite identical to that already established in the literature, but is meant to refer to all polarization-independent constructions. In the case of the Kostant-Souriau theory it includes both Kostant-Souriau prequantum bundles and metaplectic frame bundles, which we shall briefly recall in the beginning. All concepts of the Kostant-Souriau theory will henceforth be specified by the prefix KS.

Then we indicate why the fundamental structure group must be $Mp^c(2n, \mathbb{R})$ rather than $U(1) \times Mp(2n, \mathbb{R})$ as in the KS-theory. Next, we give the definition of a (generalized) prequantum bundle with structure group $Mp^c(2n, \mathbb{R})$, and derive two existence criteria for them, one in terms of the cohomology classes $\left[- \frac{1}{2\pi}\omega\right]$ and $c_1(TM, \omega)$. Also, an equivalence relation between such (generalized) prequantum bundles will be defined, and the corresponding equivalence classes turn out to be in bijection with the elements of $H^1(M, U(1))$ or the empty set.

Finally, we show how to construct (generalized) prequantum bundles from given KS-prequantum bundles and metaplectic structures. The latter are a superfluous degree of freedom in the sense that, up to equivalence, every (generalized) prequantum bundle can be obtained in this way from an arbitrary metaplectic structure.

(M,ω) will always denote a fixed 2n-dimensional symplectic manifold. All bundles have base M, and all bundle morphisms are supposed to induce the identity on M, if not stated otherwise.

Given any Lie group G, we denote by $\underline{G}$ the corresponding sheaf of (germs of) C^∞ functions on M with values in G. We use Čech cohomology with coefficients in sheaves of not necessarily abelian groups, referring to [10] , [30] .

1.1. Definition:

A KS-prequantum bundle (over ω) is a principal U(1)-bundle $\pi_L: \mathring{L}\to M$ equipped with a principal connection γ satisfying

(1) $$\operatorname{curv}\gamma = \pi_L^* i\omega \quad .$$

Given another KS-prequantum bundle $(\mathring{L}', \gamma')$, both will be called <u>equivalent</u> if there exists a principal bundle morphism $\Phi : \mathring{L}\longrightarrow \mathring{L}'$ such that

(2) $$\Phi^*\gamma' = \gamma \quad .$$

KS-prequantum bundles can also be viewed as hermitian complex line bundles L equipped with a (linear) connection ${}^L\nabla$ such that $\operatorname{curv}{}^L\nabla = i\omega$.

Obviously, the first (real) Chern class of every KS-prequantum bundle (over ω) is $\left[-\frac{1}{2\pi}\omega\right]$, in particular it is (the image of) an integral class. Moreover, we have the well-known existence criterion and classification [20] , [28] , [33] :

1.2. Theorem:

There exists a KS-prequantum bundle over ω if and only if the following equivalent conditions are satisfied

(3) $\left[-\frac{1}{2\pi}\omega\right]\in H^2(M, \mathbb{R})$ is an integral class

(4) the class $H^2(M, \exp)[i\omega] \in H^2(M, U(1))$ vanishes.

Equivalence of (3) and (4) is easily seen from the cohomology sequence induced by the exact sequence of groups

$$0 \longrightarrow \mathbb{Z} \xrightarrow{2\pi i} i\mathbb{R} \xrightarrow{\exp} U(1) \longrightarrow 0 \cdot$$

1.3. Theorem:

The group $H^1(M, U(1))$ operates in a simply transitive manner on the set of equivalence classes of KS-prequantum bundles over ω. In particular this set is either void or in bijection with $H^1(M, U(1))$.

Both of these theorems have proven to be physically significant. Indeed, the existence condition e.g. restricts the values of quantized spin to integer multiples of $\frac{1}{2}$ [23], [28], while the classification provides for different (Bose and Fermi) quantizations for systems composed of a number of indistinguishable subsystems [28] or for the 3-dimensional rotator [23].

Now consider a central extension of Lie groups

$$(5) \qquad 0 \longrightarrow C \xrightarrow{\lambda} \tilde{G} \xrightarrow{\rho} G \longrightarrow 0$$

and a principal G-bundle P.

1.4. Definition:

A ρ-lifting of P is a principal $\tilde{G}$-bundle $\tilde{P}$ together with a ρ-equivariant principal bundle morphism $\tilde{\rho}: \tilde{P} \longrightarrow P$.

Given another ρ-lifting $(\tilde{P}', \tilde{\rho}')$ of P, both will be called equivalent if there exists a principal bundle morphism $\phi: \tilde{P} \longrightarrow \tilde{P}'$ such that the diagram

$$(6) \qquad \begin{array}{ccc} \tilde{P} & \xrightarrow{\phi} & \tilde{P}' \\ & {\scriptstyle\tilde{\rho}}\searrow \quad \swarrow{\scriptstyle\tilde{\rho}'} & \\ & P & \end{array}$$

commutes.

Let us identify the isomorphism class [P] with the corresponding cohomology class in $H^1(M, \underline{G})$ induced by a system of transition functions of P. Further, consider the cohomology sequence induced by the sequence of sheaves of C^∞-functions corresponding to (5)

$$(7) \qquad \longrightarrow H^1(M,\underline{C}) \longrightarrow H^1(M,\underline{\tilde{G}}) \longrightarrow H^1(M,\underline{G}) \xrightarrow{\delta^1} H^2(M,\underline{C}).$$

Calling $w^\rho(P) := \delta^1[P]$ the ρ-obstruction class of P, we have the well-known existence criterion due to [10].

1.5. Theorem:

P admits a ρ-lifting if and only if the cohomology class $w^\rho(P) \in H^2(M,\underline{C})$ vanishes.

Moreover, we have the classification (see [14] and for discrete C also [12]).

1.6. Theorem:

The group $H^1(M,\underline{C})$ operates in a simply transitive manner on the set of equivalence classes of ρ-liftings of P. In particular this set is either void or in bijection with $H^1(M,\underline{C})$.

Next, consider the symplectic frame bundle $P(TM,\omega)$ of (M,ω), which is a principal $Sp(2n,\mathbb{R})$-bundle, and the two-fold covering group $Mp(2n,\mathbb{R})$ of $Sp(2n,\mathbb{R})$, called metaplectic group within geometric quantization. There is an associated exact sequence of Lie groups

$$(8) \qquad 0 \longrightarrow \mathbb{Z}_2 \hookrightarrow Mp(2n,\mathbb{R}) \xrightarrow{\sigma} Sp(2n,\mathbb{R}) \longrightarrow 0.$$

A σ-lifting $(\hat{P}(TM,\omega),\hat{\sigma})$ is called a metaplectic frame bundle, and an equivalence class of these a metaplectic structure. The obstruction class now is an element $w^{\sigma}(P(TM,\omega)) \in H^2(M,\mathbb{Z}_2)$, and classification is given by $H^1(M,\mathbb{Z}_2)$ or the empty set.

To give also an integrality criterion for existence of metaplectic frame bundles, we first discuss two important characteristic classes of $P(TM,\omega)$. Therefore, consider a symplectic almost complex structure J on TM, which bijectively corresponds to a reduction of structure group of $P(TM,\omega)$ from $Sp(2n,\mathbb{R})$ to $U(n)$.

Indeed, J determines the principal $U(n)$-bundle $P(TM,\omega,J) \subset P(TM,\omega)$. Such reductions of structure group always exist and are unique up to an isomorphism of principal $U(n)$-bundles, because $U(n) \subset Sp(2n,\mathbb{R})$ is the maximal compact subgroup. Hence it is reasonable to make the following

1.7. Definition:

The first Chern class of (TM,ω) is

$$(9) \qquad c_1(TM,\omega) := c_1(P(TM,\omega)) := c_1(P(TM,\omega,J)).$$

(This definition is valid for both real and integer Chern classes.) The second Stiefel-Whitney Class of (TM,ω) is

$$(10) \qquad w_2(TM,\omega) := w_2(P(TM,\omega,J) \times_{U(n)} O(2n)) = w_2(M)$$

where the associated bundle is formed via the inclusion $U(n) \subset O(2n)$. Note that this class really depends only on the topological space M.

Viewing first Chern classes as integer classes, it is well-known that $w_2(M) = H^2(M, \text{mod } 2)c_1(P(TM, \omega, J)$ holds. An easy proof of this statement may be deduced from the fact that $w_2(P(TM, \omega, J) \times_{U(n)} O(2n))$ and $c_1(P(TM, \omega, J))$ coincide with obstruction classes for certain lifting problems, referring to central extensions by $\mathbb{Z}_2$ and $\mathbb{Z}$, respectively [12]. Thus we have

$$(11) \quad w_2(TM, \omega) = H^2(M, \text{mod } 2)c_1(TM, \omega),$$

and from naturality of obstruction classes under change of groups, it follows

$$(12) \quad w^{\sigma}(P(TM, \omega)) = w_2(TM, \omega) = w_2(M), \quad \text{implying}$$

<u>1.8. Proposition:</u>

Metaplectic frame bundles exist if and only if $c_1(TM, \omega)$ as an integer class is divisible by 2, or equivalently, if the real class $\frac{1}{2}c_1(TM, \omega)$ is integral.

<u>1.9. Corollary:</u>

Consider the case $M = P^{n-1}(\mathbb{C})$, then metaplectic frame bundles exist if n is even.

<u>Proof</u>: Indeed, the first Chern class of $P^{n-1}(\mathbb{C})$ is n-times the positive generator of $H^2(P^{n-1}(\mathbb{C}), \mathbb{Z})$, cf. e.g. [31].

The complex manifold $P^{n-1}(\mathbb{C})$ arises in energy surface quantization of the n-dimensional harmonic oscillator as follows:

Consider the normalized Hamiltonian $H(q,p) := \frac{1}{2}(q^2 + p^2)$ and the energy surfaces $\Sigma_E := H^{-1}(E) \subset \mathbb{R}^{2n} = \mathbb{C}^n$ for any $E > 0$, which are presymplectic manifolds with the presymplectic form given by $\omega|\Lambda^2 T\Sigma_E$, and this form has 1-dimensional kernel **tangent to the classical orbits.** The quotient spaces with respect to this foliation coincide with $P^{n-1}(\mathbb{C})$ for every energy value $E > 0$. The presymplectic form on Σ_E induces a symplectic form ω_E on $P^{n-1}(\mathbb{C})$. In particular, $\omega_E = E \cdot \omega_1$ and the class $\left[-\frac{1}{2\pi}\omega_1\right]$ is integral, moreover it is just the image of a generator of $H^2(P^{n-1}(\mathbb{C}), \mathbb{Z})$.

Now the energy surfaces will be specialized to the Schrödinger energy values $E = N + \frac{n}{2}$, $N \in \mathbb{N}_0$, then $\left[-\frac{1}{2\pi}\omega_E\right]$ is integral if and only if n is even, otherwise it is half-integral. Therefore, KS-prequantum bundles over $\omega_{N+\frac{n}{2}}$ exist if n is even. In virtue of corollary 1.9., thus both metaplectic frame bundles and KS-prequantum bundles exist if n is even, and both do not exist if n is odd ($n > 1$).

However, KS-prequantum bundles and metaplectic frame bundles are only auxiliary structures to define a complex line bundle Q which via the sheaf of its (germs of) covariant constant sections along a polarization F determines a quantizing Hilbert space. Since [26], it is a dogma in geometric quantization that the first Chern Class of Q has to satisfy

$$(13) \qquad c_1(Q) = \left[-\frac{1}{2\pi}\omega\right] + \frac{1}{2}c_1(TM,\omega).$$

Obviously, the right hand side of this relation has to be an integral class, but it is not necessary that $\left[-\frac{1}{2\pi}\omega\right]$ and $\frac{1}{2}c_1(TM,\omega)$ are seperately integral. If both of them are half-integral, as in the above example for odd n, then complex line bundles satisfying (13) exist, too. The problem one has to worry about is the existence of additional structures on such a line bundle, in particular that of (partial) connections compatible with ω in some sense.

To motivate our approach, let us first rewrite the K-prequantization data:

Given a KS-prequantum bundle $(\mathring{L}, \gamma)$ and a metaplectic frame bundle $(\hat{P}(TM,\omega), \hat{\sigma})$, consider the Whitney product $\mathring{L} \times_M \hat{P}(TM,\omega)$, which is a principal $U(1) \times Mp(2n, \mathbb{R})$-bundle. Define a principal bundle morphism

$$(14) \qquad \begin{aligned} \Sigma : \mathring{L} \times_M \hat{P}(TM,\omega) &\longrightarrow P(TM,\omega) \\ (l,p) &\longrightarrow \hat{\sigma}(\hat{p}) \ , \end{aligned}$$

and a 1-form

$$(15) \qquad \begin{aligned} \Gamma : T(\mathring{L} \times_M \hat{P}(TM,\omega)) &\longrightarrow i\,\mathbb{R} \\ u &\longrightarrow \gamma \circ Tpr_1(u) . \end{aligned}$$

Σ makes $\mathring{L} \times_M \hat{P}(TM,\omega)$ a lifting of $P(TM,\omega)$, while Γ is something like a connection form. It is equivariant (in this case this means invariant), but not normalized in the usual way, $i\mathbb{R} = LU(1)$ being only a direct summand of the Lie algebra of the structure group $U(1) \times Mp(2n,\mathbb{R})$.

<u>1.10. Definition:</u>

The triple $(\mathring{L} \times_M \hat{P}(TM,\omega), \Sigma, \Gamma)$ will be called an <u>extended KS-prequantum bundle</u>.

Indeed, all polarization-independent data are contained in this triple. Our generalized prequantum bundle will be defined by complete analogy, except that the structure group $U(1) \times Mp(2n, \mathbb{R})$ is replaced by a more natural one.

Consider an irreducible Weyl system, i.e. an irreducible unitary representation of the Weyl-Heisenberg group $Heis(2n,\mathbb{R})$ on some Hilbert space $\mathcal{H}$. Recall that $Heis(2n,\mathbb{R})$ is the product set $U(1) \times \mathbb{R}^{2n}$ with composition given by

$$(c,x)\cdot(c',x') := (c\cdot c'\cdot e^{i\,\omega(x,x')}, x + x').$$

Since in the finite-dimensional case all irreducible Weyl systems are equivalent, it is possible to choose one which is technically helpful, e.g. the Schrödinger or the Bargmann-Segal representation.

We are only interested in automorphisms of the given Weyl system W inducing symplectomorphisms on $\mathbb{R}^{2n}$. These are characterized by the following result of A. Weil [32].

<u>1.11. Theorem:</u>

Let $Mp^c(2n,\mathbb{R})$ denote the subgroup of the normalizator of $W(Heis(2n,\mathbb{R})$ in the unitary operators on $\mathcal{H}$, consisting of all elements which induce symplectomorphisms on $\mathbb{R}^{2n}$. This yields a central extension

$$(16) \qquad 0 \longrightarrow U(1) \hookrightarrow Mp^c(2n,\mathbb{R}) \xrightarrow{\ \rho\ } Sp(2n,\mathbb{R}) \longrightarrow 0$$

where ρ is uniquely determined by

$$(17) \qquad \tilde{g} \circ W(c,x) \circ \tilde{g}^{-1} = W(c, \rho(\tilde{g})x)$$
$$(c \in U(1),\ x \in \mathbb{R}^{2n},\ \tilde{g} \in Mp^c(2n,\mathbb{R})) \quad .$$

In addition, there exists a unique character

$$\eta : Mp^c(2n,\mathbb{R}) \longrightarrow U(1)$$

such that $\eta | U(1)$ is the **squaring map**. Its kernel $Mp(2n,\mathbb{R}) := \mathrm{Ker}\,\eta$ is a two-fold covering of $Sp(2n,\mathbb{R})$.

The construction of the metaplectic group $Mp(2n,\mathbb{R})$ has also been given by [25], [24]. It is the precise analogue of the orthogonal **spinor group** $Spin(2n)$ and has been introduced to geometric quantization by Blattner, Kostant and Sternberg [3], [19]. In the same way, the group $Mp^c(2n,\mathbb{R})$ is the analogue of the group known [2] as $Spin^c(2n)$.

Unfortunately there is some terminological confusion. A. Weil has called $Mp^c(2n,\mathbb{R})$ the metaplectic group. However, we shall follow the custom in geometric quantization to call $Mp(2n,\mathbb{R})$ the metaplectic group, and propose the name <u>toroplectic group</u> for $Mp^c(2n,\mathbb{R})$. (Note that the small letter "c" should not been read as "complex" because $Mp^c(2n,\mathbb{R})$ is a real Lie group, projecting to the symplectic group on a real symplectic vector space.)

From theorem 1.11. we get a commutative diagram of Lie groups with exact rows and columns, where U(1) and $\mathbb{Z}_2$ are centrally imbedded:

$$
\begin{array}{ccccccccc}
 & & 0 & & 0 & & 0 & & \\
 & & \downarrow & & \downarrow & & \downarrow & & \\
0 & \longrightarrow & \mathbb{Z}_2 & \hookrightarrow & Mp(2n,\mathbb{R}) & \xrightarrow{\sigma} & Sp(2n,\mathbb{R}) & \longrightarrow & 0 \\
 & & \xi \downarrow & & \big| & & \big\| & & \\
0 & \longrightarrow & U(1) & \hookrightarrow & Mp^c(2n,\mathbb{R}) & \xrightarrow{\varrho} & Sp(2n,\mathbb{R}) & \longrightarrow & 0 \\
 & & .2 \big| & & \eta \downarrow & & \downarrow & & \\
0 & \longrightarrow & U(1) & = & U(1) & \longrightarrow & 0 & & \\
 & & \downarrow & & \downarrow & & & & \\
 & & 0 & & 0 & & & &
\end{array} \tag{18}
$$

U(1) being centrally imbedded in $Mp^c(2n,\mathbb{R})$, we may identify

$$Mp^c(2n,\mathbb{R}) = U(1) \times_{\mathbb{Z}_2} Mp(2n,\mathbb{R}) \tag{19}$$

where the right hand side denotes the quotient of the product $U(1) \times Mp(2n,\mathbb{R})$ with respect to the diagonal $\mathbb{Z}_2$-subgroup. To explain this identification, consider the map

$$
\begin{array}{rccc}
\tau : & U(1) \times Mp(2n,\mathbb{R}) & \longrightarrow & Mp^c(2n,\mathbb{R}) \\
 & (c,\hat{g}) & \longrightarrow & c\cdot\hat{g} \quad ,
\end{array} \tag{20}
$$

which is a group morphism in virtue of centrality of U(1). Now τ factorizes along the canonical surjection

$$
\begin{array}{ccc}
U(1) \times Mp(2n,\mathbb{R}) & \longrightarrow & U(1) \times_{\mathbb{Z}_2} Mp(2n,\mathbb{R}) \\
(c,\hat{g}) & \longrightarrow & [c,\hat{g}] \quad ,
\end{array}
$$

thus the class $[c,\hat{g}]$ is mapped to $c\cdot\hat{g} \in Mp^c(2n,\mathbb{R})$. Conversely, $\tilde{g} \in Mp^c(2n,\mathbb{R})$ corresponds to the class $[c,\hat{g}]$, where $\hat{g} \in Mp(2n,R)$ is any element satisfying $\sigma(\hat{g}) = \varrho(\tilde{g})$, and $c := \tilde{g}\hat{g}^{-1}$.

In particular, this implies that the Lie algebra of $Mp^c(2n,\mathbb{R})$ splits as a direct sum of ideals:

$$
\begin{aligned}
LMp^c(2n,\mathbb{R}) &= LU(1) \oplus LMp(2n,\mathbb{R}) \\
&= i\mathbb{R} \oplus Sp(2n,\mathbb{R}) \quad .
\end{aligned} \tag{21}
$$

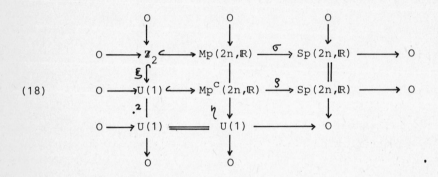

Finally let

$$(22) \qquad \mu : LMp^c(2n,\mathbb{R}) \longrightarrow i\mathbb{R}$$

denote the projection to the first summand, which is equivariant with respect to the adjoint action of $Mp^c(2n,\mathbb{R})$ due to centrality of $U(1)$.

1.12. Definition:

Let $\tilde{P}$ be a principal $Mp^c(2n,\mathbb{R})$-bundle with principal operation $\tilde{R}$. A 1-form

$$\tilde{\gamma} : T\tilde{P} \longrightarrow i\mathbb{R}$$

is said to be a μ-pseudoconnection on $\tilde{P}$ iff it satisfies

$$(23) \qquad \tilde{\gamma} \circ T\hat{R}_{\tilde{g}} = Ad_{\tilde{g}}^{-1} \circ \tilde{\gamma} \qquad (\tilde{g} \in Mp^c(2n,\mathbb{R}))$$

$$(24) \qquad \tilde{\gamma}(\underline{z}) = \mu(z) \qquad (z \in LMp^c(2n,\mathbb{R})),$$

where $\underline{z}$ is the vertical vector field on $\tilde{P}$ corresponding to z.

U(1) being central in $Mp^c(2n,\mathbb{R})$, the equivariance condition (23) just means invariance, i.e. is equivalent to

$$(25) \qquad \tilde{\gamma} \circ T\tilde{R}_{\tilde{g}} = \tilde{\gamma} \ .$$

We can also define $(1-\mu)$-pseudoconnections taking values in $sp(2n,\mathbb{R})$, clearly then only an equivariance condition like (23) may be used for the definition. In most respects, pseudoconnections behave like ordinary connections. In particular, a μ-pseudoconnection $\tilde{\gamma}$ possesses a curvature 2-form

$$(26) \qquad \begin{aligned} \operatorname{curv} \tilde{\gamma} &: \Lambda^2 T\tilde{P} \longrightarrow i\mathbb{R} \\ \operatorname{curv} \tilde{\gamma} &:= d\tilde{\gamma} + \tfrac{1}{2}[\tilde{\gamma},\tilde{\gamma}] = d\tilde{\gamma} \ . \end{aligned}$$

The local representations of μ-pseudoconnections are also similar to those of ordinary connections. The Christoffel symbols $\tilde{\gamma}_i := \tilde{s}_i^* \tilde{\gamma}$ of $\tilde{\gamma}$ according to a principal coordinate representation of $\tilde{P}$ inducing local sections $\tilde{s}_i$ and transition functions $\tilde{g}_{ij}$ satisfy

$$(27) \qquad \begin{aligned} \tilde{\gamma}_j &= Ad_{\tilde{g}_{ij}}^{-1} \circ \tilde{\gamma}_i + \mu \circ Tl_{\tilde{g}_{ij}}^{-1} \circ T\tilde{g}_{ij} \\ &= \tilde{\gamma}_i + \mu \circ Tl_{\tilde{g}_{ij}}^{-1} \circ T\tilde{g}_{ij}, \end{aligned}$$

where l_h denotes left multiplication with $h \in Mp^c(2n,\mathbb{R})$.

Conversely, given $\tilde{\gamma}_i$ satisfying (27), then there exists a unique μ-pseudoconnection $\tilde{\gamma}$ having Christoffel symbols $\tilde{\gamma}_i$.

Moreover, the $d\,\gamma_i + \frac{1}{2}[\tilde{\gamma}_i, \tilde{\gamma}_i] \;\; = d\,\tilde{\gamma}_i$ locally represent the curvature of $\tilde{\gamma}$ in the same sense.

1.13. Definition:

A prequantum $Mp^c(2n,\mathbb{R})$-bundle (over ω) is a principal $Mp^c(2n,\mathbb{R})$-bundle $\tilde{\pi} : \tilde{P} \longrightarrow M$ together with a ρ-equivariant principal bundle morphism

(28) $$\tilde{\rho} : \tilde{P} \longrightarrow P(TM,\omega)$$

and a μ-pseudoconnection $\tilde{\gamma}$ satisfying

(29) $$\operatorname{curv} \tilde{\gamma} \;=\; \pi^* i\,\omega \,.$$

(In particular, $(\tilde{P},\tilde{\rho})$ is a ρ-lifting of $P(TM,\omega)$.)

Given another prequantum $Mp^c(2n,\mathbb{R})$-bundle $(\tilde{P}', \tilde{\rho}', \tilde{\gamma}')$, both will be called equivalent if there exists a principal bundle morphism $\phi : \tilde{P} \longrightarrow \tilde{P}'$ such that the diagram

(30) $$\begin{array}{ccc} \tilde{P} & \xrightarrow{\;\;\phi\;\;} & \tilde{P}' \\ & {\scriptstyle\tilde{\rho}}\searrow \quad \swarrow{\scriptstyle\tilde{\rho}'} & \\ & P(TM,\omega) & \end{array}$$

is commutative, and

(31) $$\phi^*\tilde{\gamma}' = \tilde{\gamma}$$

holds.

By analogy to theorems 1.2, 1.5., we have the existence criterion.

1.14. Theorem:

A prequantum $Mp^c(2n,\mathbb{R})$-bundle over ω exists if and only if the cohomology class

$$H^2(M,\exp)[i\omega] \cdot H^2(M,\xi)\,w_2(M) \in H^2(M,U(1))$$

vanishes.

Proof: We shall only sketch how to get the result.

First, assume that $(\tilde{P}, \tilde{\rho}, \tilde{\gamma})$ is a prequantum $Mp^c(2n,\mathbb{R})$-bundle. By a suitable choice of principal coordinate representations of $\tilde{P}$ and $P(TM,\omega)$, the latter are represented by transition functions satisfying $\rho \circ \tilde{g}_{ij} \;=\; g_{ij}$. Moreover, it is possible to choose cochains (in general not cocycles!) c_{ij}, $\hat{g}_{ij}$ taking values in U(1), Mp(2n,R), respectively, with

(32) $$\tilde{g}_{ij} = c_{ij} \cdot \hat{g}_{ij} .$$

Since $\tilde{g}_{ij}$ is a cocycle and U(1) centrally imbedded, it follows

(33) $$e = (\partial c)_{ijk} \cdot (\xi \circ (\partial \hat{g})_{ijk}).$$

Next, relation (27), in virtue of the Ad-stable direct sum decomposition (21), may be rewritten as

(34) $$\tilde{\gamma}_j - \tilde{\gamma}_i = Tl_{c_{ij}}^{-1} \circ Tc_{ij}.$$

Moreover, we can write $c_{ij} = \exp z_{ij}$, then (34) becomes

(35) $$\tilde{\gamma}_j - \tilde{\gamma}_i = dz_{ij}.$$

Then define

(36) $$i\check{\omega}_{ijk} := (\partial z)_{ijk} .$$

The curvature condition (29) may be expressed as

(37) $$d\tilde{\gamma}_i = i\omega ,$$

and this together with (35) means that $i\check{\omega}_{ijk}$ just represents the Čech cohomology class corresponding to the de Rham class $[i\omega]$. Finally, we get

(38) $$(\partial c)_{ijk} = \exp i\check{\omega}_{ijk},$$

which inserted in (33), and using (12), proves the cohomology class in question to be trivial.

The converse is proved by nearly reversing the argument. One starts with $i\check{\omega}_{ijk}$, z_{ij}, and $\tilde{\gamma}_i$ satisfying (35), (36) and (37), thus $H^2(M,\exp)[i\omega]$ is represented by $\exp i\check{\omega}_{ijk}$. Then write

(39) $$c'_{ij} := \exp z_{ij},$$

implying the analogue of relation (34). Next, the class $H^2(M,\xi)w_2(M)$ in virtue of (12) is represented by $\xi \circ (\partial \hat{g})_{ijk}$, where $\hat{g}_{ij}$ is a cochain related to the transition functions of $P(TM,\omega)$ by $\sigma \circ \hat{g}_{ij} = g_{ij}$. Due to the vanishing assumption on the cohomology class of interest, we have

(40) $$(\exp i\check{\omega}_{ijk})(\xi \circ (\partial \hat{g})_{ijk}) = (\partial a)_{ijk}$$

for some $U(1)$-valued cochain a_{ij}. Defining

(41) $$c_{ij} := a_{ij}^{-1}\, c'_{ij} \quad ,$$

and using relation (32) to define $\tilde{g}_{ij}$, the latter turns out to be a cocycle in virtue of centrality of $U(1)$, allowing to construct a principal $Mp^c(2n,\mathbb{R})$-bundle $\tilde{P}$. Moreover, (32) implies $\rho \circ \tilde{g}_{ij} = g_{ij}$, so that a ρ-equivariant principal bundle morphism $\tilde{\rho} : \tilde{P} \longrightarrow P(TM,\omega)$ may also be constructed.

At last, the $\tilde{\gamma}_i$ now turn out to satisfy the compatibility relation (27), thus they are the Christoffel symbols of a μ-pseudoconnection $\tilde{\gamma}$ on $\tilde{P}$, the curvature of which is $\tilde{\pi}^* i\omega$ due to (37).

Next we give an integrality criterion equivalent to the above vanishing criterion.

<u>1.15. Proposition:</u>

Prequantum $Mp^c(2n,\mathbb{R})$-bundles over ω exist if and only if the (real) cohomology class

$$\left[-\frac{1}{2\pi}\omega\right] + \frac{1}{2}c_1(TM,\omega) \qquad \text{is integral.}$$

<u>Proof:</u> Consider the commutative diagram of groups with exact rows

(42) $$\begin{array}{ccccccccc} 0 & \longrightarrow & \mathbb{Z} & \xrightarrow{2\cdot} & \mathbb{Z} & \xrightarrow{\text{mod } 2} & \mathbb{Z}_2 & \longrightarrow & 0 \\ & & \downarrow & & \downarrow \tfrac{1}{2}\cdot & & \downarrow \xi & & \\ 0 & \longrightarrow & \mathbb{Z} & \hookrightarrow & \mathbb{R} & \xrightarrow{e^{2\pi i \cdot}} & U(1) & \longrightarrow & 0 \end{array} .$$

From the resulting cohomology diagram and relations (11), (12), it follows immediately that the class $H^2(M,\exp)[i\omega] \cdot H^2(M,\xi)w_2(M)$ vanishes if the class $\left[\frac{1}{2\pi}\omega\right] + \frac{1}{2}c_1(TM,\omega)$ is integral. Since $c_1(TM,\omega)$ is always integral, this condition is further equivalent to integrality of the desired cohomology class.

Due to this result, using prequantum $Mp^c(2n,\mathbb{R})$-bundles resolves all problems in the prequantization of energy surfaces of the harmonic oscillator. To derive the classification, we need the following

<u>1.16. Lemma:</u>

Let $(\tilde{P}^k, \tilde{\rho}^k, \tilde{\gamma}^k)$, $k = 1,2$, be two prequantum $Mp^c(2n,\mathbb{R})$-bundles over ω. Then the following two assertions are equivalent:

(i) $(\tilde{P}^1, \tilde{\rho}^1, \tilde{\gamma}^1)$ is equivalent to $(\tilde{P}^2, \tilde{\rho}^2, \tilde{\gamma}^2)$ via some principal bundle morphism $\phi : \tilde{P}^1 \longrightarrow \tilde{P}^2$.

(ii) $\tilde{P}^1$, $\tilde{P}^2$, $P(TM, \omega)$ admit principal coordinate representations over a contractible open covering $(U_i)_{i \in I}$ of M, such that for each $i \in I$, the diagram

(43)
$$\begin{array}{ccccccc}
U_i \times \tilde{G} & \xrightarrow{\tilde{\psi}^1_i} & \tilde{P}^1_i & & \tilde{P}^2_i & \xleftarrow{\tilde{\psi}^2_i} & U_i \times \tilde{G} \\
\downarrow {\scriptstyle 1 \times \rho} & & & {\scriptstyle \tilde{\rho}^1} \searrow \quad \swarrow {\scriptstyle \tilde{\rho}^2} & & & \downarrow {\scriptstyle 1 \times \rho} \\
U_i \times G & \xrightarrow{\psi_i} & & P(TM,\omega)_i & & \xleftarrow{\psi_i} & U_i \times G
\end{array}$$

is commutative, the Christoffel symbols of $\tilde{\gamma}^1$ and $\tilde{\gamma}^2$ coincide and the corresponding transition functions of $\tilde{P}^1$ and $\tilde{P}^2$ are related by

(44)
$$\tilde{g}^2_{ij} = b_i \cdot g^1_{ij} \, b_j^{-1}$$

for some (locally constant!) U(1)-valued cochain b_i .

Proof: We only remark that the b_i are the mapping transformations of the principal bundle morphism ϕ.

1.17. Theorem:

The group $H^1(M, U(1))$ operates in a symply transitive manner on the set of equivalence classes of prequantum $Mp^c(2n,\mathbb{R})$-bundles over ω . In particular, this set is either void or in bijection with $H^1(M,U(1))$.

Proof: Again we only sketch the argument.
Given a prequantum $Mp^c(2n,\mathbb{R})$-bundle $(\tilde{P}, \tilde{\rho}, \tilde{\gamma})$ and a class $[a] \in H^1(M,U(1))$, choose first principal coordinate representations as in one half of diagram (43). From the corresponding transition functions $\tilde{g}_{ij}$ of $\tilde{P}$, the transition functions of a new principal $Mp^c(2n,\mathbb{R})$-bundle $\tilde{P}^a$ will be defined by

(45)
$$\tilde{g}^a_{ij} := a_{ij} \cdot \tilde{g}_{ij} \quad ,$$

and then $\tilde{P}^a$ carries a unique lifting morphism $\tilde{\rho}^a : \tilde{P}^a \longrightarrow P(TM, \omega)$ since composition of $\tilde{g}^a_{ij}$ with ρ yields the transition functions of $P(TM, \omega)$. The μ -pseudoconnection $\tilde{\gamma}^a$ on $\tilde{P}^a$ is defined by the same Christoffel symbols as $\tilde{\gamma}$, thus it obviously satisfies the curvature condition.

Thus $(\tilde{P}^a, \tilde{\rho}^a, \tilde{\gamma}^a)$ is a prequantum $Mp^c(2n,\mathbb{R})$-bundle. By repeated application of the preceding lemma, it is easily shown that $H^1(M,U(1))$ operates freely on the set of equivalence classes of prequantum

$Mp^c(2n,\mathbb{R})$-bundles.

To prove transitivity, let $(\tilde{P}^k, \tilde{\varsigma}^k, \tilde{\gamma}^k)$, k = 1,2, be as in 1.16., and consider principal coordinate representations making diagram (43) commutative for each $i \in I$. Let $\tilde{g}^k_{ij}$ be the corresponding transition functions, then define

$$(46) \qquad a_{ij} := \tilde{g}^2_{ij} \cdot (\tilde{g}^1_{ij})^{-1} \quad ,$$

which is a U(1)-valued cocycle, since U(1) is central. Finally, due to curv $\tilde{\gamma}^1$ = curv $\tilde{\gamma}^2$, it can be shown that the principal coordinate representations in addition can be chosen to yield coinciding Christoffel symbols for $\tilde{\gamma}^1$ and $\tilde{\gamma}^2$. This forces a_{ij} to be locally constant, and now $[a] \in H^1(M, U(1))$ is the desired cohomology class mapping the equivalence class of $(\tilde{P}^1, \tilde{\varsigma}^1, \tilde{\gamma}^1)$ to that of $(\tilde{P}^2, \tilde{\varsigma}^2, \tilde{\gamma}^2)$.

Hence the transition from KS-prequantum bundles to prequantum $Mp^c(2n,\mathbb{R})$-bundles (absorbing metaplectic structures) preserves the highly desired classification of prequantizations by $H^1(M,U(1))$.

We conclude this section with the construction of prequantum $Mp^c(2n,\mathbb{R})$-bundles from KS-prequantum bundles and metaplectic frame bundles. The KS-data will be considered already in the form of extended KS-prequantum bundles, as in definition 1.10.

Since $\mathring{L} \times_M \hat{P}(TM,\omega)$ is a principal $U(1) \times Mp(2n,\mathbb{R})$-bundle, we may use the group morphism τ (20) to form the associated principal $Mp^c(2n,\mathbb{R})$-bundle

$$(47) \qquad \tilde{P} := (\mathring{L} \times_M \hat{P}(TM,\omega)) \times_\tau Mp^c(2n,\mathbb{R}).$$

Next, it is easily seen that the principal bundle morphism Σ (14) factorizes along the canonical principal bundle morphism

$$(48) \qquad \begin{array}{rcl} \tilde{\tau} : \mathring{L} \times_M \hat{P}(TM,\omega) & \longrightarrow & \tilde{P} \\ (l,\hat{p}) & \longmapsto & [(l,\hat{p}), e] \end{array} \quad ,$$

yielding a ς-equivariant principal bundle morphism

$$(49) \qquad \tilde{\varsigma} : \tilde{P} \longrightarrow P(TM,\omega) \quad ,$$

i.e. such that

$$(50) \qquad \Sigma = \tilde{\varsigma} \circ \tilde{\tau}$$

holds. The last relation shows $\tilde{\varsigma}$ to be unique, because $\tilde{\tau}$ is an epimorphism. Up to a natural identification, $\check{P}$ may be viewed as a quotient of $\mathring{L} \times_M \hat{P}(TM,\omega)$ with respect to the free diagonal action of $\mathbb{Z}_2$ in almost the same way as for the structure groups.

Finally, the 1-form Γ (15) factorizes along the vector bundle morphism $T\tilde{\tau}$, yielding a 1-form

(51) $\tilde{\gamma} : T\tilde{P} \longrightarrow i\mathbb{R}$

with

(52) $\Gamma = \tilde{\gamma} \circ T\tilde{\tau}$,

because $T\tilde{\tau}$ (as a morphism over $\tilde{\tau}$) is fiberwise isomorphic. Since Γ arises from an ordinary connection with values in $i\mathbb{R}$, it is easily seen that $\tilde{\gamma}$ is a μ-pseudoconnection. The curvatures then are related by

(53) $\text{curv}\,\gamma \circ \Lambda^2 T\text{pr}_1 = \text{curv}\,\tilde{\gamma} \circ \Lambda^2 T\tilde{\tau}$,

implying that the curvature condition (29) is satisfied. Therefore, $(\tilde{P}, \tilde{\xi}, \tilde{\gamma})$ is a prequantum $Mp^c(2n,\mathbb{R})$-bundle over ω.

1.18. Definition:

Given a KS-prequantum bundle $(\mathring{L}, \gamma)$ and a metaplectic frame bundle $(\hat{P}(TM,\omega), \hat{\sigma})$, the prequantum $Mp^c(2n,\mathbb{R})$-bundle $(\tilde{P}, \tilde{\xi}, \tilde{\gamma})$ constructed above is said to be the amalgamation of the former.

We only remark that the process of building amalgations is compatible with the operations of first cohomology groups on equivalence classes established in theorem 1.3., 1.6. and 1.17. Clearly, in general, there are more KS-data than prequantum $Mp^c(2n,\mathbb{R})$-bundles (up to equivalence), thus amalgamation does not necessarily induce an injective map on equivalence classes.

1.19. Proposition:

Let $(\tilde{P}, \tilde{\xi}, \tilde{\gamma})$ be a prequantum $Mp^c(2n,\mathbb{R})$-bundle and $(\hat{P}(TM,\omega), \hat{\sigma})$ a metaplectic frame bundle. Then the former is equivalent to an amalgamation of the latter and a suitable KS-prequantum Bundle $(\mathring{L}, \gamma)$.

From the classification theorems, it is obvious that the amalgamation procedure is surjective on the level of equivalence classes, when KS-data exist. However, note that metaplectic structures may be chosen completely arbitrary.

2. Symplectic connections

From now on, polarizations will be taken into account, i.e. involutive Lagrangian vector subbundles of the complexified tangent bundle $T^{\mathbb{C}}M$. Having done prequantization, quantization depends on a polarization F fixing the representation space, and in addition on an auxiliary polarization G with

(54) $$F \oplus G = T^{\mathbb{C}}M$$

which is needed to construct maps from functions on phase space to operators, cf. e.g. [11],[18].

As a first step, we assign a torsion-free symplectic connection ∇ to the pair (F,G). ∇ extends Bott's partial connection in a certain sense. The latter is known [22] to be important in geometric quantization. (It is often used in more or less implicit fashion.)

We show a uniqueness property of ∇ and derive some properties of its curvature. Finally, we briefly discuss the case, where (M,ω) is a Kähler manifold with complex structure given by $F = \overline{G}$.

Polarizations being always $\mathbb{C}$-vector bundles, we need a slightly extended notion of ordinary connection. Intermediately, we also use partial connections, cf. [15],[21].

2.1 Definition:

A (linear) connection ∇ on $T^{\mathbb{C}}M$ is called symplectic if

(55) $$\mathbb{L}_X(\omega(Y,Z)) = \omega(\nabla_X Y, Z) + \omega(Y, \nabla_X Z)$$
$$(X,Y,Z \in \Gamma(T^{\mathbb{C}}M)).$$

Symplectic manifolds, due to $d\omega = 0$, admit torsion-free symplectic connections [16], but these are not unique. Here we investigate a particular type of them. First, F being involutive, we may consider Bott's partial connection [4] along F, acting as follows

(56) $$\nabla^F : \Gamma(T^{\mathbb{C}}M/F) \longrightarrow \Gamma(\mathrm{Hom}_{\mathbb{C}}(F, T^{\mathbb{C}}M/F))$$
$$\nabla^F_X \tau_* Y := \tau_*[X,Y] \qquad (X \in \Gamma(F),\ Y \in \Gamma(T^{\mathbb{C}}M)),$$

where $\tau : T^{\mathbb{C}}M \longrightarrow T^{\mathbb{C}}M/F$ denotes the natural surjection. By dualization, it yields another partial connection

(57) $$\nabla^{F*} : \Gamma((T^{\mathbb{C}}M/F)^*) \longrightarrow \Gamma(\mathrm{Hom}_{\mathbb{C}}(F, (T^{\mathbb{C}}M/F)^*))$$
$$\nabla^{F*}_X \beta = \mathbb{L}_X \beta \qquad (X \in \Gamma(F),\ \beta \in \Gamma((T^{\mathbb{C}}M/F)^*)).$$

Since F is Lagrangian, the vector bundle isomorphism ω^b maps

$F \subset T^{\mathbb{C}}M$ to $(T^{\mathbb{C}}M/F)^* \subset T^{\mathbb{C}}M^*$, and therefore allows to transfer ∇^{F*} to a partial connection on F itself:

$$\begin{aligned} &\nabla^{FF} : \Gamma(F) \longrightarrow \Gamma(\mathrm{Hom}_{\mathbb{C}}(F,F)) \\ (58)\qquad &\nabla^{FF}_X Y := \omega^{\#}_{*} \mathbb{L}_X \omega^{b}_{*} Y \qquad (X,Y \in \Gamma(F)). \end{aligned}$$

(57) or (58) has already been used in geometric quantization to define a notion of covariant constancy for geometric objects over F or $(T^{\mathbb{C}}M/F)^*$, in particular for half-forms over F or $(T^{\mathbb{C}}M/F)^*$.

Now denote by pr_F, pr_G the projections from $T^{\mathbb{C}}M$ to F,G with kernels G,F respectively, and consider the vector bundle isomorphism $\phi: T^{\mathbb{C}}M/F \longrightarrow G$ satisfying $pr_G = \phi \circ \tau$. Then ϕ allows to transfer ∇^F to a partial connection on G:

$$\begin{aligned} (59)\qquad &\nabla^{FG} : \Gamma(G) \longrightarrow \Gamma(\mathrm{Hom}_{\mathbb{C}}(F,G)) \\ &\nabla^{FG}_X Y := pr_G[X,Y] \qquad (X \in \Gamma(F),\ Y \in \Gamma(G)). \end{aligned}$$

Next, due to the direct sum decomposition (54), the partial connections (58) and (59) may be combined to yield a rule for covariant derivation of sections in $T^{\mathbb{C}}M$ along sections in F. Interchanging the rôles of F and G, we get a rule for covariant derivation of sections in $T^{\mathbb{C}}M$ along sections in G in the same manner. From both cases together, we then obtain the desired connection on $T^{\mathbb{C}}M$:

$$(60)\qquad \nabla : \Gamma(T^{\mathbb{C}}M) \longrightarrow \Gamma(\mathrm{Hom}_{\mathbb{C}}(T^{\mathbb{C}}M, T^{\mathbb{C}}M)).$$

<u>2.2. Definition:</u>

The connection ∇ constructed above is called the <u>bilagrangian connection associated to F and G.</u>

By construction, it satisfies for every $X \in \Gamma(T^{\mathbb{C}}M)$

$$(61)\qquad \nabla_X \Gamma(F) \subset \Gamma(F)$$

and

$$(62)\qquad \nabla_X \Gamma(G) \subset \Gamma(G).$$

<u>2.3. Lemma:</u>

∇ is torsion-free and symplectic.

<u>Proof:</u> Vanishing of the torsion is proved using involutivity of F

and G.

2.4. Theorem:

There exists a unique torsion-free symplectic connection ∇ on $T^{\mathbb{C}}M$ satisfying (61) and (62).

Proof: Existence of ∇ is clear by taking the bilagrangian connection associated to F and G. Therefore, consider any connection ∇ obeying the hypothesis. Since it is torsion-free and (61), (62) holds, we have for $X \in \Gamma(F)$, $Z \in \Gamma(G)$

$$\nabla_X Z = pr_G \nabla_X Z = pr_G([X,Z] + \nabla_Z X) = pr_G[X,Z] = \nabla_X^{FG} Z .$$

Inserting this into relation (55), we get for $Y \in \Gamma(F)$ that $\nabla_X Y = \nabla_X^{FF} Y$ holds, too. Then the assertion follows by doing the same with F and G interchanged.

2.5. Lemma:

The curvature of the bilagrangian connection (associated to F and G) satisfies

$$\text{curv}\,\nabla|F \wedge F = 0 = \text{curv}\,\nabla|G \wedge G.$$

Proof: The definition of ∇ implies

$$\text{curv}\,\nabla|F \wedge F = (\text{curv}\,\nabla^{FF}|\,F \wedge F) \oplus (\text{curv}\,\nabla^{FG}|\,F \wedge F),$$

and both terms on the right hand side arise from the curvatures of the partial Bott connection ∇^F and its dual ∇^{F*}. Now these curvatures are zero (due to the Jacobi identity). The other part of the assertion follows in the same way.

Now recall (cf. e.g. [18])

2.6. Definition:

F and G are called <u>Heisenberg related</u> if for every $m \in M$ there exists an open neighborhood U and C^∞-functions $q_i, p_i : U \longrightarrow \mathbb{C}$, i=1,...,n, satisfying

(63) $$F_U = \langle \{ \omega^\# dq_i \mid i=1,\dots,n \} \rangle$$

(64) $$G_U = \langle \{ \omega^\# dp_i \mid i=1,\dots,n \} \rangle$$

and

(65) $$\{ q_i, p_j \} = \delta_{ij} ,$$

with respect to Poisson brackets.

2.7. Proposition:

F and G are Heisenberg related if and only if the bilagrangian connection ∇ associated to them is flat.

Proof: When F and G are Heisenberg related, consider the functions q_i, p_i as above. Since curv ∇ is a 2-form, it is sufficient to compute it on the vector fields $\frac{\partial}{\partial q_i} = \omega^{\#} dp_i$ and $\frac{\partial}{\partial p_i} = -\omega^{\#} dq_i$, which span $T^{\mathbb{C}}U$ in virtue of (63), (64) and (54). Now the assertion is easily proved, using that the Lie brackets between these vector fields vanish due to (65).

Conversely, suppose ∇ to be flat. By parallel transport of suitable base vectors of $T^{\mathbb{C}}_m M$, vector fields X_i, Y_i will be obtained, which are defined in some open neighborhood U of m, covariant constant along $T^{\mathbb{C}}U$, and span F_U, G_U, respectively. ∇ being torsion-free, all their mutual Lie brackets vanish. Hence there exist q_i, $p_i \in C^{\infty}(U,\mathbb{C})$ such that $X_i = -\frac{\partial}{\partial p_i}$, $Y_i = \frac{\partial}{\partial q_i}$, and then (63), (64), (65) may be obtained by choosing symplectically conjugated initial base vectors.

For the remainder of the section, let (M, ω, J) be a Kähler manifold, and $F = T^{1,0}M$, $G = T^{0,1}M$. Then we first have

2.8. Lemma:

The bilagrangian connection associated to F and G coincides with the Levi-Civita connection (according to the Riemannian metric induced by ω,J).

Proof: Since F and G are the eigenbundles of J, relations (61), (62) together are equivalent to $\nabla J = 0$. Therefore, symplecticity of ∇ (55) implies ∇ to be Riemannian, too. ∇ being torsion-free, it has to be the Levi-Civita connection.

Moreover, in this case, ∇ preserves also the hermitian structure. In other terms, the corresponding principal connection α on $P(TM,\omega)$ restricts to a principal connection on $P(TM,\omega,J) \subset P(TM,\omega)$. Hence the 2-form κ given by

$$(66) \qquad \kappa(X,Y) := -i\,\mathrm{tr}(\mathrm{curv}\,\nabla(X,Y)|T^{1,0}M) \qquad (X,Y \in \Gamma(T^{\mathbb{C}}M))$$

is related to the first Chern class $c_1(TM,\omega)$ by

$$(67) \qquad c_1(TM,\omega) = -\left[\frac{1}{2\pi}\kappa\right].$$

Then it follows [17] that

(68) $$\kappa(X,Y) = \mathrm{Ric}(X,JY) ,$$

where Ric denotes the Ricci tensor corresponding to ∇ and the Riemannian metric.

Later, it is desirable to know whether κ is harmonic with respect to the Laplace-Beltrami operator arising from ∇ and the Riemannian metric. (Note that ω is always harmonic due to $\nabla\omega = 0$.)

2.9. Lemma:

κ is harmonic if and only if the scalar curvature (corresponding to ∇ and the Riemannian metric) is locally constant.

Proof: See e.g. [34].

In particular, this result applies when the curvature of ∇ is covariant constant, and this, in turn, holds for all (hermitian) symmetric spaces [17].

For $M = P^{N-1}(\mathbb{C})$, it is even more easy to see that κ is harmonic. In fact, $P^{n-1}(\mathbb{C})$ has constant holomorphic sectional curvature [17], implying

(69) $$\kappa = \frac{1}{2}\, nc\,\omega ,$$

where $c \in R$, $c > 0$ is just the holomorphic sectional curvature.

3. Construction of quantum bundles

In this section, given a prequantum $Mp^c(2n,\mathbb{R})$-bundle $(\tilde{P}, \tilde{\xi}, \tilde{\gamma})$ and a pair of polarizations F,G as in the previous section, we are going to construct a complex line bundle Q equipped with an ordinary connection ${}^Q\nabla$ and a suitable scalar product $\langle .,. \rangle$, such that its (real) first Chern class is

(70) $$c_1(Q) = \left[-\frac{1}{2\pi}\omega\right] + \frac{1}{2}c_1(TM,\omega) .$$

This will be carried out very shortly, leaving out details and proofs.

From $(Q, {}^Q\nabla, \langle .,. \rangle)$, the representation space will arise in precisely the same manner as in the Kostant-Souriau theory.

Finally, these bundles Q together with their additional structures will be compared with the corresponding objects yielded by the Czyz theory, finding coincidence whenever the assumption of lemma 2.9. is valid.

The construction of $(Q, {}^Q\nabla, \langle .,. \rangle)$, in principle, will be performed similar to that of the half-form bundle $\Lambda_{1/2}F$ (with its additional structures) in the KS-theory, where $\Lambda_{1/2}F$ arises from the metaplectic frame bundle by reducing the structure group and then building the associated line bundle via a suitable representation. In particular, this procedure guarantees that Q (including the additional structures) will be the tensor product of the KS-prequantum bundle L and $\Lambda_{1/2}F$, if $(\tilde{P}, \tilde{\xi}, \tilde{\gamma})$ is an amalgamation of KS-data.

The need for reducing the structure group arises, because $Mp(2n,\mathbb{R})$ is simple, and $Mp^c(2n,\mathbb{R})$ possesses only η as non-trivial representation on $\mathbb{C}$. Therefore, metaplectic frame bundles have no non-trivial associated complex line bundles, while $\tilde{P}$ only has the associated complex line bundle $\tilde{P} \times_\eta \mathbb{C}$. However, the latter in general does not satisfy (70), but $c_1(\tilde{P} \times_\eta \mathbb{C}) = \left[- \frac{1}{\pi}\omega\right]$ instead.

The subgroups of $Mp(2n,\mathbb{R})$ and $Mp^c(2n,\mathbb{R})$ admitting the desired one-dimensional representations are inverse images of a subgroup of $Sp(2n,\mathbb{R})$ depending on the type of the polarization F.

Consider the sesquilinear form κ on $T^{\mathbb{C}}M$ given by

(71) $$\kappa(X,Y) := -2i\,\omega(\bar{X},Y) \qquad (X,Y \in T^{\mathbb{C}}M) .$$

3.1. Definition:

A Lagrangian vector subbundle $F \subset T^{\mathbb{C}}M$ is said to be positive, positive definite, negative, negative definite if the restriction of κ to F has the corresponding property. F is called real if the restriction of κ to F is zero.

Finally, F is called homogenous of type (f^0, f^+, f^-) if there exist vector subbundles F^0, F^+, F^- of F with ranks f^0, f^+, f^-, repectively, such that

(72) $$F = F^0 \oplus F^+ \oplus F^- ,$$

and the restrictions of κ to F^0, F^+, F^- are zero, positive definite, negative definite, respectively.

Similar definitions will be imposed for Lagrangian vector subspaces of $\mathbb{C}^{2n}$.

Obviously, $F^0 = F \cap \bar{F}$, and F is real if $F = \bar{F}$. In the following, we only consider positive polarizations F; then F is homogeneous of type $(f^0, f^+, 0)$ if $F \cap \bar{F}$ is a vector bundle, i. e. of constant rank. Then recall [22]

3.2. Proposition:

For any positive Lagrangian vector subbundle $F \subset T^{\mathbb{C}}M$

(73) $$c_1(TM,\omega) = c_1(F)$$

holds.

(If negative polarizations are considered, some signs have to be changed, e.g. in (73).)

By complexification of its elements, $Sp(2n,\mathbb{R})$ operates on $\mathbb{C}^{2n}$, thus also on the set of Lagrangian vector subspaces of $\mathbb{C}^{2n}$. Since the operation of $Sp(2n,\mathbb{R})$ on $\mathbb{C}^{2n}$ commutes with complex conjugation, this operation preserves the type of Lagrangian vector subspaces.

3.3. Proposition:

The set of Lagrangian vector subspaces of $\mathbb{C}^{2n}$ of some fixed type is a homogeneous space of $Sp(2n,\mathbb{R})$.

Let $\mathbb{F} \subset \mathbb{C}^{2n}$ be some Lagrangian vector subspace of type $(f^0, f^+, 0)$. Its isotropy group will be denoted by $Sp_{\mathbb{F}}$. The latter is a semi-direct product of two subbundle groups.

Choose $\mathbb{F}$ such that $\mathbb{F}^0 = \mathbb{C}^{f^0}$, $\mathbb{F}^+ \oplus \overline{\mathbb{F}^+} = \mathbb{C}^{2f^+}$, and a direct complement of $\mathbb{F} + \overline{\mathbb{F}}$ in $\mathbb{C}^{2n}$, which will be identified with $\mathbb{C}^{f^0}$. In this decomposition, the subgroups consist of the matrices

$$(74) \qquad \begin{pmatrix} A & 0 & 0 \\ 0 & C & 0 \\ 0 & 0 & (A^t)^{-1} \end{pmatrix} \quad \text{resp.} \quad \begin{pmatrix} 1 & B & D \\ 0 & 1 & E \\ 0 & 0 & 1 \end{pmatrix}$$

subject to the conditions

$$(75) \qquad A \in GL(f^0,\mathbb{R}) \quad , \quad C \in U(f^+) \quad ,$$

$$(76) \qquad E = -jB^t \quad , \quad BE = D - D^t \quad ,$$

where j is the canonical complex structure on $\mathbb{R}^{2f^+}$, and $.^t$ denotes **transposition**. The second subgroup is **obviously nilpotent**. It is trivial when $\mathbb{F}$ is positive definite. The first subgroup operates on the second one. It equals $GL(n,\mathbb{R})$ when $\mathbb{F}$ is real, and equals $U(n)$ when F is positive definite.

Since $Sp_{\mathbb{F}}$ by definition leaves $\mathbb{F}$ stable, restriction to $\mathbb{F}$ defines a group morphism

$$(77) \qquad \text{ind}: \; Sp_{\mathbb{F}} \longrightarrow GL(\mathbb{F},\mathbb{C}) \quad .$$

3.4. Theorem:

Lagrangian vector subbundles $F \subset T^{\mathbb{C}}M$ which are homogeneous of type $(f^0,f^+,0)$ bijectively correspond to reductions P_F of the symplectic frame bundle $P(TM,\omega)$ from structure group $Sp(2n,\mathbb{R})$ to $Sp_{\mathbb{F}}$.

Moreover, the vector bundle $P_F \times_{\mathrm{ind}} \mathbb{F}$ is naturally isomorphic to F, while the principal $GL(\mathbb{F},\mathbb{C})$-bundle $P_F \times_{\mathrm{ind}} GL(\mathbb{F},\mathbb{C})$ is naturally isomorphic to the (whole) frame bundle $P(F)$ of F. This also implies the existence of a natural isomorphism between the complex line bundles

(78) $$\Lambda^n F \cong P_F \times_{\det \circ \mathrm{ind}} \mathbb{C} \quad .$$

Now the bilagrangian connection ∇ associated to F and G determines a principal connection on the symplectic frame bundle

(79) $$\alpha : T^{\mathbb{C}} P(TM,\omega) \longrightarrow Sp(2n,\mathbb{R})$$

which, in virtue of (61), induces a principal connection on P_F

(80) $$\alpha_F : T^{\mathbb{C}} P_F \longrightarrow L\,Sp_{\mathbb{F}} \quad .$$

Next, defining $Mp^c_{\mathbb{F}} := \rho^{-1}(Sp_{\mathbb{F}}) \subset Mp^c(2n,\mathbb{R})$, there is a commutative diagram of Lie groups with exact rows

(81) $$\begin{array}{ccccccccc} 0 & \longrightarrow & U(1) & \hookrightarrow & Mp^c_{\mathbb{F}} & \xrightarrow{\rho_{\mathbb{F}}} & Sp_{\mathbb{F}} & \longrightarrow & 0 \\ & & \downarrow & & \downarrow & & \downarrow & & \\ 0 & \longrightarrow & U(1) & \hookrightarrow & Mp^c(2n,\mathbb{R}) & \xrightarrow{\rho} & Sp(2n,\mathbb{R}) & \longrightarrow & 0 \end{array}$$

For $Mp_{\mathbb{F}} := \sigma^{-1}(Sp_{\mathbb{F}}) \subset Mp(2n,\mathbb{R})$ we get a similar diagram.

Then the groups $Mp^c_{\mathbb{F}}$, $Mp_{\mathbb{F}}$, $Sp_{\mathbb{F}}$ are related to each other in precisely the same way as $Mp^c(2n,\mathbb{R})$, $Mp(2n,\mathbb{R})$, $Sp(2n,\mathbb{R})$. In particular, there is an analogue of diagram (18), the morphisms being denoted as in (18) with an additional subscript $\mathbb{F}$. Also we may identify

(82) $$Mp^c_{\mathbb{F}} = U(1) \times_{\mathbb{Z}_2} Mp_{\mathbb{F}} \quad ,$$

(83) $$LMp^c_{\mathbb{F}} = i\mathbb{R} \oplus LSp_{\mathbb{F}} \quad .$$

Denote by $\mu_{\mathbb{F}} : LMp^c_F \longrightarrow i\mathbb{R}$ the first projection.

By analogy to the procedure for the structure groups, we define $\tilde{P}_F := \tilde{\rho}^{-1}(P_F) \subset \tilde{P}$, getting a commutative diagram of principal bundles

(84) $$\begin{array}{ccc} \tilde{P}_F & \hookrightarrow & \tilde{P} \\ {\scriptstyle \tilde{\rho}_F}\downarrow & & \downarrow{\scriptstyle \tilde{\rho}} \\ P_F & \hookrightarrow & P \end{array} \quad .$$

Now the μ-pseudoconnection $\tilde{\gamma}$ on $\tilde{P}$ restricts to a $\mu_{\mathbb{F}}$-pseudoconnection $\tilde{\gamma}_F$ on $\tilde{P}_F$, which then can be augmented to yield an ordinary connection as follows:

$$(85)\qquad \begin{aligned} \tilde{\beta}_F &: T^{\mathbb{C}}\tilde{P}_F \longrightarrow LMp_{\mathbb{F}}^{c} \\ \tilde{\beta}_F &:= \tilde{\gamma}_F + \alpha_F \circ T\tilde{\zeta}_F \quad . \end{aligned}$$

Having completed the procedure of reducing the structure group, we turn to the association of a complex line bundle with connection to $(\tilde{P}_F, \tilde{\beta}_F)$. To this aim, we first have to define a suitable one-dimensional representation of $Mp_{\mathbb{F}}^{c}$.

3.5. Theorem:

There exists a unique group morphism χ making the diagram with exact rows

$$(86)\qquad \begin{array}{ccccccccc} 0 & \longrightarrow & \mathbb{Z}_2 & \hookrightarrow & Mp_{\mathbb{F}} & \xrightarrow{\sigma_{\mathbb{F}}} & Sp_{\mathbb{F}} & \longrightarrow & 0 \\ & & \downarrow & & \chi\downarrow & & \downarrow & \text{det} \circ \text{ind} & \\ 0 & \longrightarrow & \mathbb{Z}_2 & \longrightarrow & \mathring{\mathbb{C}} & \xrightarrow{\cdot^{2}} & \mathring{\mathbb{C}} & \longrightarrow & 0 \end{array}$$

commutative. Moreover, χ satisfies

$$(87)\qquad \overline{\chi(\hat{g})}\cdot\chi(g) = |\det\circ\operatorname{ind}\circ\sigma_{\mathbb{F}}(\hat{g})| \qquad (\hat{g}\in Mp_{\mathbb{F}}) \ .$$

Due to (82), the desired representation now can be defined as follows:

$$(88)\qquad \begin{aligned} \chi^{c} : Mp_{\mathbb{F}}^{c} &\longrightarrow \mathring{\mathbb{C}} \\ [c,\hat{g}] &\longrightarrow c\cdot\chi(\hat{g}) . \end{aligned}$$

This immediately implies

$$(89)\qquad (\chi^{c}(\tilde{g}))^{2} = \eta_{\mathbb{F}}(\tilde{g})\cdot(\det\circ\operatorname{ind}\circ\rho_{\mathbb{F}}(\tilde{g})) \qquad (\tilde{g}\in Mp_{\mathbb{F}}^{c}),$$

due to commutativity of (86), and (87) shows that

$$(90)\qquad \overline{\chi^{c}(\tilde{g})}\cdot\chi^{c}(\tilde{g}) = |\det\circ\operatorname{ind}\circ\rho_{\mathbb{F}}(\tilde{g})| \quad (\tilde{g}\in Mp_{\mathbb{F}}^{c}).$$

Thus consider the complex line bundle

$$(91)\qquad Q := \hat{P}_F \times_{\chi^{c}} \mathbb{C},$$

and the linear connection ${}^{Q}\nabla$ on Q associated to $\tilde{\beta}_F$.

3.6. Theorem:

The connection ${}^{Q}\nabla$ satisfies

$$(92)\qquad \operatorname{curv}{}^{Q}\nabla = i\omega + \frac{1}{2}\operatorname{tr}(\operatorname{curv}\nabla|F)$$

and the first (real) Chern class of Q is given by (70).

Proof: Both relations can be deduced from relation (89). Finally, in virtue of (90), and since the bundle of densities of weight 1 over $F \cap \bar{F}$ up to a natural isomorphism is

(93) $$|\lambda|_1 (F \cap \bar{F}) = P(F) \times_{|\det|} \mathbb{C} ,$$

we obtain a sesquilinear bundle morphism (compatible with the connections)

(94) $$\langle \cdot , \cdot \rangle \ : \ Q \times_M Q \longrightarrow |\Lambda|_1 (F \cap \bar{F}),$$

the desired (local) scalar product. It is an ordinary hermitian structure on Q when F is positive definite.

3.7. Definition:

$(Q, {}^Q\nabla, \langle .,. \rangle)$ is called the quantum bundle corresponding to the prequantum $Mp^c(2n,\mathbb{R})$-bundle $(\hat{P}, \tilde{\zeta}, \tilde{\gamma})$, the polarization F and the symplectic connection ∇.

This assignation of quantum bundles to prequantum $Mp^c(2n,\mathbb{R})$-bundles (polarizations F, G being fixed) is compatible with the respective equivalence relations and with the operations of $H^1(M,U(1))$ on equivalence classes, when F is positive definite. Moreover, inequivalent prequantum $Mp^c(2n,\mathbb{R})$-bundles determine inequivalent quantum bundles.

We stress that there is a natural isomorphism $Q = L \otimes \Lambda_{1/2}F$, when a KS-prequantum bundle $(\dot{L}, \gamma)$ and a metaplectic frame bundle $(\hat{P}(TM, \omega), \hat{\sigma})$ exist. In this case, we can define a principal $Mp_{\mathbb{F}}$-bundle $\hat{P}_F := \sigma^{-1}(PF) \subset \hat{P}(TM, \omega)$, and the half-form bundle then is given by $\Lambda_{1/2}F := \hat{P}_F \times_\chi \mathbb{C}$. The connection ${}^Q\nabla$ and the scalar product are also obtained by tensoring the corresponding objects on L and $\Lambda_{1/2}F$.

Even in this case, using the bilagrangian connection ∇ has the advantage of yielding an ordinary connection on $\Lambda_{1/2}F$ and thus on $L \otimes \Lambda_{1/2}F$. Conversely, if we do not wish Q to be equipped with an ordinary connection (e.g. since the latter depends on the auxiliary polarization G), similar constructions as those leading to our quantum bundles can be performed with ∇, α, α_F, $\tilde{\beta}_F$, and ${}^Q\nabla$ being only partial connections along F.

We conclude this section by comparing our quantum bundles with those of Czyz. Therefore, let (M, ω, J) be a compact Kähler manifold. Czyz [7], [8] obtains his quantum bundles via the harmonic representative of the cohomology class at the right hand side of (70). In more detail, there is a unique harmonic 2-form ω_{eff} satisfying

$$(95) \qquad \left[-\frac{1}{2\pi}\omega_{eff}\right] = \left[-\frac{1}{2\pi}\omega\right] + \frac{1}{2}c_1(TM,\omega) .$$

3.8. Definition:

A Czyz quantum bundle is a KS-prequantum bundle over ω_{eff}. (We should add that Czyz also considers more general polarizations F, where M and the leaves of $(F \cap \overline{F}) \cap TM$ need not necessarily be compact. For the real part F^O of F, he essentially refers to the Kostant-Souriau theory.)

To apply our approach, consider the polarizations $T^{1,0}M$ and $T^{0,1}M$. In virtue of (67), we get another representative for the right hand side of (95), namely $-\frac{1}{2\pi}\omega + \frac{1}{2}\left(-\frac{1}{2\pi}\kappa\right)$. Due to theorem 3.6. and the definition (66) of κ, the curvature of the connections on our quantum bundles is equal to $i\left(\omega + \frac{1}{2}\kappa\right)$. In addition, the scalar products (94) are hermitian structures compatible with the connections, implying that our quantum bundles in the special case of interest are KS-prequantum bundles over $\omega + \frac{1}{2}\kappa$.

Now suppose κ to be harmonic, then $\omega + \frac{1}{2}\kappa$ is harmonic, too, hence

$$(96) \qquad \omega_{eff} = \omega + \frac{1}{2}\kappa .$$

Thus we finally obtain

3.9. Proposition:

On a compact Kähler manifold with locally constant scalar curvature, the set of equivalence classes of Czyz quantum bundles coincides with the set of equivalence classes of quantum bundles according to definition 3.7. (with respect to all equivalence classes of prequantum $Mp^c(2n,\mathbb{R})$-bundles).

4. Quantization of dynamical variables according to ordering rules

Throughout this section, we consider a quantum bundle $(Q, {}^Q\nabla, \langle .,. \rangle)$ satisfying (92). We may forget how it has been constructed, but we still use the polarizations F and G, and the bilagrangian connection ∇. The representation space now is obtained from the sheaf $\underline{Q}_F$ of germs of sections in Q, which are covariant constant along F. We restrict ourselves to construct operators on this representation space which are localizable in phase space.

Consider functions on phase space depending polynomially on one set of canonical variables (associated to F) and in an arbitrary manner on the others. We are going to define maps from these functions to differential operators on Q, which induce endomorphisms of the sheaf $\underline{Q}_F$. There is no unique way to get such a map. To select one, it is necessary to have a prescription of ordering multiplication operators and differentiations, quite similar as in more conventional quantum mechanics.

Hence we first consider ordering rules for quantum mechanics over $M = \mathbb{R}^{2n}$, using the Schrödinger (position) representation, and give some important examples. Having connections on $T^{\mathbb{C}}M$ and Q to our disposition, these rules may be generalized to yield maps from the above special type of functions to differential operators on Q. In general, it is not easy to compute whether the differential operators in the image leave the sheaf $\underline{Q}_F$ stable.

Let $M = \mathbb{R}^{2n}, Q = M\times\mathbb{C}, F = \left\{\frac{\partial}{\partial p_i} \mid i=1,..,n\right\}$ and $G = \left\{\frac{\partial}{\partial q_i} \mid i=1,..,n\right\}$ with the usual position and momentum variables q_i, p_i. Then the associated bilagrangian connection ∇ turns out to be the standard connection. Since curv $\nabla = 0$, relation (92) states that curv ${}^Q\nabla = i\omega$. As usual, this relation is satisfied by setting ${}^Q\nabla = d - i\sum_i p_i dq_i$. We shall later refer to these choices as to the <u>flat standard</u> situation.

Consider the functions on $\mathbb{R}^{2n}$ of the form

$$(97) \qquad f = \sum_{0\le|i|\le h} p^i \Phi_i \ ,$$

where the Φ_i are the functions depending only on the $q_1,\dots,q_n$, but not on the $p_1,\dots,p_n$ and $p^i := p_1^{i_1}\dots p_n^{i_n}$ for a multi-index $i \in (\mathbb{N}_o)^n$. $h\in\mathbb{N}_o$ is called the <u>order of f</u>.

Denote the set of these functions by $C^h_F(\mathbb{R}^{2n})$. Remark that $C^o_F(\mathbb{R}^{2n}) \subset \dots \subset C^h_F(\mathbb{R}^{2n}) \subset C^{h+1}_F(\mathbb{R}^{2n})\dots$ and that $C^o_F(\mathbb{R}^{2n})$ is the set of all functions depending only on the position variables q_i.

In the following, all indices not specified explicitly will be multi-indices; note the definitions $\binom{j}{k} := \binom{j_1}{k_1} \dots \binom{j_n}{k_n}$, $\Pi(j) := j_1\dots j_n$ and $\underline{1} := (1,\dots,1)$.

<u>4.1. Definition:</u>

<u>An ordering rule</u> (in the above context) is a map from $\bigcup_{h=0}^{\infty} C^h_F(\mathbb{R}^{2n})$ to the set of differential operators (on the trivial line bundle

over R^n) of the following form, f being given by (97)

$$(98)\quad f \longrightarrow \hat{f} := \sum_{0 \leq |j+k| \leq h} (-i)^{|j+k|} c_{|j|} \binom{j+k}{j} \left(\frac{\partial^{|j|} \phi_{j+k}}{\partial q^j} \right)_* \frac{\partial^{|k|}}{\partial q^k} ,$$

according to some sequence $c \in \mathbb{R}^{\mathbb{N}_o}$, called that of <u>fundamental coefficients</u> of the ordering rule.

The ordering rule is said to be <u>normalized</u> if $c_o = 1$, and <u>symmetric</u> (on $C_F^h(\mathbb{R}^{2n})$) iff $\hat{f}$ is formally self-adjoint for every real-valued $f \in C_F^h(\mathbb{R}^{2n})$.

For an arbitrary normalized ordering rule, we have

$$(99)\qquad \hat{q}^i = (q^i)_* , \qquad \hat{p}^i = (-i)^{|i|} \frac{\partial^{|i|}}{\partial q^i} ,$$

thus we only consider normalized ordering rules in the sequel. Using the Leibniz formula, (98) may be written with another family of coefficients $b \in \mathbb{R}^{(\mathbb{N}_o)^n \times (\mathbb{N}_o)^n}$ in the form

$$(100)\qquad f \longrightarrow \hat{f} = \sum_{0 \leq |s+t| \leq h} (-i)^{|s+t|} b_s^{s+t} \binom{s+t}{s} \frac{\partial^{|s|}}{\partial q^s} (\Phi_{s+t})_* \frac{\partial^{|t|}}{\partial q^t} .$$

The families of coefficients then correspond via

$$(101)\qquad c_{|j|} = \sum_{0 \leq r \leq k} b_{j+r}^{j+k} \binom{k}{r}, \quad b_s^{s+t} = \sum_{0 \leq u \leq t} (-1)^{|u|} c_{|s+u|} \binom{t}{u} .$$

Note that the ordering rules do not exhaust the possible maps (100) with arbitrary coefficients b, due to the special requirements on the fundamental coefficients c. Obviously, an ordering rule is symmetric on $C_F^h(\mathbb{R}^{2n})$ if $b_s^{s+t} = b_t^{s+t}$ $(0 \leq |s+t| \leq h)$.

4.2. Examples:

Some more or less familiar (normalized) ordering rules will be listed in the following table:

Names	$c_{\|j\|}$	b_s^{s+t}	Symmetric on $C_F^h(\mathbb{R}^{2n})$ for	References
standard normal	$\delta_{\|j\|,0}$	$\delta_{s,0}$	$h = 0$	[1], [5], [6], [29]
antistandard antinormal	1	$\delta_{s,s+t}$	$h = 0$	[1], [5], [6], [29]
Rivier symmetrization	$\frac{1}{2}\delta_{\|j\|,0} + \frac{1}{2}$	$\frac{1}{2}\delta_{s,0} + \frac{1}{2}\delta_{s,s+t}$	$h \in \mathbb{N}_0$	[5], [6], [29]
Born-Jordan (extended)	$\frac{1}{\pi(j+\underline{1})}$	$\frac{1}{\pi(s+t+\underline{1})}\binom{s+t}{s}^{-1}$	$h \in \mathbb{N}_0$	[6], [29]
Weyl (-McCoy) symmetrization	$2^{-\|j\|}$	$2^{-\|s+t\|}$	$h \in \mathbb{N}_0$	[1], [5] [6], [29]
Kostant	$1 - \frac{1}{2}\|j\|$	see (101)	$h \leq 2$	[18]

Apart from the first two ones, these ordering rules agree with each other on $C_F^1(\mathbb{R}^{2n})$. In fact, we have

<u>4.3. Lemma:</u>

A normalized ordering rule, which is symmetric on $C_F^1(\mathbb{R}^{2n})$, is uniquely determined on $C_F^1(\mathbb{R}^{2n})$, and its fundamental coefficients satisfy

(102) $$c_0 = 1 \quad , \quad c_1 = \frac{1}{2} \ .$$

<u>Proof:</u> Since the fundamental coefficients $c_{|j|}$ only depend on $|j|$, this is an immediate consequence of (101).

Now we leave the flat standard situation, considering arbitrary data as indicated in the beginning of this section. First, the above set of functions $C_F^h(\mathbb{R}^{2n})$ will be generalized by a definition due to Kostant [18]:

<u>4.4. Definition:</u>

Let $\underline{C}_F^0 \subset \underline{C}$ be the subsheaf of (germs of) functions over M given by

$$\underline{C}_F^0 := \left\{ g \in \underline{C}^\infty \mid \bigwedge_{X \in \Gamma(F)} \mathbb{L}_X g = 0 \right\} .$$

For each $k \in N_0$, define a subsheaf $\underline{C}_F^k \subset \underline{C}^\infty$ by

$$\underline{C}_F^k \;:=\; \Big\{ f \in \underline{C}^\infty \;\Big|\; \bigwedge_{g_1,\dots,g_{k+1} \in \underline{C}_F^0} \mathrm{ad}\, g_1 \dots \mathrm{ad}\, g_{k+1}(f) \;=\; 0 \Big\} \;,$$

where ad refers to the Poisson bracket, and let finally

$$C_F^k(M) \;:=\; \Gamma\,(M, \underline{C}_F^k) \quad .$$

The differential operators on Q assigned to a function $f \in C_F^h$ (M) now will be the composition of several maps. Denote by

$$\check{\nabla}^{j+\ell} \;:\; C^\infty(M) \longrightarrow \Gamma(\vee^{j+\ell}(T^{\mathbb{C}}M)^*) \quad ,$$

$$\check{\nabla}^{j} \;:\; \Gamma(\vee^{j+\ell}G) \longrightarrow \Gamma(\vee^{j}(T^{\mathbb{C}}M)^* \otimes \vee^{j+\ell}G)$$

the $j+\ell$ -fold resp. j-fold symmetrized covariant derivatives induced by the bilagrangian connection ∇ $(j,\ell \in \mathbb{N}_o)$. Further, consider the maps

$$\vee^{j+\ell}(pr_G\, \omega^{\#}) \;:\; \vee^{j+\ell}(T^{\mathbb{C}}M)^* \longrightarrow \vee^{j+\ell}G \quad ,$$

$$\vee^{j} inj_G^* \otimes 1 \;:\; \vee^{j}(T^{\mathbb{C}}M)^* \otimes \vee^{j+\ell}G \longrightarrow \vee^{j}G^* \otimes \vee^{j+\ell}G \quad ,$$

where $inj_G : G \hookrightarrow T^{\mathbb{C}}M$ is the natural injection, finally let

$$\check{tr}^{j} \;:\; \vee^{j}G^* \otimes \vee^{j+\ell}G \longrightarrow \vee^{\ell}G$$

be the j-fold symmetrized contraction map. Using this, f determines a section

$$(103) \qquad X_f^{j,\ell} := \check{tr}^{j}((\vee^{j} inj_G^* \otimes 1)_* \circ \check{\nabla}^{j} \circ \vee^{j+\ell}(pr_G\, \omega^{\#}) \circ \check{\nabla}^{j+\ell} f) \in \Gamma(\vee^{\ell}G) .$$

At last, let

$$^{Q}\check{\nabla}^{\ell} \;:\; \Gamma(Q) \longrightarrow \Gamma(\vee^{\ell}(T^{\mathbb{C}}M)^* \otimes Q)$$

be the ℓ-fold symmetrized covariant derivative according to the connection $^{Q}\nabla$ on Q. Then the desired map is obtained by

$$(104) \qquad f \longrightarrow \hat{f} \;:=\; \sum_{0 \le |j+\ell| \le h} (-i)^{|j+\ell|}\, c_{|j|}\, \frac{1}{j!\,\ell!}\; X_f^{j,\ell} \lrcorner\, {}^{Q}\check{\nabla}^{\ell}$$

In general, the operators in the image of this map are not endomorphisms of the sheaf $\underline{Q}_F$. We are able to prove this only by imposing some restrictions, either on the polarizations F,G or on the function f.

4.5. Theorem:

The map given by (103), (104) coincides with (98) in the flat standard case.

In the general case, assume that F and G are Heisenberg related. Then (103), (104) define a map from $C_F^h(M)$ to the set of endomorphisms of $\underline{Q}_F$ for arbitrary $h \in \mathbb{N}_o$.

By inserting the fundamental coefficients $c_{|j|} = 1 - \frac{1}{2}|j|$ of the Kostant ordering rule, the above theorem implies a result of Kostant [18]. Note that we do not make any use of intertwining operators to prove theorem 4.5.

4.6. Theorem:

For arbitrary polarizations F,G, (103) and (104) define a map from $C_F^1(M)$ to the set of endomorphisms of $\underline{Q}_F$ if $c_o = 1$ and $c_1 = \frac{1}{2}$.

Proof: We only remark that the result heavily relies on relation (92).

This shows the importance of relation (92) concerning the curvature of the connection ${}^Q\nabla$ on a quantum bundle Q. Thus we have a new motivation for the dogma (13), (70) on the first Chern class of Q, which is a consequence of (92) when (M, ω, J) is a Kähler manifold and $F = T^{1,0}M$, $G = T^{0,1}M$.

Acknowledgements:

I want to express my gratitude to R. Schrader, M. Forger, D. Krausser and H. Scheerer for helpful discussions. Also, I thank the Studienstiftung des deutschen Volkes for financial support.

References:

1. G.S. Agarwal, E. Wolf Calculus for functions of noncommuting operators and general phase space methods in quantum mechanics I Phys. Rev. D 2 (1970) 2161-2186
2. M.F. Atiyah, R. Bott, A. Shapiro Clifford modules, Topology 3, Suppl. 1 (1964) 3-38
3. R. Blattner Quantization and representation theory in: Harmonic analysis and homogeneous spaces, ed. C.C. Moore, AMS Proc. Symp. Pure Math. 26 (1973)
4. R. Bott in: Bott, Gitler, James Lectures on algebraic and differential topology, Springer Lect. Notes in Math. 279 (1972)
5. K.E. Cahill, R.J. Glauber Ordered expansions in boson amplitude operators, Phys. Rev. 177 (1969) 1857-1881
6. L. Cohen Generalized phase space distribution functions Journ. Math. Phys. 7 (1966) 781-786
7. J. Czyz On some approach to geometric quantization in: Diff. geom. meth. in math. phys. II, Proc. Bonn 1977 Springer Lect. Notes in Math. 676 (1978)

8. J. Czyz On geometric quantization and its connections with the Maslov theory, Univ. of Warsaw reprint (1977), to app. in Rep.Math.Phys.

9. M. Forger, H. Hess Universal metaplectic structures and geometric quantization, FU Berlin reprint (1978), to app. in Comm. Math. Phys.

10. J. Fraenkel Cohomologie non abélienne et espaces fibrés Bull. Soc. Math. France 85 (1957) 135-220

11. K. Gawedzki Fourier-like kernels in geometric quantization (thesis) Diss. Math. 128 (1976) 1-83

12. W. Greub, H.-R. Petry On the lifting of structure groups in: Diff. geom. meth. in math. phys. II, Proc. Bonn 1977 Springer Lect. Notes in Math. 676 (1978)

13. H. Hess forthcoming thesis (FU Berlin)

14. H. Hess, D. Krausser Lifting classes of principal bundles I FU/TU Berlin reprint (1977), to app. in Rep. Math. Phys.

15. F.W. Kamber, P. Tondeur Foliated bundles and characteristic classes Springer Lect. Notes in Math. 493 (1975)

16. S. Kobayashi, T. Nagano On a fundamental theorem of Weyl-Cartan on G-structures, Journ. Math. Soc. Japan 17 (1965) 84-101

17. S. Kobayashi, K. Nomizu Foundations of differential geometry II Interscience (1969)

18. B. Kostant On the definition of quantization in: Colloq. Int. du CNRS, Géom. sympl. et phys. math., Aix-en-Provence 1974, ed. CNRS (1976)

19. B. Kostant Symplectic spinors in: Conv. di geom. simpl. e fisica mat., INDAM Rome 1973, Symp. Math. 14 , Academic Press (1974)

20. B. Kostant Quantization and unitary representations in: Lect. in mod. analysis and appl. III, ed. C.T. Taam, Springer Lect. Notes in Math. 170 (1970)

21. J.H. Rawnsley On the cohomology groups of a polarization and diagonal quantization, Trans. Am. Math. Soc. 230 (1977) 235-255

22. J.H. Rawnsley On the pairing of polarizations Comm. Math. Phys. 58 (1978) 1-8

23. P. Renouard Variétés symplectiques et quantification (thesis) Paris, Orsay (1969)

24. I.E. Segal Transforms for operators and symplectic automorphisms over a locally compact abelian group, Mad. Scand. 13 (1963) 31-43

25. D. Shale Linear symmetries of free boson fields
Trans. Am. Math. Soc. 103 (1962) 149-167

26. H.J. Simms Geometric quantization of energy levels in the Kepler problem, in: Conv. di geom. simpl. e fisica mat., INDAM Rome 1973, Symp. Math. 14, Academic Press (1974)

27. D.J. Simms Metalinear structures and a geometric quantization of the harmonic oscillator, in: Colloq- Int. du CNRS, Géom. sympl. et phys. math., Aix-en-Provence 1974, ed. CNRS (1976)

28. J.-M. Souriau Structure des systèmes dynamiques
Dunod, Paris (1970)

29. J. Tolar Quantization and deformation theory In: Group theoretical methods in physics, Proc. Tübingen 1977, Springer Lect. Notes in Phys. 79 (1978)

30. I. Vaisman Cohomology and differential forms Dekker (1973)

31. R.O. Wells (jr) Differential anal sis on complex manifolds
Prentice-Hall (1973)

32. A. Weil Sur certains groupes d'opérateurs unitaires
Acta Math. 111 (1964) 143-211

33. A. Weil Variétés kähleriennes Herrmann, Paris (1958)

34. K. Yano Differential geometry on complex and almost complex spaces Pergamon Press (1965)

Further applications of geometric quantization

by

Jedrzej Śniatycki

Department of Mathematics and Statistics

The University of Calgary

Calgary, Alberta

Abstract.

I. Galilei invariant quantization

The geometric quantization scheme enables one to quantize time dependent dynamics without introducing any reference frame. The resulting theory is equivalent to the one described by a time-dependent Schroedinger equation.

II. Quantization of non-relativistic dynamics with spin

Geometric quantization of a classical model of a particle with spin in an external electromagnetic field yields the Pauli theory of spin.

For details see ref. [6] .

References

1) B. Kostant, "Line Bundles and Prequantized Schroedinger Equation", Colloquium on Group Theoretical Methods in Physics, C.N.R.S., Marseille, 1972. Proceedings, pp. IV 1-22.

2) P. Renuard, Variétés Symplectiques et Quantification, Thèse, Orsay, 1969.

3) D.J. Simms, "On the Schroedinger equation given by geometric quantization," Lecture Notes in Mathematics, Vl. 676, pp. 351 - 356, Springer Verlag, Berlin, 1978.

4) J.-M. Souriau, Structure des Systèmes Dynamiques, Dunod, Paris, 1970.

5) J. Sniatycki and W.M. Tulczyjew, "Canonical Formulation of Newtonian Dynamics, " Ann. Inst. Henri Poincaré, vol. 16 (1972) pp. 23-27.

6) J. Sniatycki, Geometric Quantization and Quantum Mechanics, Springer Verlag, New York, 1980.

General Vector Field Representations of Local Heisenberg Systems

F.B. Pasemann

Institut für Theoretische Physik
TU Clausthal
3392, Clausthal, Federal Republic of Germany

Abstract:

A quantization procedure is proposed starting with the Lie algebra $\mathbb{G}$ of infinitesimal symmetries of a system and a $\mathbb{G}$-action on a principle bundle $\mathcal{P}$. For quasi-complete $\mathbb{G}$-actions the constructed vector field operators are essentially skew adjoint and can be interpreted as canonical momentum observables of a local Heisenberg system. Integrability of general vector field representations of local Heisenberg systems to unitary representations of corresponding Heisenberg systems is discussed.

1. Introduction

There are several quantization methods which start from a given symmetry of a system. As an example, systems of imprimitivity [1] are induced by a given transitive group action on a manifold. Since most of these quantization procedures are based on global symmetries, they are not appropriate if one is interested in the description of phenomena, depending on local symmetries only. They are also unsuitable for the description of systems in a configuration space which might not be a "simple" manifold (like $\mathbb{R}^n$) but a manifold with singularities, "forbidden domains", etc.

The aim of this paper is to introduce slight generalizations of these methods into two directions. We start with the action of a finite dimensional Lie algebra $\mathbb{G}$, representing the infinitesimal symmetry of a system, on a principal bundle $\mathcal{P}$. Since non trivial base manifolds are allowed, on one hand local symmetries of the system, which are inextendable to global ones, may be considered. On the other hand, since the formalism is adapted to bundle structures, the description of fields is changed from functions on manifolds to sections in bundles, thus giving the possibility to incorporate in a clear geometric way the internal symmetries of a system into the quantization procedure.

As an example of this method general vector field representations [2,3] of systems, called "local Heisenberg systems" by Snellman [4], are discussed. These systems have been introduced by Snellman as a generalization of imprimitivity systems [1] or, equivalently, of Heisenberg systems [5,6] . Here, as usual, the so called vector field operators of the Lie algebra representation correspond to the momentum observables associated with the infinitesimal symmetries of the system, whereas the functions with compact support on the base manifold M represent the canonical localization observables.

In section 2 there is a brief review of the construction of general vector field representations of a finite dimensional Lie algebra $\mathbb{G}$. These representations depend an a $\mathbb{G}$-action on a principle bundle $\mathcal{P}$ over M with structure group K, on a representation ς of this structure group K and on the choice of a volume form Ω on M.

In section 3 general vector field representations of local Heisenberg systems are defined. The question, when a general vector field representation of a local Heisenberg system is integrable to a unitary representation of a corresponding Heisenberg system is answered by a proposition using an infinitesimal version of Palais' local G-actions [7].

We use the following notation:

$\mathcal{F}(M,F)$	the set of functions on M with values in a space F,
$\mathcal{F}_c(M)$	the set of functions on M with compact supporrt
$\mathfrak{X}(M)$	the $\mathcal{F}(M)$-module of vector fields on M,
$\theta(X)$	the Lie derivative with respect to X $\in \mathfrak{X}(M)$,
${}^X\varphi_t$	the local diffeomorphisms on M associated with X $\in \mathfrak{X}(M)$,
γ^*	the pull-back with respect to the map $\gamma: M \longrightarrow N$,

$Sec_c(\mathcal{H}, N)$ the $\mathcal{J}(M)$-module of sections in a vector bundle $\mathcal{H}$ over M with compact support in $N \subset M$.

2. General vector field representations

For the construction of general vector field representations (gvr) of a finite dimensional Lie algebra $\mathbb{G}$ we need the following ingredients [3]:

1.) An oriented manifold M with volume form Ω.
2.) A finite dimensional complex vector space F with hermitian inner product $\langle .,. \rangle$.
3.) A representation ρ of a compact connected Lie group K in F preserving the inner product $\langle .,. \rangle$.
4.) A principal bundle $\mathcal{P} = (P, \pi, M, K)$ over M with a ρ-associated vector bundle $\mathcal{H}^\rho = (H, \bar{\pi}, M, F)$, i.e. we have the commuting diagram

$$\begin{array}{ccc} P \times F & \xrightarrow{\quad q \quad} & H \\ \downarrow{\scriptstyle pr_1} & & \downarrow{\scriptstyle \bar{\pi}} \\ P & \xrightarrow[\quad \pi \quad]{} & M \end{array}$$

and a linear isomorphism $q_p: \quad F \longrightarrow H_{\pi(p)}$ on the fibres defined by $q_p(z) := q(p,z)$, $p \in P$, $z \in F$.

5.) A Hilbert space $\mathcal{L}^2(\mathcal{H}^\rho, \Omega)$ of sections in $\mathcal{H}^\rho$ with compact support, where the inner product is given by

$$(\sigma_1, \sigma_2) := \int_M \langle \sigma_1(m), \sigma_2(m) \rangle \Omega \quad , \quad \sigma_1, \sigma_2 \in Sec_c(\mathcal{H}^\rho).$$

6.) A $\mathbb{G}$-action $(\hat{\phi}, \phi)$ on $\mathcal{P}$, i.e. $\hat{\phi}, \phi$ are Lie algebra homomorphisms such that the following diagram commutes:

$$\begin{array}{ccc} & & \mathfrak{X}_K(P) \\ & \overset{\hat{\phi}}{\nearrow} & \downarrow{\scriptstyle \pi_*} \\ \mathbb{G} & \xrightarrow[\quad \phi \quad]{} & \mathfrak{X}(M) \end{array}$$

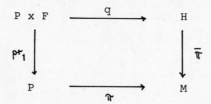

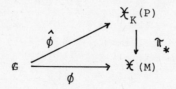

where $\mathfrak{X}_K(P)$ denotes the Lie algebra of K-invariant vector fields on P

Furthermore we need the notion of a quasi-complete $\mathbb{G}$-action and the definition of the determinant function of a vector field with respect to the given volume form Ω.

<u>Definition</u>: Let ${}^X\varphi$ denote the flow of the vector field $X \in \mathfrak{X}(M)$ and let $M_t^X \subset M$ be given s.t. ${}^X\varphi_t: M_t^X \longrightarrow M_{-t}^X$, $t \in \mathbb{R}$, is a diffeomorphism. Then $X \in \mathfrak{X}(M)$ is called <u>quasi-complete</u> iff the set $E_t^X := M - M_t^X$, $t \in \mathbb{R}$, is of measure zero with respect to the measure μ^Ω induced by Ω, [8].
A $\mathbb{G}$-action $(\hat{\phi}, \phi)$ on $\mathcal{P}$ is called <u>quasi-complete</u> iff $\phi(x) \in \mathfrak{X}(M)$, $x \in \mathbb{G}$, is quasi-complete.

It is just this property of quasi-completeness, which makes the whole method sensitive for some sort of singularity structure of the base manifold M. To get an impression of what quasi-completeness means, lets have a look at the following examples:

Let $M = \mathbb{R}^2 - \{\vec{0}\}$. Then

1.) the vector field $X = \frac{\partial}{\partial \varphi}$ is complete,

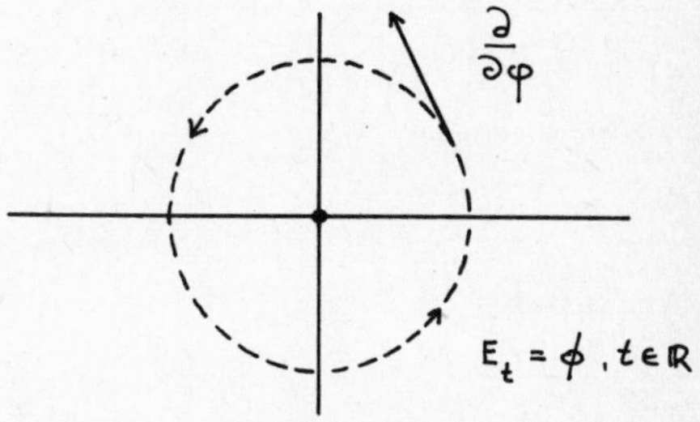

2.) the vector field $Y = \frac{\partial}{\partial x}$ is quasi-complete but not complete

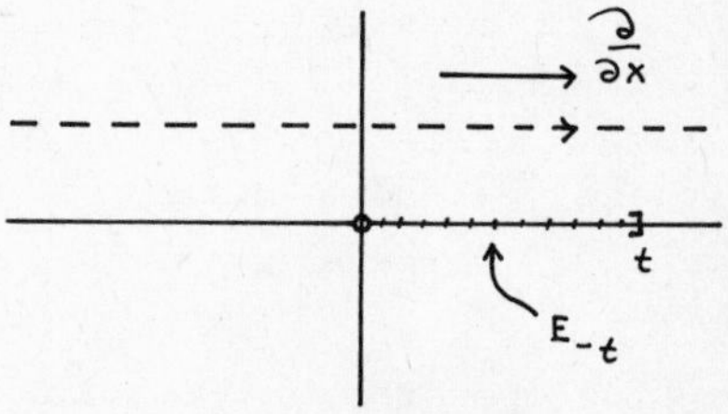

3.) the vector field $Z = \frac{\partial}{\partial r}$ is not complete and not quasi-complete.

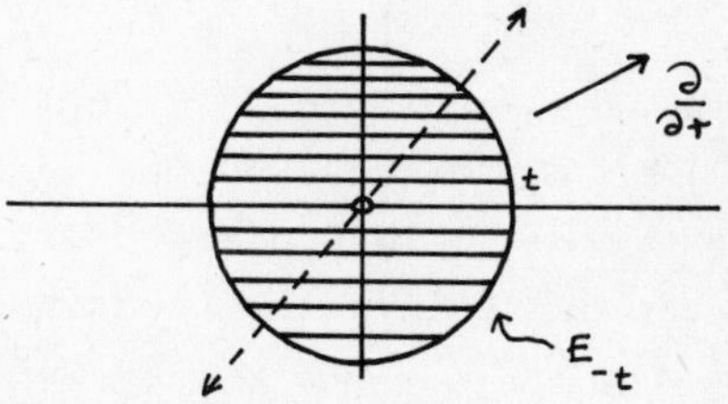

Now, as a last step, we have the following

<u>Definition</u>: The <u>determinant function</u> $\det_t^{\Omega} X$ of $X \in \mathfrak{X}(M)$ with respect to the volume form Ω on M is defined by

$$\det_t^{\Omega} X \cdot \Omega := {}^X\varphi_t^* \Omega$$

where ${}^X\varphi_t^*\Omega$ is the pull back of Ω under the local diffeomorphism ${}^X\varphi_t$. (We use the same symbol Ω for the volume form Ω and its restriction $\Omega | M_{-t}^X$ to M_{-t}^X).

We observe, that $\left.\frac{d}{dt}\right|_{t=0} \det_t^{\Omega} X$ equals the divergence $\operatorname{div}^{\Omega} X$ of $X \in \mathfrak{X}(M)$ with respect to Ω, [8].

A linear isomorphism $q^{\#}$: $\operatorname{Sec}(\mathcal{H}^{\rho}) \longrightarrow \mathcal{F}_K(P,F)$, where $\mathcal{F}_K(P,F)$ denotes the K-invariant functions on P with values in F, is given by

$$\sigma^{\#}(p) \;=\; (q^{\#}(\sigma))(p) \;:=\; q_p^{-1} \circ \sigma \circ \pi(p), \quad p \in P, \; \sigma \in \operatorname{Sec}(\mathcal{H}^{\rho}).$$

Using this isomorphism, we give the following definition of operators.

<u>Definition</u>: Let $(\hat{\phi},\phi)$ be a quasi-complete $\mathbb{G}$-action on $\mathcal{P}$.
The <u>vector field operator</u> D(x) on $\operatorname{Sec}_c(\mathcal{H}^{\rho}) \subset \mathcal{L}^2(\mathcal{H}^{\rho},\Omega)$ associated with $(\hat{\phi},\phi)$ is defined by

$$(D(x)\sigma)^{\#} := \Theta(\hat{\phi}(x))\sigma^{\#} + \tfrac{1}{2}(\pi^*\operatorname{div}^{\Omega}\phi(x))\,\sigma^{\#}, \quad x \in \mathbb{G}, \; \sigma \in \operatorname{Sec}_c(\mathcal{H}^{\rho}),$$

and the <u>flow operator</u> $U_t(x)$ on $\operatorname{Sec}_c(\mathcal{H}^{\rho}, M_t^X)$ associated with $(\hat{\phi},\phi)$ is defined by

$$(U_t(x)\sigma)^{\#}(p) := \begin{cases} (\pi^* \det_t^{\Omega}\phi(x))^{1/2} ({}^{\hat{\phi}(x)}\varphi_t^* \sigma^{\#})(p), & \pi(p) \in M_t^{\phi(x)} \\ 0, & \pi(p) \in E_t^{\phi(x)}. \end{cases}$$

For these operators associated with a given quasi-complete $\mathbb{G}$-action on $\mathcal{P}$ we have the following results [3] :

1. The flow operator $U_t(x)$, $x \in \mathbb{G}$, is isometric, and since $\operatorname{Sec}_c(\mathcal{H}^{\rho}, M_t^{\phi(x)})$, $t \in \mathbb{R}$, is dense in $\mathcal{L}^2(\mathcal{H}^{\rho},\Omega)$ for a quasi-complete $\mathbb{G}$-action on $\mathcal{P}$, $U_t(x)$ extends to an unitary operator on $\mathcal{L}^2(\mathcal{H}^{\rho},\Omega)$ denoted by the same symbol.
2. The vector field operator D(x), x $\mathbb{G}$, is essentially skew adjoint on the dense invariant domain $\operatorname{Sec}_c(\mathcal{H}^{\rho})$.
3. The map D: $x \longrightarrow D(x)$, $x \in \mathbb{G}$, is a Lie algebra homomorphism. It depends on the $\mathbb{G}$-action $(\hat{\phi},\phi)$, the representation ρ and the volume form Ω , i.e. $D = D(\hat{\phi}, \rho, \Omega)$.

Definition: The skew adjoint representation $D(\hat{\phi}, \varrho, \Omega)$ associated with a given quasi-complete $\mathbb{G}$-action $(\hat{\phi},\phi)$ on $\mathcal{P}$ is called a general vector field representation (gvr) of $\mathbb{G}$ in $\mathcal{L}^2(\mathcal{H}^\varrho,\Omega)$.

In [3] it was shown that two gvrs $D(\hat{\phi}, \varrho, \Omega)$, $D(\hat{\phi}, \varrho, \Omega')$ depending on different volume forms Ω and Ω' as well as two gvrs $D(\hat{\phi}, \varrho, \Omega)$, $D(\hat{\phi}, \varrho', \Omega)$ depending on two unitarily equivalent representations ϱ and ϱ' of the structure group K are unitarily equivalent.

We now want to apply these gvrs to local Heisenberg systems.

3. General vector field representations of local Heisenberg systems

Let us first recall the notions of Heisenberg systems and their representations [6].

Definition: A Heisenberg system (Hs) is a triple $(\mathbb{A},G,\varphi)$ where $\mathbb{A}$ is a $*$-algebra, G a Lie group and φ: $G \longrightarrow {}^*\mathrm{Aut}(\mathbb{A})$ a Lie group homomorphism.

Let $\mathfrak{H}$ be a Hilbert space and let $\mathbb{L}(\mathfrak{H})$ denote the set of linear operators in $\mathfrak{H}$. A representation (Q,U) of a Hs $(\mathbb{A},G,\varphi)$ is given by a $*$-algebra morphism $Q: \mathbb{A} \longrightarrow \mathbb{L}(\mathfrak{H})$ and a unitary representation U of G in $\mathfrak{H}$, such that

$$Q(\varphi_g(f)) = U(g)\, Q(f)\, U^{-1}(g), \quad g \in G,\ f \in \mathbb{A}.$$

Then, following Snellman [4], local Heisenberg systems and their representations are given by

Definition: A local Heisenberg system (lHs) is a triple $(\mathbb{A},\mathbb{G},\phi)$ with $\mathbb{A}$ a $*$-algebra, $\mathbb{G}$ a finite dimensional Lie algebra and ϕ: $\mathbb{G} \longrightarrow {}^*\mathrm{Der}(\mathbb{A})$ a Lie algebra homomorphism into the $*$-derivations of $\mathbb{A}$.

Let $\mathfrak{H}$ be a Hilbert space and let $\mathbb{L}(\mathfrak{H})$ denote the set of linear operators in $\mathfrak{H}$. A representation (Q,D) of a lHs $(\mathbb{A},\mathbb{G},\phi)$ is given by a Lie algebra homomorphism $D: \mathbb{G} \longrightarrow \mathbb{L}(\mathfrak{H})$ such that

$$(Q(f))^* = Q(f^*),$$
$$[D(x), Q(f)] = Q(\phi(x)f), \qquad f \in \mathbb{A},\ x \in \mathbb{G},$$

on some dense invariant domain $\Delta \subset \mathfrak{H}$.

In the following let us choose the algebra $\mathcal{J}_c(M)$ of real valued functions on M with compact support, provided with an appropriate family of semi-norms, so that $\mathcal{J}_c(M)$ is a subalgebra of the $*$-algebra of complex valued functions on M. Since the vector fields on M are $*$-derivations of the algebra $\mathcal{J}_c(M)$ we will consider lHs of the form

$(\mathcal{S}_c(M), \mathbb{G}, \phi)$ where $\phi\colon \mathbb{G} \longrightarrow \mathfrak{X}(M)$ is a representation of $\mathbb{G}$ by vector fields on M.

Furthermore let us consider a physical context, where there is given a principal bundle $\mathcal{P}$ over M, M representing for instance the configuration space of a system. Let there be given a quasi-complete $\mathbb{G}$-action $(\hat{\phi}, \phi)$ on $\mathcal{P}$ and let $\mathcal{H}^\rho$ be a vector bundle ρ-associated to $\mathcal{P}$. Let $\mathfrak{H} = \mathcal{L}^2(\mathcal{H}^\rho, \Omega)$ denote the Hilbert space of sections in $\mathcal{H}^\rho$, constructed as in the last section.

Then the map $Q\colon \mathcal{S}_c(M) \longrightarrow \mathbb{L}(\mathfrak{H})$ given by

$$Q(f)\sigma \ := \ f\cdot\sigma, \quad f \in \mathcal{S}_c(M), \ \sigma \in \mathcal{L}^2(\mathcal{H}^\rho, \Omega),$$

and the gvr $D = D(\phi, \rho, \Omega)$ of $\mathbb{G}$ associated with the $\mathbb{G}$-action $(\hat{\phi}, \phi)$ on $\mathcal{P}$ defines a representation (Q,D) of the lHs $(\mathcal{S}_c(M), \mathbb{G}, \phi)$ in $\mathcal{L}^2(\mathcal{H}^\rho, \Omega)$.

Proof: Essentially we have to show, that

$$[D(x), Q(f)] \ = \ Q(\phi(x)f), \quad x \in \mathbb{G}, \quad f \in \mathcal{S}_c(M).$$

Writing $\hat{X} := \hat{\phi}(x) \in \mathfrak{X}_K(P)$ and $X := \phi(x) \in \mathfrak{X}(M)$ and using the isomorphism $q^{\#}\colon \mathrm{Sec}_c(\mathcal{H}^\rho) \longrightarrow \mathcal{S}_K(P)$ we get for $\sigma \in \mathrm{Sec}_c(\mathcal{H}^\rho)$

$$\begin{aligned}(D(x)Q(f)\sigma)^{\#} \ &= \ \theta(\hat{X})((\pi^{*}f)\cdot\sigma^{\#}) + \tfrac{1}{2}\,\pi^{*}(f\cdot \mathrm{div}^{\Omega} X)\cdot\sigma^{\#} \\ &= \ (\pi^{*}f)\cdot\theta(\hat{X})\,\sigma^{\#} + \sigma^{\#}\theta(\hat{X})\,(\pi^{*}f) + \tfrac{1}{2}\,\pi^{*}(f\cdot \mathrm{div}^{\Omega} X)\cdot\sigma^{\#}\end{aligned}$$

where $\theta(\hat{X})$ denotes the Lie derivative with respect to the vector field $\hat{X} \in \mathfrak{X}(P)$. On the other hand we have

$$(Q(f)D(x)\sigma)^{\#} \ = \ (\pi^{*}f)\left[\theta(\hat{X})\sigma^{\#} + \tfrac{1}{2}\,\pi^{*}\mathrm{div}^{\Omega} X\cdot\sigma^{\#}\right] \ \Longrightarrow$$

$$([D(x), Q(f)]\sigma)^{\#} = \ \theta(\hat{X})\,(\pi^{*}f)\cdot\sigma^{\#}$$

The representation $(Q,D) \ = \ D(\hat{\phi}, \rho, \Omega)$ is called the gvr of the lHs $(\mathcal{S}_c(M), \mathbb{G}, \phi)$ in $\mathcal{L}^2(\mathcal{H}^\rho, \Omega)$ associated with the $\mathbb{G}$-action $(\hat{\phi}, \phi)$.

Since the map D is a Lie algebra homomorphism, the following relations for the operators Q(f) and D(x) hold:

$$\begin{aligned}[][Q(f_1), Q(f_2)] \ &= \ 0, && f_1, f_2 \in \mathcal{S}_c(M) \\ [D(x), Q(f)] \ &= \ Q(\phi(x)f), && f \in \mathcal{S}_c(M) \\ [D(x), D(y)] \ &= \ D([x,y]), && x,y \in \mathbb{G} \\ \alpha D(x) + \beta D(y) \ &= \ D(\alpha x + \beta y), && x,y \in \mathbb{G}, \ \alpha,\beta \in \mathbb{R}\end{aligned}$$

i.e. they give the usual commutation relations.

Let there be given a representation of a lHs. Then it is interesting to know if this representation is integrable to a unitary representation of a corresponding Hs. By integrability of a representation (Q,D) of a lHs $(\mathbb{A},\mathbb{G},\phi)$ in the Hilbert space $\mathfrak{H}$ we understand, that there exists a representation (Q,U) of a Hs $(\mathbb{A},G,\varphi)$ in $\mathfrak{H}$ such that the two diagrams commute:

$$\begin{array}{ccc} \mathbb{G} & \xrightarrow{\;D\;} & \$(\Delta) \\ {\scriptstyle \exp}\downarrow & & \downarrow{\scriptstyle \mathrm{Exp}} \\ G & \xrightarrow[\;U\;]{} & \mathbb{U}(\mathfrak{H}) \end{array} \qquad\qquad \begin{array}{ccc} \mathbb{G} & \xrightarrow{\;\phi\;} & {}^{*}\mathrm{Der}(A) \\ {\scriptstyle \exp}\downarrow & & \downarrow{\scriptstyle \widetilde{\exp}} \\ G & \xrightarrow[\;\varphi\;]{} & {}^{*}\mathrm{Aut}(A) \end{array}$$

where $\$(\Delta)$ denotes the Lie algebra of essentially skew adjoint operators on the common dense invariant domain $\Delta \subset \mathfrak{H}$ and $\mathbb{U}(\mathfrak{H})$ denotes the group of unitary operators on $\mathfrak{H}$ and G is a connected Lie group with Lie algebra $\mathbb{G}$.

Concerning the integrability of gvr's, we have the

<u>Proposition</u>: Let $(\hat{\phi},\phi)$ be a quasi-complete $\mathbb{G}$-action on $\mathcal{P}$. Then the associated gvr $(Q,D)(\hat{\phi},\varrho,\Omega))$ of a lHs $(\mathcal{S}_c(M),\mathbb{G},\phi)$ in $\mathcal{L}^2(\mathcal{H}^\varrho,\Omega)$ is integrable to a representation (Q,U) of a Hs $(\mathcal{S}_c(M),G,\varphi)$ iff $(\hat{\phi},\phi)$ is essentially G-maximal with respect to the representation ϱ of the structure group K.

The geometric property of essential G-maximality of a Lie algebra homomorphism $\hat{\phi}$: $\mathbb{G} \longrightarrow \mathfrak{X}(P)$, which controls the integrability of gvr's, is equivalent to the so called "loop criteria" discussed in [2]. It corresponds to an infinitesimal version of Palais' maximal local G-action [7]. Here we give the following definition:

Let L(G) denote the set of all finite ordered r-tupels $\bar{x}_r = (x_1,\dots,x_r)$, $x_i \in \mathbb{G}$, given by

$$L(G) := \left\{ \bar{x}_r = (x_1,\dots,x_r) \;\middle|\; x_i \in \mathbb{G},\ \prod_{i=1}^{r} \exp x_i = e \in G \right\} .$$

Let $(\hat{\phi},\phi)$ be a $\mathbb{G}$-action on $\mathcal{P}$. Then write $\hat{\phi}(x_i) = \hat{X}_i$ and $\phi(x_i) = X_i$ for short and $\hat{\bar{X}}_r$ for the ordered r-tupel $\hat{\bar{X}}_r = (\hat{X}_1,\dots,\hat{X}_r)$ of vector fields on P. The subset $L(G,\hat{\phi},p) \subset L(G)$ for $p \in P$, is given by

$$L(G,\hat{\phi},p) := \left\{ \bar{x}_r \,\middle|\, \bar{x}_r \in L(G),\ {}^{\hat{\bar{X}}_r}\varphi_1(p) \text{ exists} \right\},\ p \in P,$$

where ${}^{\hat{\bar{X}}_r}\varphi_t(p)$ denotes the combination of local diffeomorphisms corresponding to the r-tupel $\hat{\bar{X}}_r$ of vector fields:

$${}^{\hat{\bar{X}}_r}\varphi_t(p) := {}^{\hat{X}_r}\varphi_t \circ \dots \circ {}^{\hat{X}_1}\varphi_t(p) \quad , \quad p \in P.$$

<u>Definition</u>: A $\mathbb{G}$-action $(\hat{\phi},\phi)$ on $\mathcal{P}$ is called <u>essentially G-maximal</u> with respect to the representation ϱ of the structure group K of $\mathcal{P}$ iff

$$\bar{x}_r \in L(G,\hat{\phi},p) \Longrightarrow {}^{\hat{\bar{X}}_r}\varphi_1(p) = p\cdot\ker\varrho \quad , \quad p \in P.$$

We now can give the

<u>Proof</u> of the proposition: Let $\bar{x}_r \in L(G)$. Let $(\hat{\phi},\phi)$ be the $\mathbb{G}$-action on $\mathcal{P}$. Let $M_1^{\bar{X}_r} \subset M$ denote the open set of points $m \in M$ for which the combination of local diffeomorphisms ${}^{\bar{X}_r}\varphi_t$ corresponding to the r-tupel $\bar{X}_r = (\phi(x_1),\dots,\phi(x_r))$ of vector fields on M is defined. Since $(\hat{\phi},\phi)$ is quasi-complete, the set $E_1^{\bar{X}_r} = M - M_1^{\bar{X}_r}$ is of measure zero and $\mathrm{Sec}_c(\mathcal{H}^\varrho, M_1^{\bar{X}_r})$ is dense in $\mathcal{L}^2(\mathcal{H}^\varrho,\Omega)$.

i) Let $(\hat{\phi},\phi)$ be essentially G-maximal. Then we have to show that $\mathrm{Exp}\, D(\hat{\phi},\varrho,\Omega)(x_r) = \mathbb{1}$ on the dense domain $\mathrm{Sec}_c(\mathcal{H}^\varrho, M_1^{\bar{X}_r})$.
By Stone's theorem, on $\mathrm{Sec}_c(\mathcal{H}^\varrho, M_1^{\bar{X}_r})$ we have

$$\mathrm{Exp}\, D(\hat{\phi},\varrho,\Omega)(x) = U_1(\hat{\phi},\varrho,\Omega)(x) \quad , \quad x \in \mathbb{G},$$

where $U_1(x)$ is the flow operator $U_t(x)$ for t=1.
Then for $\bar{x}_r \in L(G,\hat{\phi},p)$ essential G-maximality of $(\hat{\phi},\phi)$ means

$${}^{\hat{\bar{X}}_r}\varphi_1(p) = p\cdot\ker\varrho \quad \text{and} \quad {}^{\bar{X}_r}\varphi_1(m) = m, \quad \pi(p) = m \in M,$$

so that ${}^{\hat{\bar{X}}_r}\varphi_1^{\#}\sigma^{\#} = \sigma^{\#}$, $\sigma \in \mathrm{Sec}_c(\mathcal{H}^\varrho, M_1^{\bar{X}_r})$ and $\det_1^\Omega \bar{X}_r(m) = 1$,
where $\det_1^\Omega \bar{X}_r$ is given by

$$\det{}_1^\Omega \bar{X}_r\, \Omega = {}^{\bar{X}_r}\varphi_1^{\#}\Omega .$$

Then for $\bar{x}_r \in L(G,\hat{\phi},p)$, $\sigma \in \mathrm{Sec}_c(\mathcal{H}^\varrho, M_1^{\bar{X}_r})$,

$$(\mathrm{Exp}\circ D(\hat{\phi}, \varsigma, \Omega)(\bar{x}_r)\sigma)^{\#} = (U_1(\hat{\phi}, \varsigma, \Omega)(\bar{x}_r)\sigma)^{\#}$$
$$= (U_1(x_r)\circ \ldots \circ U_1(x_1)\sigma)^{\#}$$
$$= (\pi^{*}\det{}_1^{\Omega}\bar{X}_r)\cdot {}^{\hat{\bar{X}}_r}\varphi_1^{*}\sigma^{\#} = \sigma^{\#} .$$

Since the map $\widetilde{\exp}$: $X \longmapsto {}^{X}\varphi_1^{*}$ assigns to every vector field $X \in \mathcal{X}(M)$ (derivation in $\mathcal{J}_c(M)$) the pull back ${}^{X}\varphi_1^{*}$ (automorphism of $\mathcal{J}_c(M)$) with respect to the corresponding local diffeomorphism ${}^{X}\varphi_1$, essential G-maximality of the $\mathbb{G}$-action $(\hat{\phi}, \phi)$ induces commutativity of the second diagram.

ii) The reversed direction of the proof is obvious.

4. Concluding remarks

The quantization procedure given by general vector field representations of local Heisenberg systems may have promising applications to the physical situations where there is given a principal bundle $\mathcal{P}$ over a manifold M as a natural starting point.

Then states of a system are described by sections (with compact support) in a vector bundle $\mathcal{H}^{\varsigma}$, ς-associated to $\mathcal{P}$, spanning the Hilbert space $\mathcal{L}^2(\mathcal{H}^{\varsigma}, \Omega)$ with respect to the volume form Ω on M. Given a finite dimensional Lie algebra $\mathbb{G}$ of infinitesimal symmetries of the system and a quasi-complete $\mathbb{G}$-action on $\mathcal{P}$, the corresponding vector field operators together with the functions on M with compact support then give the usual commutation relations for the canonical momentum and localization observables.

Especially new results may be derived with this quantization procedure if the base manifold M has a non-trivial topology so that twisted fields [9] viewed as sections in non-trivial vector bundles may be considered.

General vector field representations are applicable for example in the context of geometric quantization [10] where one starts with a given quantizing bundle over M [11] and the Lie algebra action on this bundle is given via the connection on this principal bundle and the symplectic form on the base manifold M. But the interpretation of this quantization procedure is quite different to that of pre-quantization.

In other contexts physically motivated structures have to be used in an analous way. Applications of general vector field representations based on gauge principal bundles [12] will be given elsewhere.

References:

1. G.W. Mackey, Bull. Amer. Math. Soc. 69, 628 (1963)
2. H.D. Doebner, H.-E. Werth, J. Math. Phys., 20, 1011 (1979)
3. F.B. Pasemann, General vector field representations - a geometric tool for quantization, IAEA, IC 78/48 (1978)
4. H. Snellman, Ann. Inst. Henri Poincaré A 24, 393 (1976)
5. I.E. Segal, Duke Math. J. 18, 221 (1951)
6. J. Dixmier, Algèbres d'operateur. In Proc. Int. Summer School Phys. Varenna 1968, Ed. R. Jost, Academic Press, New York 1970
7. R.S. Palais, Mem. of the Amer. Math. Soc. 22, (1957)
8. R. Abraham, J.E. Marsden, Foundation of Mechanics Benjamin, Reading 1978
9. C.J. Isham, Proc. R. Soc. Lond. A 362, 383, (1978)
10. B. Kostant, Quantization and unitary representations, Lecture Notes in Mathematics Vol. 170, Springer, New York 1970
11. J.-E. Werth, Ann. Inst. Henri Poincaré A 25, 165 (1976)
12. W. Drechsler, M.E. Meyer, Fiber Bundle Techniques in Gauge Theories, Lecture Notes in Physics Vol. 67, Springer, New York 1977

Aspects of Relativistic Quantum Mechanics on Phase Space

S. Twareque Ali

Department of Mathematics
University of Toronto
Toronto, Canada

Abstract

Recent work on formulating relativistic quantum mechanics on stochastic phase spaces is described. Starting with a brief introduction to the mathematical theory of stochastic spaces, an account is given of non-relativistic quantum mechanics on stochastic phase space. The relativistic theory is introduced by constructing certain classes of representations of the Poincaré group on phase space, obtaining thereby both the classical and the quantum dynamics. Applications to the Dirac equation are discussed, and an alternative 2-component equation for a charged spin-1/2 particle, interacting with an external electromagnetic field is studied.

1. Introduction

The scope of the present paper is somewhat outside the ken of differential geometry or geometric quantization per se. However, it is at one with the other papers, in the sense that it also approaches quantum mechanics (both non-relativistic as well as relativistic) from the point of view of its possible representations on phase space. As in the case of geometric quantization, the aim here also is to put both classical and quantum mechanics on a common phase space, and thereby to understand better the relationship between the two theories. Some aspects of the work reported here overlap in spirit though not in the formal apparatus used, with that of Bayen et al. (1). Likewise the non-relativistic portion of the present theory is in some sense a 'rigorization' of the work of Agarwal and Wolf (2).

The problem to which we address ourselves is the following: Classical statistical mechanics is formulated on the phase space Γ. The observables are functions of the coordinates $\underline{q}$ and momenta $\underline{p}$, while the states are normalized probability measures over phase space. Quantum mechanics, on the other hand, is formulated over a Hilbert space. The question now arises as to whether quantum mechanical states could also be represented as probability measures over phase space, and the quantum observables as functions (or in general distributions) over Γ.

It is clear that in view of the Heisenberg uncertainty principle operating in quantum mechanics, $\underline{q}$ and $\underline{p}$ cannot be interpreted as referring simultaneously to the exact position and momentum of the particle. In other words, if we are to insist on the positivity (cf. also Ref. 3, in this connection) of the probability measures corresponding to quantum mechanical states, we have to adopt a stochastic interpretation for the simultaneous assignment of the position $\underline{q}$ and momentum $\underline{p}$ of a particle. We have, therefore, to define the appropriate sense in which this can be done.

The next question is how much of these non-relativistic phase space considerations can be extended to the relativistic domain. In particular, would such an approach solve any of the standard problems of relativistic quantum mechanics, such as for example, the non-existence of a Hilbert space in the coordinate representation, consisting only of states with positive energy whose square moduli also represent positive definite probability densities? In the context of the present approach the answer to this question is in the affirmative. In addition, a relativistic phase space approach also deepens our understanding of the problem of a quantum mechanical particle interacting with an electromagnetic field (4), overcoming many of the difficulties of the old theory, such as instability of the positive energy states against small perturbations. We will attempt to present in this paper an outline of some of the recent developments in this direction.

The rest of this report is organized as follows. In Section 2 we outline briefly some ideas about generalized and smooth observables in quantum mechanics, and mention how these observables lend an additional structure to quantum mechanics and lead to the concept of stochastic value spaces. The idea of the amount of information carried by an observable is also mentioned here. In Section 3 we use the notions introduced in Section 2 to motivate and interpret the phase space representations of non-relativistic quantum mechanics, which we introduce at this point. In particular, we see how the notion of a stochastic phase space arises naturally in quantum mechanics. We also present some examples of practical computations using phase space methods. In Section 4 we extend the phase space representations to the relativistic regime. We do this by looking at certain induced representations of the Poincaré group on Hilbert spaces of functions on phase space. Representations are obtained both for classical as well as quantum mechanical systems. In the classical case we derive a relativistic generalization of the Liouville equation. The irreducible quantum mechanical representations are shown to be equivalent to the standard ones describing particles having mass $m > 0$ and spin j (= 0, 1/2, 1, ...). At this point, we achieve an interesting unification of classical and quantum mechanics, both relativistic and non-relativistic, in the sense that they are all described on the same Hilbert space consisting of phase space functions. In Section 5 we briefly mention the problem of the Dirac equation on phase space, and as an alternative, a 2-component equation of a spin-1/2 particle interacting with an external electromagnetic field, treated also on phase space. The case of a spin-0 particle is collaterally examined. Finally, we end in Section 6 with a recapitulation of some of the ways in which the present framework for looking at relativistic quantum mechanics is a definite advantage over the traditional treatments. We also emphasize here that it is only in the context of stochastic phase spaces, as introduced in Section 3, that an adequate physical interpretation can be given to our treatment of quantum dynamics on phase space. Also, it is only against a background of such stochastic phase spaces that some of the problems of relativistic quantum mechanics find a resolution here.

2. Generalized Observables in Quantum Mechanics

Throughout this section we consider a model of quantum mechanics based upon a separable Hilbert space $\mathcal{H}$. Ignoring superselection rules, the observables of the system are self adjoint operators A on $\mathcal{H}$, while the (pure) states are the normalized vectors in $\mathcal{H}$. An observable A can equally well be described by the associated spectral measure. In this case, we shall refer to its spectrum as its <u>value space</u>. Looked at in this way, the concept of an observable has an immediate generalization (5,6). Let X be a locally compact space, and $\mathcal{B}(X)$ the set of all Borel sets (generated by the topology) of X. Let $E \longrightarrow a(E)$ be a normalized positive operator valued (POV) measure on $\mathcal{B}(X)$, i.e., $\forall E \in \mathcal{B}(X), a(E) \in \mathcal{L}(\mathcal{H})^+$ (= set of bounded positive operators on $\mathcal{H}$) and $a(E)$ satisfies

(i) $a(\bigcup_i E_i) = \sum_i a(E_i)$ for $E_i \cap E_j = \phi$, $i \neq j$,

(ii) $a(X) = \mathbb{1}$.

The sum in (ii) is assumed to converge weakly. Then a is called a <u>generalized observable</u>, with value space X, the idea being that a measurement of the observable a on a given physical state yields results lying in the sets $E \in \mathcal{B}(X)$.

This generalization of the notion of an observable becomes very useful if we wish, for example, to describe the outcomes of experiments such as those which try to localize a particle in a region of three dimensional Euclidean space, or more generally to a region of six dimensional phase space. At other times, we shall find such observables extremely useful for describing the outcomes of imprecise measurements.

Let μ be a Borel measure also defined on $\mathcal{B}(X)$. We say (7) that the observable a is <u>smooth</u> with respect to μ if a and μ both have the same null sets, i.e., $(\phi, a(E)\psi) = 0, \forall \phi, \psi \in \mathcal{H}$ iff $\mu(E) = 0$. It is clear that in this case a is actually smooth with respect to a whole measure class equivalent to μ. Given a generalized observable a we can construct the measure class with respect to which it is smooth by means of the following lemma:

<u>Lemma 2.1</u>

Let a be a generalized observable and $\{\phi_n\}$ an orthonormal set of basis vectors in $\mathcal{H}$. Let ν be the probability measure

$$\nu(E) = \sum_{n=1}^{\infty} \frac{1}{2^n} (\phi_n, a(E)\phi_n) \tag{2.1}$$

Then a is smooth with respect to ν.

<u>Proof:</u>

Clearly $a(E)=0 \Longrightarrow \nu(E)=0$. Let $\nu(E) = 0$. Then, since the sum on the right hand side of (2.1) is composed of positive terms only,

$$(\phi_n, a(E)\phi_n) = 0 \quad \forall n \in \mathbb{N}. \tag{2.2}$$

Next, $a(E)$ being a positive operator, $a(E) = B^2$, where B is the square root of $a(E)$. Hence (2.2) implies

$$\|B\phi_n\|^2 = 0 \quad , \quad \forall n \quad , \tag{2.3}$$

and since $\{\phi_n\}$ is a basis set, we get $B = 0$, so that $a(E) = 0$. Q.E.D.

Let us next introduce the idea of the amount of information carried by a generalized observable. Let a_1 and a_2 be two generalized observables. Then a_1 is said to give more information than a_2, (written $a_2 \subset a_1$) if for any two vectors ϕ and $\psi \in \mathcal{H}$, the equality $(\phi, a_1(E)\phi) = (\psi, a_1(E)\psi)$, $\forall E$ implies $(\phi, a_2(E)\phi) = (\psi, a_2(E)\psi)$, $\forall E$ (cf. also (8)). In other words, if $a_2 \subset a_1$ then using a_2 alone we cannot distinguish between states which are indistinguishable using a_1 alone. If both $a_2 \subset a_1$ and $a_1 \subset a_2$ then a_1 and a_2 are said to be informationally equivalent.

Consider next the case where the quantum mechanical system has a symmetry group $\mathcal{G}$ acting on it. We shall assume that $\mathcal{G}$ is locally compact and has a representation $g \longrightarrow U_g$ by unitary operators on $\mathcal{H}$. If a is a normalized POV measure, then $a(E)$, U_g are said to form a POV system of imprimitivity (POVI), if $\mathcal{G}$ acts transitively on X (in the sense of Mackey (9)) and

$$U_g^* a(E) U_g = a(g^{-1}[E]) \tag{2.4}$$

where $g^{-1}[E]$ is the translate of the set E by g. In the case $a(E) = P(E)$, a projection valued measure then $P(E)$, U_g are said to form a projection valued system of imprimitivity (PVSI). Suppose μ is a Borel measure on X which is invariant with respect to the action of $\mathcal{G}$. We mention a result in the next lemma, the proof of which is similar to that of Lemma 2 in Appendix A of (10).

Lemma 2.2

Let $a(E)$, U_g be a POVSI and μ a (quasi-) invariant measure on X. Then a is smooth with respect to μ.

We proceed now to introduce the concept of a stochastic value space. Earlier in this section we had introduced the concept of a value space X in relation to the (commutative) generalized observable $a(E)$. Let us associate to each point $x \in X$ a probability measure ν_x also defined on $\mathcal{B}(X)$. Let

$$\hat{X} = \{ (x, \vartheta_x) \mid x \in X \} . \qquad (2.5)$$

We transfer the Borel structure of X to $\hat{X}$ via the association $x \longrightarrow \vartheta_x$, and call $\hat{X}$, thus equipped with a Borel structure, a stochastic value space. The physical justification for this nomenclature is that we are often concerned with value spaces X for which the points $x \in X$ are randomly distributed with probabilities given by ϑ_x. Thus for example, if we wish to measure the position of a particle, we can only determine it up to a certain degree of precision (depending upon the apparatus used), and the observed position itself becomes a random variable distributed according to the confidence function of the apparatus in question. Similarly, for the stochastic value space $\hat{X}$, if the system is observed to lie in a small region surrounding x, then the probability that it actually lies in some other region E of the value space is $\vartheta_x(E)$.

We may now state the most important result of this section.

Theorem 2.1.

Let $a(E)$, U_g be a POVSI and $P(E)$, U_g a PVSI on $\mathcal{H}$, with respect to the action of the group $\mathcal{G}$, and let a and P be informationally equivalent. If for all $E, F \in \mathcal{B}(X)$, $a(E)$ commutes with $a(F)$, then a determines a unique stochastic value space which is Borel isomorphic to X.

The proof to this theorem is exactly similar to the proof of Theorem 2 in (10), and will be omitted here. We only mention how the space $\hat{X}$ is obtained: Since $P(E)$, U_g is a PVSI, by Mackey's imprimitivity theorem (9) the Hilbert space $\mathcal{H}$ is isomorphic to $\mathcal{K} \otimes L^2(X, \mu)$, where, as before, μ is the (quasi-) invariant measure on X and $\mathcal{K}$ is a Hilbert space carrying a unitary representation of a subgroup H of $\mathcal{G}$ (for which $X = \mathcal{G}/H$). On $\mathcal{K} \otimes L^2(X, \mu)$ the operators $a(E)$ act as

$$(a(E)\phi)(x) = \vartheta_x(E)\,\phi(x) , \qquad (2.6)$$

where ϕ is in $\mathcal{K} \otimes L^2(X, \mu)$ and the ϑ_x's are Borel measures, determined by a at each point x in X, which satisfy

$$\vartheta_{[x]g}(E) = \vartheta_x([E]g^{-1}) , \qquad (2.7)$$

$\forall g \in \mathcal{G}$ and $E \in \mathcal{B}(X)$. The stochastic value space $\hat{X}$ then consists of the ordered pairs (x, ϑ_x). In the same spirit, if δ_x is the delta measure concentrated at x, then the PSVI $P(E), U_g$ determines the stochastic value space $\hat{X}$ which consists of ordered pairs (x, δ_x).

In conventional quantum mechanics, for the system in a given state ϕ, $|\phi(x)|^2$ is the probability density of finding it at the point x, if measurements of x can be made absolutely precise. Such measurements lead to the operators $P(E)$ given by

$$(\phi, P(E)\phi) = \int_E |\phi(x)|^2 \, d\mu(x) \tag{2.8}$$

On the other hand, in the present approach, since imprecise measurements are allowed, we obtain in addition the operators

$$(\phi, a(E)\phi) = \int_E \vartheta_0 * |\phi|^2(x) \, d\mu(x) \tag{2.9}$$

where $\vartheta_0 * |\phi|^2$ is the convolution of the measure ϑ_0 with $|\phi(x)|^2$. Comparing Eq. (2.9) with (2.8) we see that the operators $a(E)$ represent measurements which are averages (in a definite sense) of the results of measurements represented by the sharp operators $P(E)$. This immediately points up the additional richness of structure in the present approach.

3. Phase Space Representations of Non-Relativistic Quantum Mechanics

The generalized observables introduced in the last section are not very useful if one adheres to the usual coordinate space or momentum space representation of non-relativistic quantum mechanics. In this case, replacing the standard momentum (or position) observable by its stochastic counterpart leads to no new physical result (11), since the action of the underlying Galilei group implies their informational equivalence. However, if one is looking for a phase space representation of quantum mechanics, i.e., a representation in which the wave functions ψ are functions of the phase space variables $\underline{q}$ and $\underline{p}$, then generalized observables play a crucial role. The reason for this is that in talking about phase space theories one is dealing, in some sense, with the simultaneous measurement of the non-commuting operators Q and P, and this is not possible while remaining within limits of conventional quantum mechanics.

In this section, we shall construct phase space representations of non-relativistic quantum mechanics. More precisely, we shall start with the set of all states ϱ of the system, i.e., with the set of all density matrices on $\mathcal{H}$, and map them linearly onto positive functions $f(\underline{q},\underline{p})$ of the phase space variables satisfying

$$\int_\Gamma f(\underline{q},\underline{p})\, d^3\underline{q}\, d^3\underline{p} = 1 \tag{3.1}$$

Dual to this, we shall also construct a linear map which will take the observables A of the system which are of interest here (viz, the set of all bounded self adjoint operators and a certain class of unbounded and symmetric operators on $\mathcal{H}$) onto a family of real tempered distributions $F(\underline{q},\underline{p})$ in such a way that the relationship

$$\mathrm{tr}[A\varrho] = \int_\Gamma F(\underline{q},\underline{p})\, f(\underline{q},\underline{p})\, d^3\underline{q}\, d^3\underline{p} \tag{3.2}$$

will hold. This will complete the analogy with classical statistical mechanics.

There is, however, one point of departure from the classical treatment and its quantum counterpart as presented here. Classically, if we have a state ν^{cl}, which is a probability measure defined on the Borel sets $\mathcal{B}(\Gamma)$ of the six dimensional phase space

$$\Gamma = \{ (\underline{q},\underline{p}) \mid \underline{q} \in \mathbb{R}^3,\ \underline{p} \in \mathbb{R}^3 \} \quad , \tag{3.3}$$

then, for any $\Delta \in \mathcal{B}(\Gamma)$, $\nu^{cl}(\Delta)$ is the probability of finding the system localized (sharply) in the region Δ of phase space. Furthermore, the <u>marginal distributions</u>

$$\bar{\nu}^{cl}(\Delta_1) = \nu^{cl}(\Delta_1 \times \mathbb{R}^3) , \tag{3.4a}$$

$$\hat{\nu}^{cl}(\Delta_2) = \nu^{cl}(\mathbb{R}^3 \times \Delta_2) \tag{3.4b}$$

where $\Delta_1, \Delta_2 \in \mathcal{B}(\mathbb{R}^3)$, represent respectively, the probabilities of the system being localized sharply in the region Δ_1 of configuration space and the region Δ_2 of momentum space. We cannot of course expect to have marginality conditions similar to (3.4a) and (3.4b) satisfied in the quantum mechanical case also, precisely because $\underline{p}$ and $\underline{q}$ cannot be measured simultaneously with absolute precision any more. Hence, if ϱ is a density matrix and f its phase space representative, $f(\underline{q},\underline{p})$ cannot (if we insist upon the positivity of f) be considered as being the probability density for finding the system at the sharp point ($\underline{q}$, $\underline{p}$) in phase space. Thus, if f is a positive (semi-definite) function representing ρ, the marginality conditions (3)

$$\int_{\mathbb{R}^3} f(\underline{q},\underline{p})\, d^3\underline{p} = \langle \underline{q} | \rho | \underline{q} \rangle \,, \tag{3.5a}$$

$$\int_{\mathbb{R}^3} f(\underline{q},\underline{p})\, d^3\underline{q} = \langle \underline{p} | \rho | \underline{p} \rangle \,, \tag{3.5b}$$

will not be satisfied. In Eqs. (3.5) the quantities $\langle \underline{q}|\rho|\underline{q}\rangle$ and $\langle \underline{p}|\rho|\underline{p}\rangle$ have their usual meanings. For example, in the Schrödinger representation (spinless particle), if ρ is a pure state corresponding to a wave function $\psi(\underline{q})$, so that $\rho = |\psi\rangle\langle\psi|$ then

$$\langle \underline{q} | \rho | \underline{q} \rangle = |\psi(\underline{q})|^2 \tag{3.6a}$$

and,

$$\langle \underline{p} | \rho | \underline{p} \rangle = |\tilde{\psi}(\underline{p})|^2 \tag{3.6b}$$

$\tilde{\psi}$ being the Fourier transform of ψ. On the other hand, if we do not insist on getting the 'sharp' marginality conditions (3.5), and if we replace the phase space Γ by a stochastic phase space $\hat{\Gamma}$ in the sense of the preceding section, then we may reasonably expect $f(\underline{q},\underline{p})$ to satisfy the modified marginality conditons

$$\int_{\mathbb{R}^3} f(\underline{q},\underline{p})\, d^3\underline{p} = \int_{\mathbb{R}^3} \chi_{\underline{q}}(\underline{q}') \langle \underline{q}' | \rho | \underline{q}' \rangle\, d^3\underline{q}' \,, \tag{3.7a}$$

$$\int_{\mathbb{R}^3} f(\underline{q},\underline{p})\, d^3\underline{q} = \int_{\mathbb{R}^3} \hat{\chi}_{\underline{p}}(\underline{p}') \langle \underline{p}' | \rho | \underline{p}' \rangle\, d^3\underline{p}' \,. \tag{3.7b}$$

In Eqs. (3.7), χ (resp. $\hat{\chi}$) is a probability measure on $\mathbb{R}^3$ centred at $\underline{q}$ (resp. $\underline{p}$), so that $\hat{\Gamma}$ is the set of all quadruples

$$\hat{\Gamma} = \{ ((\underline{q},\chi_{\underline{q}}), (\underline{p},\hat{\chi}_{\underline{p}})) \mid (\underline{q},\underline{p}) \in \Gamma \} \,. \tag{3.8}$$

This is what, as we will now show, does indeed occur. In fact, the probability measures $\chi_{\underline{q}}$ and $\hat{\chi}_{\underline{p}}$ (which are to be interpreted here as confidence functions) are also completely determined by the representation in question.

Let us begin with the Schrödinger representation for a massive spinless quantum mechanical particle. Thus $\mathcal{H} = L^2(\mathbb{R}^3, d^3\underline{x})$ and we take the Weyl group $\mathcal{W}$ as the underlying symmetry group of the system. $\mathcal{W}$ has the unitary representation $(\underline{q},\underline{p}) \longrightarrow U(\underline{q},\underline{p})$, where

$$U(\underline{q},\underline{p}) = \exp(\tfrac{i}{\hbar}\,\underline{p}\cdot\underline{Q})\,\exp(-\tfrac{i}{\hbar}\,\underline{q}\cdot\underline{P}) \tag{3.9}$$

with

$$(Q_j\psi)(\underline{x}) = x_j\,\psi(\underline{x}) \;,\quad j=1,2,3\;, \tag{3.10a}$$

$$(P_j\psi)(\underline{x}) = -i\hbar\frac{\partial}{\partial x_j}\psi(\underline{x})\;,\; j=1,2,3\;. \tag{3.10b}$$

Let $\mathcal{T}(\mathcal{H})$ denote the set of all trace-class operators on $\mathcal{H}$, $\mathcal{T}(\mathcal{H})^+$ its positive cone, $M(\Gamma)$ the set of all bounded measures on the phase space Γ , and $M(\Gamma)^+$ its positive cone.

Definition 3.1

A phase space representation of quantum mechanics is a positive, linear map

$$\pi : \rho \in \mathcal{T}(\mathcal{H}) \longrightarrow \mu_\rho \in M(\Gamma) \tag{3.11}$$

which satisfies

(i) $$\int_\Gamma d\mu_\rho(\underline{q},\underline{p}) = \operatorname{tr}\rho \qquad \forall\, \rho \in \mathcal{T}(\mathcal{H})^+ \tag{3.12}$$

(ii) $$\mu_{\rho_1} \neq \mu_{\rho_2} \quad \text{iff} \quad \rho_1 \neq \rho_2 \tag{3.13}$$

(iii) $$\pi\left[U^*(\underline{q},\underline{p})\,\rho\,U(\underline{q},\underline{p})\right] = [\mu_\rho]_\gamma \quad , \tag{3.14}$$

where for any $\gamma = (\underline{q},\underline{p}) \in \Gamma$, $\Delta \in \mathcal{B}(\Gamma)$,

$$([\mu_\rho]_\gamma)(\Delta) = \mu_\rho(\Delta - \gamma) \tag{3.15}$$

The following two theorems completely describe all phase space representations of non-relativistic quantum mechanics and their relationship to stochastic phase spaces. The proofs are fairly involved and lengthy and may be found in (10).

Theorem 3.1

Every phase space representation of non-relativistic quantum mechanics is completely determined by a unique element $g \in \mathcal{J}(\mathcal{H})^+$, satisfying $\mathrm{tr}\, g = 1$ via the following association

$$\rho \in \mathcal{J}(\mathcal{H}) \longrightarrow \mu_\rho^g \in M(\Gamma) \tag{3.16}$$

$$d\mu_\rho^g(\underline{q},\underline{p}) = \frac{1}{\hbar^3}\, \mathrm{tr}\left[U(\underline{q},\underline{p})\, g\, U^*(\underline{q},\underline{p})\, \rho\right] d^3\underline{q}\, d^3\underline{p} \quad . \tag{3.17}$$

Since according to Eq. (3.17), the measure μ_ρ^g which represents ρ on phase space always happens to be absolutely continuous with respect to the Lebesgue measure $d^3\underline{q}\, d^3\underline{p}$ on Γ we shall write the association implied in (3.16) as

$$\rho \longrightarrow \rho_g(\underline{q},\underline{p}) = \hbar^{-3}\, \mathrm{tr}\left[U(\underline{q},\underline{p})\, g\, U^*(\underline{q},\underline{p})\, \rho\right] \quad . \tag{3.18}$$

Thus every state ρ of the system is represented on phase space by a positive (semi-definite) function ρ_g which is properly normalized, i.e,

$$\int_\Gamma \rho_g(\underline{q},\underline{p})\, d^3\underline{q}\, d^3\underline{p} = 1 \quad . \tag{3.19}$$

We call g the <u>generator</u> of the representation $\rho \longrightarrow \rho_g(\underline{q},\underline{p})$. In particular if g is of the form

$$g = |e\rangle\langle e| \qquad , e \in \mathcal{H} , \tag{3.20}$$

then the corresponding representation is said to be <u>extremal</u>.

<u>Theorem 3.2</u>

The generator g determines uniquely the stochastic phase space

$$\Gamma_g = \left\{ ((\underline{q},\chi_{\underline{q}}),(\underline{p},\hat{\chi}_{\underline{p}})) \mid (\underline{q},\underline{p}) \in \Gamma \right\} \tag{3.21}$$

where

$$\chi_0(\underline{q}) = \hbar^{-3} \langle \underline{q}|g|\underline{q}\rangle \tag{3.22a}$$

$$\hat{\chi}_0(\underline{p}) = \hbar^{-3} \langle \underline{p}|g|\underline{p}\rangle \quad ; \tag{3.22b}$$

and,

$$\chi_{\underline{q}}(\underline{q}') = \chi_0(\underline{q}'-\underline{q}) \tag{3.23a}$$

$$\hat{\chi}_{\underline{p}}(\underline{p}') = \hat{\chi}_{o}(\underline{p}'-\underline{p}) \tag{3.23b}$$

The probability densities $\rho_g(\underline{q},\underline{p})$ satisfy the marginality conditions.

$$\int_{\mathbb{R}^3} \rho_g(\underline{q},\underline{p})\, d^3\underline{p} = \int_{\mathbb{R}^3} \chi_{\underline{q}}(\underline{q}') < \underline{q}' | \rho | \underline{q}' > d^3\underline{q}' \tag{3.24a}$$

$$\int_{\mathbb{R}^3} \rho_g(\underline{q},\underline{p})\, d^3\underline{q} = \int_{\mathbb{R}^3} \hat{\chi}_{\underline{p}}(\underline{p}') < \underline{p}' | \rho | \underline{p}' > d^3\underline{p}' \quad . \tag{3.24b}$$

Let us define the spreads s_j and t_j of the functions $\chi_o(\underline{q}), \hat{\chi}_o(\underline{p})$ by

$$s_j = \left[2 \int_{\mathbb{R}^3} (q_j - \bar{q}_j)^2 \chi_o(\underline{q})\, d^3\underline{q} \right]^{\frac{1}{2}} \quad , \; j=1,2,3 \, , \tag{3.25a}$$

$$t_j = \left[2 \int_{\mathbb{R}^3} (p_j - \bar{p}_j)^2 \hat{\chi}_o(\underline{p})\, d^3\underline{p} \right]^{\frac{1}{2}} \quad , \; j=1,2,3 \, ; \tag{3.25b}$$

with,

$$\bar{q}_j = \int_{\mathbb{R}^3} q_j \chi_o(\underline{q})\, d^3\underline{q} \tag{3.26a}$$

$$\bar{p}_j = \int_{\mathbb{R}^3} p_j \hat{\chi}_o(\underline{p})\, d^3\underline{p} \quad . \tag{3.26b}$$

Then it follows from Eqs. (3.22) that $s_j t_j$ satisfy the Heisenberg uncertainty relations

$$s_j t_j \geq \hbar \quad , \; j=1,2,3 \, . \tag{3.27}$$

Finally the stochastic phase space representation determined by g gives rise to the generalized position (resp. momentum) observable a^Q (resp. a^P) defined by

$$(a^Q(\Delta_1)\psi)(\underline{x}) = \int_{\Delta_1} \chi_{\underline{x}}(\underline{x}')\, \psi(\underline{x}')\, d^3\underline{x}' \tag{3.28a}$$

$$(a^P(\Delta_2)\tilde{\psi})(\underline{k}) = \int_{\Delta_2} \hat{\chi}_{\underline{k}}(\underline{k}')\, \tilde{\psi}(\underline{k}')\, d^3\underline{k} \tag{3.28b}$$

where again $\tilde{\psi}$ is the Fourier transform of ψ . As is to be expected

a^Q (resp. a^P) is informationally equivalent to the standard (sharp) position (resp. momentum) observable, defined through the usual PV spectral measure.

To complete the description of non-relativistic quantum mechanics on phase space we construct next the map dual to π in Eq. (3.11). As mentioned earlier, this involves mapping self adjoint operators A onto distributions F, over suitable class of test functions, in a way which preserves Eq. (3.2). Let g be the representation generator introduced above. Let

$$g_W(\underline{q},\underline{p}) = \hbar^{-3/2} \, tr\left[U(\underline{q},\underline{p})g\right] , \tag{3.29}$$

and let

$$\tilde{S}_g(\mathbb{R}^6) = \{ g_W(\underline{q},\underline{p}) f(\underline{q},\underline{p}) \mid f \in S(\mathbb{R}^6) \} \tag{3.30}$$

be a space of test functions obtained by allowing f to vary over the Schwartz space $S(\mathbb{R}^6)$. We equip S_g with the natural topology induced by the mapping $f \longrightarrow g_W f$ (i.e., the topology under which this mapping is a homomorphism). Similarly, let $S_g(\mathbb{R}^6)$ be the space of test functions which are Fourier transforms of the functions in $\tilde{S}_g(\mathbb{R}^6)$, and equip $S_g(\mathbb{R}^6)$ with the natural topology induced by this transform. The next theorem, whose proof may be found in (12) summarizes all the properties of the dual map we are seeking:

<u>Theorem 3.3</u>

To every operator A on $L^2(\mathbb{R}^3, d^3\underline{x})$, which is either bounded, or symmetric with domain containing $S(\mathbb{R}^3)$, there corresponds a unique distribution $A_g(\underline{q},\underline{p})$ which satisfies

$$tr[A\varrho] = \int_\Gamma A_g(\underline{q},\underline{p}) \, \varrho_g(\underline{q},\underline{p}) \, d^3\underline{q} \, d^3\underline{p} \tag{3.31}$$

for a dense set of elements $\varrho \in \mathcal{T}(\mathcal{H})$, and more generally, whenever $\varrho_g \in S_g(\mathbb{R}^6)$. If A is symmetric, the distribution is real.

The distribution A_g may be explicitly obtained, for any given operator A , by using the techniques described in (12). For instance, if g is taken to be the density matrix corresponding to the wave function (ground state of the harmonic oscillator of mass m, with potential energy $\frac{1}{2} K x^2$ and $mK = \alpha^2$) :

$$e(\underline{x}) = \left(\frac{\alpha^2}{\pi\hbar}\right)^{\frac{3}{4}} \exp\left(-\frac{\alpha^2}{2\hbar}\underline{x}^2\right) \quad , \tag{3.32}$$

then with $A = Q_j$, the j-th position operator, we have $A_g = q_j$. Similarly let ρ be the canonical equilibrium state (at inverse temperature β)

$$\rho^{(\beta)} = e^{-\beta H} / tr\, e^{-\beta H} \quad , \tag{3.33}$$

where

$$H = \frac{1}{2}\left(\underline{P}^2 + \omega^2 \underline{Q}^2\right) \tag{3.34}$$

is the 3-dimensional oscillator Hamiltonian. Then with the same g as before,

$$\rho_g(\underline{q},\underline{p}) = \frac{1}{(\pi\hbar)^3 \sigma_1 \sigma_2} \exp\left[-\left(q^2/\sigma_1^2 + p^2/\sigma_2^2\right)\right] \quad , \tag{3.35}$$

$$\left.\begin{aligned} \sigma_1 &= \frac{\hbar}{\alpha^2} + \omega^{-1} \coth\left(\frac{\beta\omega}{2}\right) \\ \sigma_2 &= \hbar\alpha^2 + \omega \coth\left(\frac{\beta\omega}{2}\right) \end{aligned}\right\} \tag{3.36}$$

Once again, we omit all details of computation.

Having set up a formalism to transcribe both states and observables on phase space, we can also study the evolution of quantum systems on phase space, and derive the analogues of the Schrödinger equation. Much work in this direction has been reported in Refs. 12,13,14 and 15 .

4. Relativistic Quantum Mechanics on Phase Spaces

The object of this section is to extend the considerations of the last section to the relativistic domain. Since any of the phase space representations introduced in the last section can be described entirely in terms of the extremal ones, we concentrate exclusively on these representations. Let e be a unit vector in $L^2(\mathbb{R}^3, d^3\underline{x})$ and $g = |e\rangle\langle e|$ be the generator of the non-relativistic phase space representation of quantum mechanics. If ψ is any other vector in $L^2(\mathbb{R}^3, d^3\underline{x})$ and $\rho = |\psi\rangle\langle\psi|$, then clearly

$$\rho_g(\underline{q},\underline{p}) = |\psi_e(\underline{q},\underline{p})|^2 \tag{4.1}$$

where

$$\psi_e(\underline{q},\underline{p}) = \hbar^{-3/2}(e_{\underline{q},\underline{p}}, \psi) \quad , \tag{4.2}$$

$$e_{\underline{q},\underline{p}} = U(\underline{q},\underline{p})e \quad . \tag{4.3}$$

Let $L^2(\Gamma)$ be the Hilbert space of all complex valued functions on Γ which are square-integrable with respect to $d^3\underline{q}\, d^3\underline{p}$. Then it is easily verified (cf. (15)) that $\psi_e \in L^2(\Gamma)$ for all $\psi \in L^2(\mathbb{R}^3, d^3\underline{x})$ and that

$$\|\psi_e\|^2 = \|\psi\|^2 \quad . \tag{4.4}$$

Thus the mapping $\psi \longrightarrow \psi_e$ embeds $L^2(\mathbb{R}^3, d^3\underline{x})$ isometrically onto a subspace of $L^2(\Gamma)$. Let us denote this subspace of $L^2(\Gamma)$ by $L^2(\Gamma_e)$, and let $\mathbb{P}_e$ be the projection operator for which

$$L^2(\Gamma_e) = \mathbb{P}_e L^2(\Gamma) \quad . \tag{4.5}$$

If we denote the image of e in $L^2(\Gamma)$ under the isometry (4.2) also by e , then the following relations are easily verified (cf. (15)):

$$\mathbb{P}_e = \hbar^{-3}\int_\Gamma |e_{\underline{q},\underline{p}}\rangle\langle e_{\underline{q},\underline{p}}|\, d^3\underline{q}\, d^3\underline{p} \quad ; \tag{4.6}$$

$$(\mathbb{P}_e f)(\underline{q},\underline{p}) = \hbar^{-3/2}(e_{\underline{q},\underline{p}}, f) \quad ; \tag{4.7}$$

$$e_{\underline{q},\underline{p}}(\underline{q}',\underline{p}') = \hbar^{-3/2}(e_{\underline{q}',\underline{p}'}, e_{\underline{q},\underline{p}}) \quad , \tag{4.8}$$

where $f \in L^2(\Gamma)$, and we have assumed that the operators $U(\underline{q},\underline{p})$ have been mapped into the unitary operators $W(\underline{q},\underline{p})$ on $L^2(\Gamma_e)$:

$$(W(\underline{q},\underline{p})f)(\underline{q}',\underline{p}') = \exp\left[\tfrac{i}{\hbar}\underline{q}\cdot(\underline{p}-\underline{p}')\right] f(\underline{q}'-\underline{q}, \underline{p}'-\underline{p}) \tag{4.9}$$

and,

$$e_{\underline{q},\underline{p}} = W(\underline{q},\underline{p})e \tag{4.10}$$

It is clear that the operators $W(\underline{q},\underline{p})$ as given in (4.9) can be extended to the whole of $L^2(\Gamma)$, obtaining thereby a reducible representation of the Weyl group $\mathcal{W}$. The irreducible subrepresentations, which are then projected out by the projection operators of the type (4.6), give rise to extremal phase space representations of quantum mechanics.

Our approach to the relativistic problem will follow similar lines in the quantum mechanical case. However, it is advantageous in this instance to rely more on a group theoretical approach, and to study the classical problem first. Let $\mathcal{P}_+^\uparrow$ denote the Poincaré group, which is the semidirect product of $\mathcal{L}_+^\uparrow$, the proper, orthocronous Lorentz group and T^4 the abelian group of space-time translations. Let $T \otimes SO(3)$ be the subgroup of $\mathcal{P}_+^\uparrow$ consisting of time translations and spatial rotations. We shall construct first representations of $\mathcal{P}_+^\uparrow$ which are induced from unitary irreducible representations of this subgroup. These representations will describe the classical statistical mechanics of an ensemble of free particles having mass m and spin j.

Let $\{a, \Lambda\}$ denote an element in $\mathcal{P}_+^\uparrow$, and $\{(a_0, \underline{0}), R\}$ an element in $T \otimes SO(3)$. Then $\{a, \Lambda\}$ has the (Mackey) decomposition (cf. (16))

$$\{a, \Lambda\} = \{(0, \underline{a} - \frac{a_0}{c}\underline{v}), \Lambda_v\}\{(\frac{a_0}{\gamma}, \underline{0}), R\}, \qquad (4.11)$$

$$\Lambda = \Lambda_v R \quad , \qquad (4.12)$$

corresponding to the left coset space

$$\mathcal{M} = \mathcal{P}_+^\uparrow / T \otimes SO(3) \qquad (4.13)$$

of $\mathcal{P}_+^\uparrow$ modulo the subgroup $T \otimes SO(3)$. In Eqs. (4.11) and (4.12), Λ_v is a pure Lorentz boost for the velocity $\underline{v}$, R is a spatial rotation, $a = (a_0, \underline{a})$, c is the velocity of light and $\gamma = (1 - v^2/c^2)^{-\frac{1}{2}}$. In view of the nature of the right hand side of Eq. (4.11), $\mathcal{M}$ can be identified with $\mathbb{R}^3 \times \mathcal{V}^+$ ($\mathcal{V}^+$ forward 'velocity hyperboloid'),

$$u_0^2 - \underline{u}^2 = -c^2 \qquad (4.14)$$

$$u = (u_0, \underline{u}) = \gamma(c, \underline{v}) \quad . \qquad (4.15)$$

It is possible to prove (16) that the measure $d^3\underline{q}\, d^3\underline{u}$ on $\mathbb{R}^3 \times \mathcal{V}^+$ is invariant with respect to the action of $\mathcal{P}_+^\uparrow$.

In the sequel it will be more useful to parametrize $\mathcal{M}$ by elements of the phase space

$$\Gamma^{(m)} = \mathbb{R}^3 \times \mathcal{V}_m^+ \tag{4.16}$$

where $\mathcal{V}_m^+$ is the 'forward mass hyperboloid' consisting of 4-momentum vectors

$$p = (p_0, \underline{p}) = mu = (mc\gamma, m\underline{v}\gamma) \tag{4.17}$$

of particles having a fixed mass m. We shall then denote the boost Λ_v by Λ_p, and replace the measure $d^3\underline{q}\,d^3\underline{u}$ by the equivalent measure $d^3\underline{q}\,d^3\underline{p}$. Once again, we shall denote by $L^2(\Gamma^{(m)})$ the Hilbert space of functions on $\Gamma^{(m)}$ which are square integrable with respect to $d^3\underline{q}\,d^3\underline{p}$.

Consider the unitary irreducible representation $V^{j,m}$ of $T \otimes SO(3)$ defined by

$$V^{j,m}(q_0, R)\xi = \exp[-\frac{imc}{\hbar} q_0] D^j(R)\xi \tag{4.18}$$

Here j is one of the numbers 0,1/2,1, 3/2,...;$m > 0$ and ξ is a vector in a 2j+1 dimensional spinor space $\mathcal{K}^j$. The representation D^j is the usual 2j+1 dimensional unitary irreducible representation of $SO(3)$ (or rather its covering group $SU(2)$). The representation $V^{j,m}$ induces a (reducible) unitary representation $\{a,\Lambda\} \longrightarrow W^{cl}(a,\Lambda)$ of $\mathcal{P}_+^\uparrow$ on $\mathcal{K}^j \otimes L^2(\Gamma^{(m)})$ in the manner

$$(W^{cl}(a,\Lambda)f)(\underline{q},\underline{p}) = \exp[-\frac{im^2c^2}{\hbar}(\Lambda^{-1})^0{}_\nu(q-a)^\nu/(\Lambda^{-1})^0{}_\nu p^\nu]\times \\ \times D^j(\Lambda_p^{-1}\Lambda\Lambda_{\Lambda_p^{-1}})f(\{a,\Lambda\}^{-1}(\underline{q},\underline{p})) . \tag{4.19}$$

In Eq. (4.19) standard summation conventions for 4-vectors have been employed. Also, f is a vector in $\mathcal{K}^j \otimes L^2(\Gamma^{(m)})$, $q-a$ denotes the 4-vector $(-a_0, \underline{q}-\underline{a})$, and $\{a,\Lambda\}^{-1}(\underline{q},\underline{p})$ is the translate of the point $(\underline{q},\underline{p})$, treated as a coset element in $\mathcal{M}$ under the action of the group element $\{a,\Lambda\}^{-1}$. Details of the construction of the representation $W^{cl}(a,\Lambda)$ may be found in Ref. 16.

It is interesting to study the generator of the subgroup of time translations for the representation W^{cl}. Let

$$V^{cl}(t) = W^{cl}((-ct,\underline{0}), I) \tag{4.20}$$

be the one-parameter abelian subgroup of time translations, I = identity element of $SO(3)$, and

$$f(\underline{q},\underline{p},t) = (V^{ce}(t)f)(\underline{q},\underline{p}) \quad , \tag{4.21}$$

$\forall f \in \mathcal{K}^j \otimes L^2(\Gamma^{(m)})$. Writing

$$\underline{\hat{P}} = (\hat{P}^1, \hat{P}^2, \hat{P}^3) = -i\hbar \left(\frac{\partial}{\partial q^1}, \frac{\partial}{\partial q^2}, \frac{\partial}{\partial q^3}\right) , \tag{4.22}$$

we obtain the relations

$$V^{ce}(t) = \exp\left[-\frac{i}{\hbar} H^{ce} t\right] , \tag{4.23}$$

$$H^{ce} = \frac{c}{p_0}\left[m^2c^2 + \underline{p}\cdot\underline{\hat{P}}\right] . \tag{4.24}$$

For vectors f which are in the common dense domain of the operators $\hat{P}^j$, we have the equation of motion

$$\frac{\partial}{\partial t} f(\underline{q},\underline{p},t) = -\frac{ic}{\hbar p_0}\left[m^2c^2 + \underline{p}\cdot\underline{\hat{P}}\right] f(\underline{q},\underline{p},t) . \tag{4.25}$$

This last equation can be cast into a manifestly covariant form if we introduce the 4-vector operator

$$\hat{P}^\mu = (\hat{P}^0, \underline{\hat{P}}) = \left(\frac{i\hbar}{c}\frac{\partial}{\partial t}, \underline{\hat{P}}\right) ; \tag{4.26}$$

for then we have the alternative expression

$$(p_\mu \hat{P}^\mu - m^2c^2)\, f(\underline{q},\underline{p}) = 0 \tag{4.27}$$

On the other hand, if we study the time evolution of the square amplitudes

$$\rho(q,p) = \| f(\underline{q},\underline{p},t) \|^2 \tag{4.28}$$

we obtain the similar equation

$$p_\mu \hat{P}^\mu \rho(q,p) = 0 . \tag{4.29}$$

Eq. (4.29) is immediately recognized to be the relativistic generalization of the Liouville equation of classical statistical mechanics. Indeed, in the limit of $c \longrightarrow \infty$ this equation is easily seen to go over to the usual Liouville equation

$$\frac{\partial}{\partial t} \rho(\underline{q}, \underline{p}, t) = - \underline{p}/m \cdot \underline{\nabla}_q \rho(\underline{q}, \underline{p}, t) \quad , \tag{4.30}$$

one that could have been obtained by studying similar induced representations of the Galilei group. Equations of motion in the presence of electromagnetic interactions can also be studied in connection with Eq. (4.29) (cf. Ref. 4). We close our discussion of the classical case by noting that if we define a probability 4-current density

$$j^\mu(\underline{q}, t) = \int_{V_m^+} p^\mu \rho(\underline{q}, \underline{p}, t)\, p_0^{-1}\, d^3\underline{p} \tag{4.31}$$

then j^μ satisfies the equation of continuity

$$\frac{\partial}{\partial q^\mu} j^\mu = 0 \quad . \tag{4.32}$$

Also, for the integrated form of this current density

$$J^\mu(t) = \int_{\mathbb{R}^3} j^\mu(\underline{q}, t)\, d^3\underline{q} \tag{4.33}$$

we have

$$\frac{\partial J^0}{\partial t} = 0 \quad . \tag{4.34}$$

To go now to the quantum case, we start by looking at the usual irreducible representation of $\mathcal{P}_+^\uparrow$ for mass m and spin j. The Hilbert space is

$$\tilde{\mathcal{H}}_m = \mathcal{K}^j \otimes L^2(V_m^+, \frac{d^3\underline{k}}{k_0}) \quad , \tag{4.35}$$

and the representation $\{a, \Lambda\} \longrightarrow \tilde{U}(a, \Lambda)$ is given by

$$(\tilde{U}(a, \Lambda)\tilde{\varphi})(k) = \exp\left[-\frac{i}{\hbar} ka\right] D^j(\Lambda_k^{-1} \Lambda \Lambda_{\Lambda^{-1}k})\, \tilde{\varphi}(\Lambda^{-1}k) \tag{4.36}$$

$\forall \tilde{\varphi} \in \mathcal{H}$, and where we have written ka for the 4-scalar product $k_0 a_0 - \underline{k} \cdot \underline{a}$.

Let $\hat{e}$ be a complex valued function on $\mathbb{R}^3$ which satisfies

$$\int_{\mathbb{R}^3} |\hat{e}(\underline{k})|^2 d^3\underline{k} = mc \quad . \tag{4.37}$$

Let us make the further assumption that $\hat{e}$ is rotationally invariant, i.e.,

$$\hat{e}(R\underline{k}) = \hat{e}(\underline{k}) \tag{4.38}$$

for all pure rotations R in $\mathcal{P}_+^\uparrow$. The next theorem is the relativistic analogue of the isometry described by Eqs. (4.1) - (4.4) above (cf. Ref. 16).

Theorem 4.1

The mapping $\tilde{\varphi} \longrightarrow \varphi_e$ defined by

$$\varphi_e(\underline{q},\underline{p}) = \hbar^{-\frac{3}{2}} \int_{V_m^+} \overline{\hat{e}(k)} \, (\tilde{U}^*((0,\underline{q}),\Lambda_p)\tilde{\varphi})(k) \frac{d^3k}{k_0} \tag{4.39}$$

establishes a linear Hilbert space isometry between $\tilde{\mathcal{H}}_m$ and a proper subspace of $\mathcal{K}^j \otimes L^2(\Gamma^{(m)})$

Once again we denote the range of the isometry (4.39) by $\mathcal{K}^j \otimes L^2(\Gamma_e^{(m)})$ and by $\mathbb{P}_e$ the projector onto this range. Further, we denote by $\hat{e}$ the matrix valued function on $\Gamma^{(m)}$ which is obtained when $\tilde{\varphi}(k)$ in (4.39) is replaced by $\hat{e}(k)$. It follows then that $\mathbb{P}_e$ is of the form

$$(\mathbb{P}_e f)(\underline{q},\underline{p}) = \hbar^{-3/2} \int_{\Gamma^{(m)}} \hat{e}(\underline{q}',\underline{p}') \, (\tilde{U}^*((0,\underline{q}),\Lambda_{\underline{p}}) f)(\underline{q}',\underline{p}') \, d^3\underline{q}' d^3\underline{p}' \tag{4.40}$$

The image W^{qm} of the representation $\tilde{U}$ under the above isometry is now given by

$$\begin{aligned}(W^{qm}(a,\Lambda)\varphi_e)(\underline{q},\underline{p}) = D^j(\Lambda_p^{-1}\Lambda\Lambda_{\Lambda^{-1}p}) \Big(\exp\Big[-\frac{i}{\hbar c}(\Lambda^{-1})^0{}_\nu \, \times \\ \times (\underline{q}-a)^\nu \{m^2c^3/(\Lambda^{-1})^0{}_\nu p^\nu + H^{qm} - H^{ce}\}\Big]\varphi_e\Big)(\{a,\Lambda\}^{-1}(\underline{q},\underline{p}))\end{aligned} \tag{4.41}$$

where H^{qm} is the quantum mechanical Hamiltonian

$$(H^{qm}\varphi_e)(\underline{q},\underline{p}) = (c[\hat{\underline{P}}^2 + m^2c^2]^{\frac{1}{2}}\varphi_e)(\underline{q},\underline{p}) \; . \tag{4.42}$$

It is clear that the representation described in Eq. (4.41) can be extended to the whole of $\mathcal{K}^j \otimes L^2(\Gamma^{(m)})$, so that the quantum systems correspond to irreducible subsectors which are projected out by operators of the type (4.40). It is possible to show (17) that non-relativistically these subsectors go over exactly to the ones containing irreducible representations of the Weyl group $\mathcal{W}$ discussed in the beginning of this section. Consequently, they describe stochastic phase space realizations of quantum mechanics.

As in Eq. (4.21) let us define the functions $\psi_e(\underline{q},\underline{p},t)$ by

$$\psi_e(\underline{q},\underline{p},t) = \left(\exp\left[-\tfrac{i}{\hbar} H^{qm} t\right] D(\Lambda,\hat{P},p)\,\psi_e\right)(\underline{q},\underline{p}) \quad , \qquad (4.43)$$

where $D(\Lambda,\hat{P},p)$ is the operator obtained from $D^j(\Lambda_k^{-1}\cdot\Lambda_p\Lambda_{\Lambda_p^{-1}k})$ by replacing k with P. We then have the following very interesting result (16,17).

Theorem 4.2

The functions $\psi_e(\underline{q},\underline{p},t)$ satisfy the Klein-Gordon equation

$$\left(\Box_q - \frac{m^2c^2}{\hbar^2}\right)\psi_e(\underline{q},\underline{p},t) = 0 \ , \quad \Box_q = -\frac{1}{c^2}\frac{\partial^2}{\partial t^2} + \underline{\nabla}_q^2 \ . \qquad (4.44)$$

The 4-current

$$j_e^{\mu}(\underline{q},t) = \int_{V_m^+} p^{\mu}\,\|\psi_e(\underline{q},\underline{p},t)\|^2\,\frac{d^3\underline{p}}{p_0} \qquad (4.45)$$

transforms as a 4-vector under Lorentz transformations, and if $\hat{e}$ is a real function in $\tilde{\mathcal{H}}$, this current is also conserved, i.e.,

$$\frac{\partial j_e^{\mu}}{\partial q^{\mu}} = 0 \quad . \qquad (4.46)$$

The above theorem assures us that the square amplitudes $\|\psi_e\|^2$ on phase space have proper transformation properties and hence are bona-fide probability densities for finding the system localized at a stochastic phase space point. In addition, the Hamiltonian H^{qm} in (4,42) has only a positive spectrum, so that in our theory the positivity of the energy as well as of the probability densities is maintained. It is possible to study the position space localizability of systems described by the wave functions ψ_e, by looking at POVSI's associated with the representation W^{qm}. The resulting localization operators have to be interpreted in terms of the stochastic nature of the underlying space. We end this section with the remark that it is possible to transcribe the relativistic observables once again into distributions over phase space, and much of the work done in the last section for non-relativistic systems can be carried over to the relativistic domain.

5. The Dirac Equation and the External Field Problem

In this section we very briefly describe how the Dirac equation can be described on a stochastic phase space, and then discuss an alternative 2-component equation for a spin-1/2 particle interacting with an external electromagnetic field. The problem of a charged spin-zero particle is also discussed in the same setting. The work described here is developed in detail in Refs. 4, 18, 19 and 20.

To derive a phase space version of the Dirac equation, we start with the representations W^{qm} of $\mathcal{P}_+^\uparrow$ introduced in the last section for j= 1/2. Since we shall wish to incorporate parity and charge conjugation symmetries, we have to 'double' the Hilbert space of the problem to

$$\mathcal{H}_W = \mathcal{K}^{1/2} \otimes \mathcal{L}^2(\Gamma_e^{(m)}) \oplus \mathcal{K}^{1/2} \otimes \mathcal{L}^2(\Gamma_e^{(m)}) \tag{5.1}$$

As usual $\mathcal{K}^{1/2}$ is a two dimensional spinor space. Let γ_0 be a 4 x 4 matrix formed out of the 2 x 2 unit matrix $\mathbb{1}$:

$$\gamma_0 = \begin{pmatrix} \mathbb{1} & 0 \\ 0 & -\mathbb{1} \end{pmatrix} . \tag{5.2}$$

Consider the two representations $W_\pm$ of $\mathcal{P}_+^\uparrow$ (corresponding to positive and negative energies) on $\mathcal{K}^{1/2} \otimes \mathcal{L}^2(\Gamma_e^{(m)})$:

$$\begin{aligned}(W_+(a,\Lambda)\psi_e)(\underline{q},\underline{p}) = {} & D^{1/2}(\Lambda_p^{-1}\Lambda\Lambda_{\Lambda_p^{-1}})\,(\exp[-\tfrac{i}{\hbar c}(\Lambda^{-1})^0{}_\nu\,(\underline{q}-a)^\nu \times \\ & \times\{m^2c^3/(\Lambda^{-1})^0{}_\nu p^\nu + H^{qm} - H^{cl}\}]\,\psi_e)(\{a,\Lambda\}^{-1}(\underline{q},\underline{p})) \quad ;\end{aligned} \tag{5.3a}$$

$$\begin{aligned}(W_-(a,\Lambda)\psi_e)(\underline{q},\underline{p}) = {} & \tau\,\overline{D^{1/2}(\Lambda_p^{-1}\Lambda\Lambda_{\Lambda_p^{-1}})}\,\tau^{-1}(\exp[\tfrac{i}{\hbar c}(\Lambda^{-1})^0{}_\nu(\underline{q}-a)^\nu \times \\ & \times\{m^2c^3/(\Lambda^{-1})^0{}_\nu\, p^\nu + H^{qm} - H^{cl}\}]\,\psi_e)(\{a,\Lambda\}^{-1}(\underline{q},\underline{p})) \quad ,\end{aligned} \tag{5.3b}$$

where τ is the 2x2 unitary matrix

$$\tau = \begin{pmatrix} 0 & 1 \\ -1 & 0 \end{pmatrix} \tag{5.4}$$

which has the property that if $A \in SU(2)$ then $\tau A \tau^{-1}$ is a matrix which is the complex conjugate $\bar{A}$ of A. Thus in fact $\tau \overline{D^{\frac{1}{2}}} \tau^{-1} = D^{\frac{1}{2}}$ but we prefer to write Eq. (5.3b) in this particular form for future convenience. Next we see that $W_{FW} = W_+ \oplus W_-$ is a unitary (reducible) representation of $\mathcal{P}_+^\uparrow$ on $\mathcal{H}_W$. Indeed, if $\psi_W \in \mathcal{H}_\omega$,

$$(W_{FW}(a,\Lambda)\psi_W)(\underline{q},\underline{p}) = \mathcal{L}^{\frac{1}{2}}(\Lambda_p^{-1}\Lambda\Lambda_{\Lambda^{-1}p})(\exp[-\frac{i}{\hbar c}(\Lambda^{-1})^0{}_\nu \times \times (\underline{q}-a)^\nu \gamma_0 \{m^2c^3/(\Lambda^{-1})^0{}_\nu p^\nu + H^{qm} - H^{cl}\}]\psi_W)(\{a,\Lambda\}^{-1}(\underline{q},\underline{p})) \tag{5.5}$$

where $\mathcal{L}^{\frac{1}{2}}$ is the 2-fold representation,

$$\mathcal{L}^{\frac{1}{2}} = \begin{pmatrix} D^{\frac{1}{2}} & 0 \\ 0 & \tau\overline{D^{\frac{1}{2}}}\tau^{-1} \end{pmatrix} , \tag{5.6}$$

of $SU(2)$ on $\mathcal{K}^{\frac{1}{2}} \oplus \mathcal{K}^{\frac{1}{2}}$. Equation (5.5) already contains the Dirac equation in the Foldy-Wouthuysen form, i.e., in the form in which the positive and negative energy parts of the wave function ψ_W appear decoupled. To display this explicitly, let

$$H_W = \gamma_0 H^{qm} \tag{5.7}$$

and define

$$\psi_W(\underline{q},\underline{p},t) = (\exp[-\frac{i}{\hbar} H_W t]\psi_W)(\underline{q},\underline{p}) \tag{5.8}$$

Then it is straight forward to verify that

$$i\hbar \frac{\partial \psi_W}{\partial t}(\underline{q},\underline{p}) = (H_W \psi_W)(\underline{q},\underline{p}) \tag{5.9}$$

which is the Dirac equation in the Foldy-Wouthuysen representation.

The standard form of the Dirac equation can now be obtained by a Foldy-Wouthuysen type of a transformation. We note first that $D^{\frac{1}{2}}$ can be extended to the representation $D^{\frac{1}{2},0}$ of $SL(2,\mathbb{C})$ and similarly, $\overline{D^{\frac{1}{2}}}$ to $D^{0,\frac{1}{2}}$.
Let us then write

$$\mathcal{L} = \begin{pmatrix} D^{\frac{1}{2},0} & 0 \\ 0 & \tau D^{0,\frac{1}{2}}\tau^{-1} \end{pmatrix} \tag{5.10}$$

and introduce the 4-spinors ϕ_D defined as

$$\phi_D(q,p) = \mathcal{L}(\Lambda_p)\psi_W(q,p) \tag{5.11}$$

This transformation is not unitary, and hence necessitates a change in the scalar product of the underlying Hilbert space $\mathcal{H}_W$. On the space of transformed functions ϕ_D let us introduce the scalar product

$$(\phi_D, \phi_D) = \frac{1}{mc}\int_{\Gamma(m)} \langle \phi_D(\underline{q},\underline{p}) \,|\, p_\mu \tilde{\sigma}^\mu \phi_D(\underline{q},\underline{p})\rangle \, d^3\underline{q}\, d^3\underline{p} \tag{5.12}$$

where $\langle\ |\ \rangle$ represents the scalar product in $\mathcal{K}^{1/2} \oplus \mathcal{K}^{1/2}$.
The $\tilde{\sigma}$'s are the 4x4 matrices

$$\tilde{\sigma}_0 = \begin{pmatrix} 1\!\!1 & 0 \\ 0 & 1\!\!1 \end{pmatrix}, \quad \tilde{\sigma}^k = \begin{pmatrix} \sigma^k & 0 \\ 0 & -\sigma^k \end{pmatrix}, \tag{5.13}$$

and for $k = 1,2,3$, the σ^k's are the Pauli spin matrices. With the scalar product of Eq. (5.12) the ϕ_D's form a Hilbert space $\mathcal{H}_D$ which is isometric to $\mathcal{H}_W$ via the mapping (5.11).

On $\mathcal{H}_D$ we finally make the unitary transformation

$$\mathcal{F} = \frac{\hat{P}^0 + mc - \underline{\gamma}\cdot\hat{\underline{P}}}{[2\hat{P}^0(\hat{P}^0 + mc)]^{1/2}}, \tag{5.14}$$

with

$$\mathcal{F}\phi_D = \psi_D, \tag{5.15}$$

where the γ's are the usual Dirac matrices. We get thereby the Dirac equation in standard form:

$$\left(i\gamma_\mu \frac{\partial}{\partial q^\mu} - \frac{mc}{\hbar}\right)\psi_D(\underline{q},\underline{p}) = 0 \quad . \tag{5.16}$$

It is possible to introduce an electromagnetic interaction into Eq. (5.16) via a minimal coupling. Let us however, study an alternative model of a spin-1/2 particle described by a 2-component equation, and introduce a coupling there. The procedure will be a canonical quantization of a classical equation, from which it will also be clear how a minimal coupling could be introduced on phase space (cf. 4,18,19 and 20 for details).

The starting point is a classical Hamiltonian (19,20) E for a charged 'spinning' particle interacting with an electromagnetic field $F^{\mu\nu}$ with potentials A^μ at the space-time point ($\underline{x}$, t). The particle is assumed to have intrinsic electric and magnetic dipole moments $\underline{d}$ and $\underline{\mu}$ respectively, described by the dipole moment tensor $\sigma_{\mu\nu}^{cl}$ with

$$\sigma_{oj}^{cl} = -d^j \ , \ \sigma_{ij}^{cl} = \varepsilon_{ijk}\,\mu^k \ , \ i,j,k = 1,2,3 \ , \tag{5.17a}$$

$$\sigma_{\mu\nu} = -\sigma_{\nu\mu} \ , \ \mu,\nu = 0,1,2,3 \ , \tag{5.17b}$$

where ε_{ijk} is the completely antisymmetric tensor in three dimensions. Thus (20):

$$E = eA^0(\underline{x},t) + c\Big[\big(\underline{p} - \tfrac{e}{c}\underline{A}(\underline{x},t)\big)^2 + \big(mc + \tfrac{1}{2c}\sigma_{\mu\nu}^{cl}F^{\mu\nu}(\underline{x},t)\big)^2\Big]^{\frac{1}{2}} \ , \tag{5.18}$$

e being the electric charge of the particle. In (5.18) if we set $\sigma_{\mu\nu}^{cl} = 0$, we recover the classical Hamiltonian for a charged spinless particle.

To quantize (5.18) we first write it as an operator on $\mathcal{K}^{\frac{1}{2}} \otimes L^2(\Gamma^{(m)})$ by means of the substitutions

$$\underline{p} \longrightarrow \underline{\hat{P}} \ , \ A_\mu(\underline{x},t) \longrightarrow A_\mu(q) \ , \ F_{\mu\nu}(\underline{x},t) \longrightarrow F_{\mu\nu}(q) \tag{5.19a}$$

$$\underline{d} \longrightarrow f\frac{e\hbar}{2mc}\underline{\sigma} \ , \ \underline{\mu} \longrightarrow g\frac{e\hbar}{2mc}\underline{\sigma} \ , \tag{5.19b}$$

where $A_\mu(q)$ and $F_{\mu\nu}(q)$ are multiplication operators, f and g are constants and $\underline{\sigma}$ has the Pauli matrices as components. Thus on $\mathcal{K}^{\frac{1}{2}} \otimes L^2(\Gamma^{(m)})$ we have the Hamiltonian

$$H(q^0) = eA^0(q) + c\Big[\big(\underline{\hat{P}} - \tfrac{e}{c}\underline{A}(q)\big)^2 + \big(mc + \tfrac{1}{2c}\sigma_{\mu\nu}F^{\mu\nu}(q)\big)^2\Big]^{\frac{1}{2}} \tag{5.20}$$

where now

$$\sigma^{\mu\nu} = \frac{ie\hbar}{4mc}\big(g\,g^{\mu\lambda}g^{\nu k} + f\,\varepsilon^{\mu\nu\lambda\kappa}\big)(\sigma_\kappa\sigma_\lambda - \sigma_\lambda\sigma_\kappa) \ . \tag{5.21}$$

In (5.21) the metric tensor $g^{\mu\nu}$ has the non-zero components $g^{00} = -g^{11} = -g^{22} = -g^{33} = 1$, $\varepsilon^{\mu\nu\lambda\kappa}$ is the completely anti-symmetric tensor in four dimensions and σ_0 is the 2x2 unit matrix. On $\mathcal{K}^{1/2} \otimes L^2(\Gamma^{(m)})$ let $\mathbb{P}_e^{1/2}$ be the projector onto $\mathcal{K}^{1/2} \otimes L^2(\Gamma_e^{(m)})$ (cf. Eq. (4.40)). It can then be shown explicitly (20) that operators such as

$$A_e^{\mu} = \mathbb{P}_e^{1/2} A^{\mu} \mathbb{P}_e^{1/2} \tag{5.22a}$$

$$F_e^{\mu\nu} = \mathbb{P}_e^{1/2} F^{\mu\nu} \mathbb{P}_e^{1/2} \quad , \tag{5.22b}$$

etc., have the interpretation of stochastic potentials and fields, respectively, i.e., they act at smeared out points. To complete the quantization of (5.18), we restrict (5.20) by means of $\mathbb{P}_e^{1/2}$ and write on $\mathcal{K}^{1/2} \otimes L^2(\Gamma_e^{(m)})$

$$H_e(q^0) = eA_e^0 + c\left[(\hat{\underline{P}} - \frac{e}{c}\underline{A}_e)^2 + (mc + \frac{1}{2c}\sigma_{\mu\nu}F_e^{\mu\nu})^2\right]^{1/2} \tag{5.23}$$

The selfadjointness of (5.20) or (5.23) is relatively easy to prove (4). Also noteworthy is the fact that (5.23) leaves the space of positive energy solutions stable. Furthermore, the spin-0 case is obtained by dropping the $\sigma_{\mu\nu}F_e^{\mu\nu}$ term in (5.23).

The dynamics given by (5.23) can be shown to be gauge invariant (19, 20); for a central potential it has a spin-orbit interaction term and hydrogen-like bound state solutions (20). Moreover, since the interaction appears at a stochastic point, it is non-local and leads to interesting renormalization properties. Also noteworthy is the absence of any zitterbewegung effects.

6. Conclusion

To end this somewhat lengthy account of the theory of quantum mechanics on stochastic phase spaces, let us recapitulate some of its salient features. Firstly, the older theory is completely contained in the present theory. Secondly, the current theory achieves a great descriptive unification, in the sense that for a system with a given mass and spin, the classical and quantum dynamics - both relativistic and non-relativistic - are all described on the same Hilbert space. The theory also makes clear the transition from the relativistic to the non-relativistic regimes, since the corresponding representations of the Poincaré and the Galilei groups are all defined on the same space. Further, the classical limit of a quantum theory can also be very effectively studied within the present framework.

However, it is mainly in the relativistic portion of the theory, that most of its interesting advantages appear. The wave functions give rise to true probability densities on position and momentum spaces. The energy is positive definite and particle and antiparticle states appear in orthogonal subspaces, without the possibility of unwanted transitions of particles into anti-particle states, through small perturbations. The interaction with an electromagnetic field occurs at a smeared out point and this leads in turn to interesting consequences in the second quantized theory. Let us emphasize again, that it is really in the concept of a stochastic space that the advantages of the present theory are to be traced. It would certainly be worth investigating, in this same spirit, the problem of higher spin equations, of possible field theories on phase space and the Feynman path-integral approach on phase space. Work in some of these directions is currently in progress.

References

1. F. Bayen, M. Flato, C. Fronsdal, A. Lichnerowiz and D. Sternheimer, Ann. Phys. (NY), 111 (1978), 61 and 111.
2. G.S. Agarwal and E. Wolf, Phys. Rev. D2 (1970) 2161, 2187 and 2206.
3. E.P. Wigner, Perspectives in Quantum Theory, edited by W. Yourgrau and A. Vander Merwe (M.I.T. Press, Cambridge, Mass., 1971), pp. 25-36)
4. E. Prugovečki, Phys. Rev. D18 (1978), 3655.
5. G. Ludwig, 'Deutung des Begriffs "physikalische Theorie" and axiomatische Grundlegung der Hilbertraumstruktur der Quantenmechanik durch Hauptsätze des Messens', Lecture Notes in Physics 4 (Springer-Verlag, Berlin-Heidelberg-New York, 1970)
6. E.B. Davis and J.T. Lewis, Commun. Math. Phys. 17 (1970).
7. S. Gudder and C. Piron, J. Math. Phys. 12 (1971), 1583.
8. E.B. Davies, J. Funct. Anal. 6 (1970), 318.
9. G.W. Mackey, Proc. Math. Acad. Sci. USA 35 (1949), 537.
10. S.T. Ali and E. Prugovečki, J. Math. Phys. 18 (1977), 219.
11. S.T. Ali and H.D. Doebner, J. Math. Phys. 17 (1976), 1105.
12. S.T. Ali and E. Prugovečki, Internat. J. Theor. Phys. 16 (1977), 689.
13. E. Prugovečki, Physica 91A (1978), 202.
14. E. Prugovečki, Ann. Phys. (NY) 110 (1978), 102.
15. S.T. Ali and E. Prugovečki, Physica 89A (1977), 501.
16. S.T. Ali, J. Math. Phys. 20 (1979), 1385; and 21 (1980), 818.
17. E. Prugovečki, J. Math. Phys. 19 (1978), 2260.
18. E. Prugovečki, Rep. Math. Phys. 18 (1980), to appear.
19. S.T. Ali and E. Prugovečki, 'Consistent models of spin-0 and 1/2 extended particles scattering in external fields' in Proceedings of the 1979 Conference on Mathematical Methods and Applications of Scattering Theory, J.A. De Santo, A.W. Saenz and W.W. Zachary, eds., Springer-Verlag, to appear.
20. S.T. Ali and E. Prugovečki, 'Self-consistent relativistic model for extended spin-1/2 particles in external fields', to appear.

On the Confinement of Magnetic Poles

by

H.R. Petry

Institut für Theoretische Kernphysik
University of Bonn, Nussallee

0. Introduction

In this note, the properties of Hamiltonians describing electron-monopole scattering will be discussed. In the first section we review some interesting topological features connected with this problem. In section2 we present the eigenfunctions of the Schrödinger and the Dirac equation. Section 3 reviews the difficulties connected with the asymptotic scattering conditions. In section 4, a discussion of the electron vacuum states will provide a possible mechanism for the confinement of magnetic poles. Finally, section 5 yields a unified description of the case with several monopoles together with a treatment of the self-adjoint extensions of the Dirac operator.

1. Magnetic monopoles, topological preliminaries:

It has been shown [1) 2)] that the wave function of an electron in an external magnetic monopole field B, $\vec{B} = \mu\, \vec{x}/\,|\vec{x}|^3$, can no longer be regarded as an ordinary function, but has to be redefined as a section in a complex line bundle ξ over $\dot{\mathbb{R}}^3 = \mathbb{R}^3 - \{0\}$ with the following properties:

(a) ξ has a covariant derivative ∇ and a fibre metric $<,>$, such that

$$y < \sigma_1, \sigma_2 > \; = \; <\nabla_y \sigma_1, \sigma_2> \; + \; <\sigma_1, \nabla_y \sigma_2>$$

holds for arbitrary vector fields y and the sections σ_1 and σ_2;

(b) the curvature $\omega(\nabla)$ satisfies

$$\omega(\nabla) = ieB \quad .$$

(In the last equation, B is regarded as a two-form). The necessary and sufficient condition that ξ exists is simply that $eB/2\pi$ defines an integer cohomology class [1], which holds if $2e\mu \in \mathbb{Z}$ and yields the Dirac quantization of the monopole strength μ [3].

It follows directly from ref. [1] that in the more general case of

$$\vec{B} = \sum_{i=1}^{n} \mu_i (\vec{x} - \vec{x}_i)/|\vec{x} - \vec{x}_i|^3 \quad ,$$

a suitable line bundle $\tilde{\xi}$, satisfying (a) and (b), exists if

$$\mu_i e = m_i \quad ,$$

m_i being integer or half-integer. Moreover, $\tilde{\xi}$ is determined up to bundle isomorphisms which, in physical terms, correspond to gauge transformations. If $\tilde{\xi}$ is given, one can immediately write down a Schrödinger equation for sections σ,

$$(1) \qquad E\sigma = -\frac{1}{2M}\sum_{k=1}^{3} \nabla_k^2 \sigma$$

where ∇_k denotes the covariant derivative in the direction of the orthonormal basis vector fields e_k and M is the particle mass.

The corresponding Dirac equation is an equation for a column $\tilde{\sigma}$ of 4 sections $\sigma_1, \dots, \sigma_4$:

$$(2) \qquad E\tilde{\sigma} = -i\sum_{k=1}^{3} \alpha_k \nabla_k \tilde{\sigma} + \beta M \tilde{\sigma}$$

with suitable Dirac matrices α_k and β. Bilinear invariants are, of course, formed with the help of the fibre metric.

2. Explicit solutions for one monopole

The case of a single monopole, located at the origin, has been treated explicitely in the following way [1]: If $e \cdot \mu = m$, (m integer or half-integer), then ξ_m is an associated vector bundle $P \times_{d_m} \mathbb{C}$. $\pi : P \longrightarrow \mathbb{R}^3$ is a principle U(1)-bundle and the representation

$$d_m : U(1) \longrightarrow U(1) \qquad (:= z \in \mathbb{C}, \quad |z| = 1)$$

is defined by

$$d_m(g) = g^{2m} \quad .$$

By a standard theorem, there is a bijection $d_m^{\#}$ of sections in ξ_m with equivariant functions ψ on P, i.e. $\psi(q\cdot z) = z^{-2m}\psi(z)$, $(q \in P,\ z \in U(1))$.

If $H: T_{\pi(q)} \longrightarrow T_q$ is a horizontal lift of the tangent spaces, then a covariant derivative ∇ in ξ_m is given by

$$\nabla_y \sigma = d_m^{\#-1}(H(y)d_m^{\#}(\sigma))$$

where $H(y)\cdot d_m^{\#}(\sigma)$ is the ordinary derivative of $d_m^{\#}(\sigma)$ in the direction of $H(y)$.

One has to adjust H in such a way that (a) and (b) holds. Before doing this, it is useful to write down the equations (1) and (2) in terms of equivariant functions: If we set $\psi = d_m^{\#}(\sigma)$ it then follows that

$$(1') \qquad E\psi = -\frac{1}{2M}\sum_{k=1}^{3} H(e_i)^2\psi \quad ,$$

respectively

$$(2') \qquad E\tilde{\psi} = -i\sum_{k=1}^{3} i\,\alpha_k H(e_k)\tilde{\psi} + \beta M\tilde{\psi} \quad ,$$

where, in the last equation, $\tilde{\psi}$ is now a column of 4 equivariant functions.

We now sketch the construction of P and H (see ref. [1]) for further details).

Let H denote the quaternions, $\dot{H} = H - \{0\}$, and identify $\dot{\mathbb{R}}^3$ with the elements of $\dot{H}$ orthogonal to the identity element. Set $P = \dot{H}$ and define $\pi: \dot{H} \longrightarrow \dot{\mathbb{R}}^3$ by $\pi(q) = qe_3\bar{q}$. If $z \in U(1)$ has the form $z = \cos\alpha + i\sin\alpha$, define the group action by $q\cdot z = q(\cos\alpha + e_3\cdot\sin\alpha)$. If h is a vector field in $\dot{\mathbb{R}}^3$ we define its horizontal lift with the help of quaternion multiplication:

$$H(h)_q = -h(\pi(q))\cdot qe_3/2|q|^2 \quad .$$

Conditions (a) and (b) hold if we set

$$\langle\sigma_1,\sigma_2\rangle_x = \overline{((d_m^{\#}\sigma_1)(q))}\cdot(d_m^{\#}\sigma_2)(q)$$

where $q \in \pi^{-1}(x)$ is arbitrarily chosen.

The group $SU(2) = \{q \in \dot{Q};\ |q| = 1\}$ acts on P by right and left multiplication. The 3 vector fields, or angular momentum operators, induced by left and right actions, are given for arbitrary functions $\psi: P \longrightarrow \mathbb{C}$ by

$$(I_1\psi)(q) = \frac{i}{2}\frac{d}{dt}\psi(\exp(-te_1)q)\Big|_{t=0}$$

and $$(K_1\psi)(q) = \frac{i}{2}\frac{d}{dt}\psi(q\exp te_1)\Big|_{t=0},$$

respectively.

The equations (1') and (2') are invariant under the left action of SU(2) on P, which reflects the rotational invariance of our problem. For the Schrödinger equation (1') one obtains a complete set of solutions which satisfy $(\vec{I})^2\psi = j(j+1)\psi$, $I_3\psi = n\psi$ and $K_3\psi = -m\psi$, the last equation being the equivariance condition. More explicitely these solutions are

$$\psi^{Ej}_{n,m}(q) = |q|^{-1}J_\lambda(\sqrt{2ME}|q|^2)D^j_{n,-m}(q/|q|)$$

with $E>0$; $n=-j, -j+1, -n, j$; $j = |m|, |m|+1, \ldots$, and

$$\lambda = ((j+\tfrac{1}{2}) - m^2)^{1/2} .$$

J_λ is a Bessel function and $D^j_{n,-m}$ is the Wigner coefficient of an irreducible representation of SU(2) [4].

The case of the Dirac equation is more complicated because the Dirac Hamiltonian is not self-adjoint. But it turns out that it has self-adjoint extensions [5]. There are only two rotationally and CT-invariant possibilities, one of them yielding a bound state. If that possibility is rejected (because it does not appear in the classical problem) we are left with a unique self-adjoint extension. (in rejecting the bound state, we differ from ref. [5]). In order to describe a complete set of eigenstates, we identify $\mathbb{C}^4 = \mathbb{C}^2 \times \mathbb{C}^2$ and define

$$\alpha_i = \sigma_2 \times \sigma_i, \qquad \beta = 1 \times \sigma_1 .$$

Next we introduce the operator

$$D = \sum_{k=1}^{3} (\sigma_k I_k + K_3 \cdot \sigma_k \pi(q)_k) + 1$$

whose square is

$$D^2 = (\vec{I} + \tfrac{1}{2}\vec{\sigma})^2 - K_3^2 + \tfrac{1}{4} .$$

Hence D has eigenvalues $\lambda = \pm (I(I+1) - m^2 + \frac{1}{4})^{1/2}$ for

$I = |m| + \frac{1}{2}, |m| + \frac{3}{2}, \ldots$, and a vanishing eigenvalue for $I = |m| - \frac{1}{2}$.

There is, for each $I > |m| - \frac{1}{2}$ a function $\psi_{I,n}(q/|q|)$ with values in $\mathbb{C}^2$, unique up to scalar multiplication, such that

$$D\psi_{I,n} = +(I(I+1) - m^2 + \tfrac{1}{4})\psi_{I,n})$$

$$(I_3 + \tfrac{1}{2}\sigma_3)\psi_{I,n} = n\psi_{I,n})$$

$$K_3\psi_{I,n} = -m\psi_{I,n}$$

where the last equation reflects the equivariance of $\psi_{I,n}$. n is restricted to $n = -j, -j+1, \ldots, j$ and $I = |m| - \frac{1}{2}, |m| + \frac{1}{2}, \ldots$

In addition

$$(\vec{I} + \tfrac{1}{2}\vec{\sigma})^2\psi_{I,n} = I(I+1)\psi_{I,n} .$$

The functions

$$|q|^{-1}V(E) \times \Big[J_{\lambda-\frac{1}{2}}(\sqrt{E^2 - M^2}\,|q|^2) \pm$$

$$\pm i\sum_{k=1}^{3}\sigma_k\,\pi(q)_k\,J_{\lambda+\frac{1}{2}}(\sqrt{E^2 - M^2}\,|q|^2)\Big]\psi_{I,n}(q/|q|)$$

with $EV(E) = (\pm\sigma_2(E^2 - M^2)^{1/2} + \sigma_1 M)V(E)$ and $I > |m| - \frac{1}{2}$

together with the functions

$$V(E,|q|) \times \psi_{|m|-1/2,n}(q/|q|) \quad , \quad \text{with}$$

$$V(E,q) = \frac{1}{2|q|^2}\exp\left(-i\frac{m}{|m|}\sigma_2(E - \sigma_1 M)\cdot|q^2|\right)\begin{pmatrix}1 + m/|m|\\ 1 - m/|m|\end{pmatrix}$$

yield a complete set of our self-adjoint extension of the Dirac Hamiltonian. E has to satisfy $E^2 > M^2$.

3. Scattering solutions and the confinement of magnetic monopoles

The solutions we have found are not scattering solutions. Such solutions have to be defined by asymptotic conditions at large distances r. To define these conditions one would, naively, try to find a local section σ_o in ξ_m, normalized to one, which is defined outside a large sphere around the origin and formulate the asymptotic condition as:

$$\psi \sim \sigma_o(e^{i\vec{k}\vec{x}} + A(\theta)\frac{e^{ikr}}{r}), \quad r \longrightarrow \infty .$$

The choice of σ_o would correspond to a choice of a gauge at large distances. Unfortunately, such a σ_0 does not exist, because the curvature form is not exact outside any sphere, and hence, the bundle is non-trivial there and admits no non-vanishing section.

The situation is the same for several monopoles, with one exception: when the total monopole charge is zero, the curvature form $\omega(\nabla)$ is indeed exact outside a large sphere, the bundle becomes trivial there and admits a section σ_0. Hence, in the latter case, we can define the normal boundary conditions and find Möller operators

$$\Omega_{\pm}: \ L^2(\mathbb{R}^3) \longrightarrow \mathrm{Sec}(\tilde{\xi})$$

which are isometries from the free momentum space into the square integrable sections Sec $(\tilde{\xi})$. In the case of non-vanishing total monopole charge, this is impossible; the "free" states are not in $L^2(\mathbb{R}^3)$ [6]. This can be quite drastically shown in the case of a single magnetic pole:

Assume that we have isometric Möller operators $\Omega_{\pm}: L^2(\mathbb{R}^3) \rightarrow \mathrm{Sec}(\xi_m)$. They are intertwining operators for the free and the interacting Hamiltonians,

$$H\,\Omega_{\pm} = \Omega_{\pm} H_0$$

and, from the rotational invariance of our problem, they should also intertwine the angular momentum operators for the free and the interacting case, i.e. $\vec{I}\,\Omega_{\pm} = \Omega_{\pm}\vec{L}$. But we see from the explicit solutions, that the spectrum of $\vec{I}^2$ differs from the spectrum of $\vec{L}^2$. Hence we have a contradiction. Note that the situation is really different, e.g. from Coulomb scattering; there we have also difficulties with the boundary conditions, but, at least, Möller operators can be defined.

Hence, for a nonvanishing total magnetic charge there are no "free" asymptotic electron states in the normal technical sense. Nevertheless, a satisfactory treatment of the scattering problem can be formulated also in this case by changing the asymptotic conditions appropriately[8-1]. The important point is, that that the new asymptotic states are not in one-to-one correspondence with the normal free states ψ_k (free plane-waves of momentum k). This has an important effect on the vacuum energy of the electrons. Remember that the vacuum energy, according to Dirac, is equal to

$$E_0 = -\sum_k (k^2 + m^2)^{1/2} \langle \psi_k^-, \psi_k^- \rangle$$

for the free case, and can be written as

$$E_o' = -\sum_k (k^2 + m^2)^{1/2} \langle \Omega_\pm(\psi_k^-), \Omega_\pm(\psi_k^-) \rangle$$

for any perturbed Dirac equation which has no negative energy bound-states and which yields Möller operators $\Omega_\pm$. (ψ_k^- denotes a negative energy solution of momentum k). It follows that

$$E_o' - E_o = -\sum_k (k^2+m^2)^{1/2} (\langle \Omega_\pm(\psi_k^-), \Omega_\pm(\psi_k^-) \rangle - \langle \psi_k^-, \psi_k^- \rangle) = 0$$

because $\Omega_\pm$ is an isometry.

Hence there will be no change in the vacuum energy when Möller operators exist in the normal technical sense. In the case of a nonvanishing magnetic charge this latter condition does not hold and we may expect a change in the vacuum energy. Therefore, when a magnetic charge is removed to infinity from a magnetically neutral assembly of monopoles, leaving finally a magnetically charged object, it will cost energy to change the electron vacuum. We may speculate that this energy is so high, that the separation of magnetic charges actually never occurs. Whether this speculation is true can be decided by solving the Dirac equation of an electron in the field of a monopole-antimonopole pair separated by a distance d and by computing then the change of the electron vacuum energy as a function of d.

4. Mathematical aspects of bundles belonging to several monopoles

The physical picture presented in the last section is rather incomplete because the intermediate two-center problem with a monopole-antimonopole pair at a finite distance has not yet been treated analytically. In particular, we do not have a formula for the force between the poles induced by the change in the electron vacua. What could be achieved up to now is only a unified description of the underlying complex line bundles when several monopoles are present. We give this description here for the sake of completeness. Let the magnetic field B be given by

$$\vec{B} = \sum_{k=1}^{n} \mu_k (\vec{x} - \vec{x}_k)/|\vec{x} - \vec{x}_k|^3$$

with $e\mu_k = m_k$, m_k integer of half-integer, such that the Dirac quantization is fulfilled.

Let $D^j = SU(2) \longrightarrow GL(V_j)$, $V_j = \mathbb{C}^{2j+1}$ be an irreducible representation of SU(2) [4]. Identify the Lie algebra Lie(SU(2)) with $\mathbb{R}^3$, the vector product $[\,,\,]$ replacing the Lie bracket. Let d_j denote the representation of Lie(SU(2)) induced by D^j. Define

$$V = \bigotimes_{k=1}^{n} V_{|m_k|} \quad .$$

Hence, we have for every k=1,...,m a structural representation of Lie(SU(2)) in GL(V), which we also denote by $d_{|m_i|}$. For every $\vec{x} \in \overset{(n)}{\mathbb{R}}{}^3 := \{\mathbb{R}^3 - \{\vec{x}_1,\dots,\vec{x}_n\}\}$ define the linear subspace $F_{\vec{x}} \subset V$ by

$$F_{\vec{x}} = \{v \in V;\ d_{|m_k|}\ (\vec{x} - \vec{x}_k)/|\vec{x} - \vec{x}_k|\ v = -im_k v,\ \ k=1,\dots,n)$$

$F_{\vec{x}}$ is one-dimensional and

$$\tilde{V} = \bigcup_{\vec{x}} F_{\vec{x}}$$

defines the total space of a complex line bundle over $\overset{(n)}{\mathbb{R}}{}^3$. A section $\sigma : \overset{(n)}{\mathbb{R}}{}^3 \longrightarrow \tilde{V}$ is obviously a map

$\sigma : \overset{(n)}{\mathbb{R}}{}^3 \longrightarrow \tilde{V}$ which satisfies

$$d_{|m_k|}(\vec{x} - \vec{x}_k)/|\vec{x} - \vec{x}_k|\sigma(x) = -im_k\sigma(\vec{x}),\ k = 1,\dots,n.$$

If h is a vector field on $\overset{(n)}{\mathbb{R}}{}^3$, define

$$(\nabla_h\sigma)(\vec{x}) = (h\sigma)(\vec{x}) + \sum_{k=1}^{n} d_{|m_k|}([\vec{h}, \vec{x} - \vec{x}_k])/|\vec{x} - \vec{x}_k|^2\sigma(\vec{x}).$$

By straightforward computation, this is found to be a suitable covariant derivative with curvature $\omega(\nabla) = ieB$. (By the theorem mentioned in section 1, we know that any other description of the line bundles is obtained by bundle isomorphisms). Compare also with ref. [6).

Hence we know the bundles and the covariant derivative for all monopole configurations and we can study the Dirac equation (2). As we mentioned already, the Dirac Hamiltonian H is no longer self-adjoint for one monopole and we expect this to be also the case for several monopoles. For one monopole the particular self-adjoint extension which we met in section 3, can be described with the help of the operator CT which anticommutes with H: Define the space

$$D: = \{ \tilde{\sigma};\ \mathrm{supp}\, \tilde{\sigma} \subset \overset{(n)}{\mathbb{R}}{}^3 \text{ compact} \}$$

and the defect spaces

$$K_{\pm} = \{ \tilde{\sigma};\ (H \pm i)\tilde{\sigma} = 0 \};$$

By the theorem of von Neumann [11) the self-adjoint extensions of H are given by H_U with domain

$$D_U = \{ \tilde{\sigma};\ \tilde{\sigma} = \tilde{\sigma}_0 + \tilde{\sigma}_1 + U\tilde{\sigma}_1 \quad \tilde{\sigma}_0 \in D_0,\ \tilde{\sigma}_1 \in K_+ \}$$

where U is any isometry from K_+ onto K_- (provided U exists!) If $\tilde{\sigma} \in D_U$, then

$$H_U \tilde{\sigma} := H\tilde{\sigma} .$$

The particular self-adjoint extension of H which we have chosen in section 3 for one monopole is characterized by $(m \neq 0)$

$$U = \frac{m}{|m|} CT .$$

A configuration of several monopoles with total magnetic charge behaves at large distances as a single monopole with $m=Q$. Hence, if $Q \neq 0$, we expect that the self-adjoint extension which we should choose in the general case is characterized by

$$U = \frac{Q}{|Q|} CT.$$

This prescription only works if $Q \neq 0$. Nothing is known about the case of a magnetically neutral configuration of several monopoles. It is an interesting speculation, that, in this case, the defect spaces $K_{\pm}$ vanish, like in the free case.

Conclusion

The mathematical description of quantum mechanics of charged particles in the field of magnetic monopoles seems to reach a rather consistent level. As we have seen, there might even be a natural mechanism which provides the confinement of magnetic poles. In principle, this could be checked by solving the Dirac equation for an electron in the field of a monopole-antimonopole pair.

References

(1) W. Greub, H.R. Petry. J. Math. Phys. 16 (1975) 1347

(2) T.T. Wu, C.Y. Yang, Nucl. Phys. B107 (1976) 365

(3) P.A.M. Dirac, Proc. Roy. Soc. A133 (1931) 60

(4) A.R. Edmonds, Angular momentum in Quantum mechanics (Princeton U.P., Princeton N.J. 1957)

(5) A.S. Goldhaber, Phys. Rev. D16 (1977) 1815

(6) A.S. Goldhaber, Phys. Rev. 140 (1965) 1407

(7) M. Fierz, Helv. Phys. Acta 17 (1944) 27

(8) P. Banderet, Helv. Phys. Acta 19 (1946) 503

(9) K. Ford, J.A. Wheeler, Ann. Phys. 7 (N.Y.) 287

(10) Y. Kazama, C.N. Yan, A.S. Goldhaber, Phys. Rev. D15 (1977) 2287

(11) M. Reed, B. Simon: Fourier Analysis, Self-adjointness, Academic Press (N.Y.) 1975

SU(3) and SU(4) as Spectrum-generating Groups [a,b]

by

A. Bohm

Center for Particle Theory, University of Texas, Austin, Texas 78712, U.S.A.

and

R.B. Teese

Max-Planck-Institut für Physik und Astrophysik, München, Fed. Rep. of Germany

Abstract:

Using the analogy between the spectrum-generating SU(n) approach in particle physics and the dynamical group approach in atomic and molecular physics, we outline the basic ideas behind this alternative to broken-symmetry SU(n) approaches. We review various tests of dynamical SU(3) and SU(4) method, and discuss in particular two crucial tests of the fundamental assumptions.

I. Introduction

Historically there appear to be two distinct stages in the use of groups in quantum physics. The first and best known stage involves the use of groups to describe symmetry transformations. The first name which comes to mind is Wigner, but of course many other famous

a) Talk presented by R.B. Teese at the Conference on Differential-Geometric Methods in Physics, Clausthal, July 13-15, 1978

b) Research supported in part by NSF grant GF 420(O and DOE grant E(40-1) 3992.

people, such as Weyl, van der Wearden, Hund and Bargmann, also contributed. In this use of groups, one begins with the fundamental assumptions of quantum mechanics and plausible properties of the symmetry transformations, and is led to unitary representations of the covering group of the symmetry transformations, with the fundamental observables as the generators of the transformation group [1]. The space of physical states is then the unitary representation space of the symmetry; consequently, if we know the mathematical properties of this representation space, then we know the physical properties of the physical system that it describes. We may distinguish two classes of properties of these structures that are used: First, there are those properties that are used for the classification of physical states - for example, the irreducible representations and their reduction with respect to subgroups, the spectra of the generators and of elements of the enveloping algebra. Second, there are those properties that are used for the calculation of transitons between subsystems - for example, with definite assumptions about the transformation properties of the observables, the Wigner-Eckart theorem and the Clebsch-Gordan coefficients.

The second stage in the application of group theory started around 1965. Its purpose was the same as that for symmetry groups; namely, the properties of the representation served to classify the states, and assumed the transformation properties of the observables were used to calculate transitions. However, there was an essential difference: the existence of these group representations could not be derived from a symmetry transformation of the physical system. The name which was first given to this concept was "dynamical group" [2], although many other names have been used since then, such as "spectrum-generating group" [3] and "non-invariance group" [4]. We will use the names "dynamical group" and "spectrum-generating group" interchangeably.

The first application [2] of this use of groups was to the rotator. This gave a mathematical structure which can describe, for example, a diatomic molecule in a particular vibrational and electronic state. The rotator can have any integral value of angular momentum j, so its weight diagram is that of Fig. (1). The symmetry group of the rotator is $SO(3)_{J_i}$ (the subscript indicates that J_i are generators). Each line in the weight diagram corresponds to an irreducible representation of R^j of SO(3). However, if we add operators Q_i such that J_i, Q_i generate E(3), then the entire weight diagram of Fig. (1) corresponds to the irreducible representation space

$$R \xrightarrow[SO(3)]{} \sum_j \oplus R^j \qquad (1)$$

of E(3). That is, the dynamical group E(3) contains operators Q_i which transform between different irreducible representation spaces R^j of the symmetry group SO(3) $\subset$ E(3).

A question which naturally arises is then, whether or not such considerations have any application to elementary particle physics. In fact, this question was already addressed in the original paper by generalizing the rotator to a relativistic rotator [2]. However, there is another possible application, which I want to discuss today, and that is the reinterpretation of the SU(n) of particle physics in terms of dynamical groups. These groups, SU(2), SU(3),.., SU(n), where n is apparently limited only by the current experimental budgets, are well accepted as groups whose irreducible representations classify the observed particles and resonances. They have been customarily treated as symmetry groups, which, for SU(2) isospin, was a very good approximation. As experimental budgets continue to rise, though, this approximation has been getting much worse. The increase of "symmetry breaking" with increasing n will, if it has not already done so, prevent us from being able to use the Wigner-Eckart theorem. It is in an effort to save this second aspect of the usefulness of group theory that we have investigated the reinterpretation of SU(n).

II. Dynamical SU(n)

To demonstrate this approach, we shall use SU(3) as an example. For the representation space $\mathcal{H}^{SU(3)}$ we choose a basis labelled by I, I_3, Y and any other quantum numbers which may be needed, with the notation

$$|\alpha\rangle = |I, I_3, Y, \ldots\rangle \quad . \qquad (2)$$

In addition to these charges, the hadrons have properties coming from the space-time symmetry group, the poincaré group $\mathcal{P}_{P_\mu, L_{\mu\nu}}$. Consequently each hadron is also described by an irreducible representation space $\mathcal{H}$ (m,s) of $\mathcal{P}$. The basis vectors usually used for this space are the Wigner basis vectors $|p, s, s_3\rangle$, which are generalized eigenvectors of the momentum operator P_μ . Letting

$\mathcal{H}^{\alpha}$ denote the space spanned by $|\alpha\rangle$, the combination of internal and spacetime properties (according to a fundamental assumption of quantum mechanics concerning the combination of physical systems [5]) has as its space of physical states for the hadron, the direct product space $\mathcal{H}(m,s) \otimes \mathcal{H}^{\alpha}$. Such a space is represented by a dot in the weight diagram. For example, the weight diagram for the pseudoscalar meson octet is shown in Fig. (2). The space of physical states for the whole octet of Fig. (2) is

$$\mathcal{H}^{\{8\}} = \sum_{\alpha} \oplus \mathcal{H}(m,s) \otimes \mathcal{H}^{\alpha} \quad . \tag{3}$$

The basis system that is usually chosen for this direct product space is the direct product basis

$$|p s s_3 \alpha\rangle = |p s s_3\rangle \otimes |\alpha\rangle \quad . \tag{4}$$

This basis may not exist, if for example the operators whose eigenvalues are α and the mass operator cannot be simultaneously diagonal. Nevertheless, even if the basis (eq.(4)) does exist, it is not suitable if we take into account the fact that the SU(3) classification group is not a symmetry. To illustrate this, we shall compare this situation to that of the rotator in atomic physics.

For an atomic system we assume SO(3) rotational symmetry

$$[H, J_i] = 0 \tag{5}$$

where H is the Hamiltonian. For a quantum-mechanical rotator (e.g., a diatomic molecule in a definite vibrational state) the transitions between different angular momentum states take place through a triplet of operators Q_i having the property [1]

$$[Q_i, J_j] = i\,\varepsilon_{ijk} Q_k \quad . \tag{5b}$$

The Wigner-Eckart theorem may then be applied to the transition matrix elements:

$$\langle E'j'j_3' | Q_{(K)} | j_3 j E\rangle = C(j\,1\,j'; j_3\, K\, j_3')\,\langle E'j' \| Q \| j E\rangle \tag{5c}$$

where the reduced matrix elements $\langle E'j' \| Q \| j E\rangle$ do not depend upon j_3, K, j_3'.

The SU(3) which classifies the hadrons is, however, not a symmetry group. The mass operator, and therefore the 4-momentum operator cannot commute with all of the SU(3) generators E_α:

$$[P_\mu, E_\alpha] \neq 0 \quad . \tag{6a}$$

To describe weak transitions from one hadron state to another, the algebra of observables must include the weak "current" operator. For the hadronic term in K_{ℓ_3} and π_{e_3} decays, one uses a Lorentz vector operator V_α^μ with the property

$$[E_\alpha, V_\beta^\mu] = i f_{\alpha\beta\gamma} V_\gamma^\mu \tag{6b}$$

and calculates transitions (decays) using the formula

$$\langle \alpha' p' | V_\beta^\mu | p\alpha \rangle = \sum_{\gamma = F,D} C(\alpha\beta\alpha';\gamma) \langle p' \| V^\mu \| p \rangle \quad . \tag{6c}$$

Although this formula looks like the Wigner-Eckart theorem, it is not. The quantities $\langle p' \| V^\mu \| p \rangle$ depend upon the particle masses through the momenta, so they are not independent of the SU(3) indices $\alpha\beta\alpha'$. One would expect that eq. (6c) could be used as an approximation, to the extent that the mass differences in a multiplet may be neglected. For SU(2), since $m_{\pi^+}/m_{\pi^0} \simeq 1$, the approximation is very good. For SU(3), since $m_k/m_\pi \simeq 4$, the approximation is highly questionable, and for SU(4), since $m_{\chi_c}/m_k \simeq 20$, the symmetry-breaking corrections to eq. (6c) could be much larger than the effects of the Clebsch-Gordan coefficients.

It is clear from the above analogy that the problem with eq. (6c) lies in eq. (6a). In order to replace eq. (6c) by an exact equation, one must assume that the SU(3) is a symmetry of something other than the momentum. A suggestion which was made many years ago by Werle [6] is that eq. (6a) should be replaced by

$$[\hat{P}_\mu, E_\alpha] = 0 \quad , \tag{7}$$

where $\hat{P}_\mu = P_\mu M^{-1}$ is the 4-velocity operator and M is the mass operator. Actually, eq. (7) is more general than it at first appears. If we multiplied the momentum operator by a difference power of M, it would lead to the unphysical relation $[M, E_\alpha] = 0$.

Under the assumption (eq. (7)) that SU(3) is a symmetry of the velocity operator and the usual assumption

$$[L_{\mu\nu}, E_\alpha] = 0 \quad , \tag{8}$$

it is more convenient to use the velocity-Poincaré group $\hat{\mathcal{P}}_{\hat{P}_\mu, L_{\mu\nu}}$ rather than the physical spacetime symmetry $\mathcal{P}_{P_\mu, L_{\mu\nu}}$. Neither this SU(3) nor $\hat{P}$ are connected with physical symmetry tranformations. Nevertheless we may assume that $\hat{P} \otimes SU(3)$ describes the spectrum of the physical system of a hadron, with each hadron of an octet being a different state of this physical system. The space of physical states is then

$$\mathcal{H}^{[8]} = \sum_\alpha \oplus \mathcal{H}^{\hat{P}}(m=1,s) \otimes \mathcal{H}^\alpha \quad . \tag{9}$$

The physical Poincaré group $\mathcal{P}_{P_\mu, L_{\mu\nu}}$ is still represented in this space, only P_μ cannot be written in the direct product form $P_\mu \otimes \mathbb{1}$ due to eqs. (6a) and (7). As a basis for eq. (9) one chooses

$$|\hat{p} s s_3 \alpha\rangle = |\hat{p} s s_3\rangle \otimes |\alpha\rangle \tag{10}$$

where $|\hat{p} s s_3\rangle$ are generalized eigenvectors of the 4-velocity operator $\hat{P}_\mu$ which span the space $\mathcal{H}^{\hat{P}}(1,s)$. The mass operator acts only on $|\alpha\rangle$. The basis vectors (eq.(10)) may or may not be generalized eigenvectors of the 4-momentum, depending on whether or not the SU(3) lable α represents a physical mass eigenstate.

Using eq. (7) we can now write the Wigner-Eckart theorem for matrix elements of V^μ_α between the 4-velocity eigenvectors (eq.(10)):

$$\langle \hat{p}' s' s_3' \alpha' | V^\mu_\beta | \hat{p} s s_3 \alpha\rangle = \sum_{\gamma = F, D} C(\alpha\beta\alpha' ; \gamma) \langle \hat{p}' s' s_3' \| V^\mu \| \hat{p} s s_3\rangle . \tag{11}$$

The reduced matrix elements in eq. (11), unlike in eq. (6c), are SU(3) invariant functions of the SU(3) invariant 4-velocities. That is, eq. (11) is not a symmetry-limit approximation, but rather an exact relationship. Continuing with the example of K_{ℓ_3} decays, eq. (11) becomes [7]

$$\langle \pi \hat{p}_\pi | V^\mu_\beta | K \hat{p}_K\rangle = C_F(\pi\beta K)\left[(\hat{p}_K + \hat{p}_\pi)^\mu F_+(\hat{q}^2) + (\hat{p}_K - \hat{p}_\pi)^\mu F_-(\hat{q}^2)\right] \tag{12}$$

The reduced matrix elements $F_\pm(\hat{q}^2)$ are functions of the SU(3)-invariant $\hat{q}^2 = (\hat{p}_K - \hat{p}_\pi)^2$. This is to be contrasted with the conventional expression which comes from eq. (6c),

$$\langle \pi p_\pi | V_\beta^\mu | K p_K \rangle = C_\mp(\pi\beta K)\left[(p_K+p_\pi)^\mu f_+^{\pi K}(q^2) + (p_K-p_\pi)^\mu f_-^{\pi K}(q^2)\right] \quad (13)$$

in which the formfactors $f_\pm^{\pi K}(q^2)$ depend upon the masses and are therefore not SU(3)-invariants.

We thus see that the basic idea behind dynamical SU(n) is really very simple: Expression (6c) cannot be correct because of the mass differences, and it may not even be an acceptable approximation. By using the Werle relation, one obtains a formula which is in principle exact, and which can be tested. However, the task of testing this idea is not simple. Conventional formulae found in textbooks cannot be used, since they were derived under the assumption that the masses are SU(n)-invariant. Quantities like the partial decay rates must be completely rederived, starting from the basic principles of quantum mechanics. The result of such rederivations is in general that the new formula differs from the conventional one by a factor (suppression factor), which is a well-defined function of the hadron mass ratios and differences [8]. The exact form of this function depends not only upon the Werle assumption [7] but also upon the assumptions made about the transformation property of the transiton operators (weak and electromagnetic currents).

Before going on to discuss some applications of these ideas, we will quote van Dam and Biedenharn, who have independently studied the idea that some groups in particle physics should commute with the 4-vleocity rather than 4-momentum operator. Referring to their "dynamical stability group of P_μ/M ", before they knew of the spectrum-generating SU(3) results, they wrote [9], "We suggest that the concept of a dynamical stability group is the proper concept to replace the unworkable concept of a global Lie group symmetry in relativistic quantum mechanics." We hope that the present results lend support to their suggestion.

III. Applications

Spectrum-generating SU(3) and SU(4) have been applied to five processes which involve no more than one hadron. Possible applications involving multi-hadron states have been discussed by Kielanowski [10]. Since this talk is a review, we will only briefly describe these five processes, and refer to the original literature for details.

1.) <u>$V \to e\bar{e}$</u>: The leptonic decays of vector mesons ($\rho, \omega, \phi, \psi, ..$) do not fulfill the ordinary quark model predictions which come from SU(4) symmetry (with mass differences taken into account in the phase space). Instead, they fulfill Yennie's empirical rule [11]. With a suitable assumption for the electromagnetic current operator within dynamical SU(4), this rule can be derived [7,12] .

2.) <u>$V \to P\gamma$</u> : The radiative decays of vector mesons ($\rho, \omega, \phi, \psi, K^{o*}, D^*$) into pseudoscalar mesons (π, η, K^o, η_c, D) are difficult to explain with ordinary SU(3) and SU(4). In particular, the decays of new particles $\psi \to X(2.8)\gamma$ and $D^{o*} \to D^o \gamma$ are strongly suppressed experimentally, although there is no principle such as the OZI rule to forbid them. Within dynamical SU(4), this suppression is explained by the large hadron mass differences, since suppression factors arise naturally in this framework [13].

3.) <u>Hyperon Magnetic Moments</u>: Neither SU(3) symmetry nor quark model predictions [14] (including different "masses" of the quarks) can fit the present experimental magnetic moments. A suggestion factor gives a slight improvement [15], but the magnetic moments remain an unsolved problem.

4.) <u>Baryon Semileptonic Decays</u>: The Cabibbo model (discussed below) can be applied to these decays. Most fits of this model to the data only use that part of the data which can be expressed as rates or ratios of leading form-factors, and such fits appear to be acceptable [16]. However, when all of the data is compared, the fit looks rather poor [17]. The use of dynamical SU(3) techniques improves the fit, but the results are still not really satisfactory [18]. This is an area which is not yet well understood and requires further experimental and theoretical work.

5.) <u>Weak Leptonic and Semileptonic Meson Decays:</u> We will discuss this in some detail.

The original motivation for developing the spectrum-generating SU(3) approach was to explain the Cabibbo supression (defined below) as a consequence of symmetry breaking. That is, the Cabibbo angle was to be determined as a function of hadron masses. Using the weak transition operator (hadronic current) it was in fact possible to obtain the Cabibbo suppression for some processes. However, it was not possible to find suitable assumptions which would give the correct supression for all processes. Therefore, the Cabibbo angle had to be accepted as a phenomenological constant that cannot be expressed as a suppression factor. At present, we consider the dynamical SU(n) approach to be a correction of the Cabibbo model, which takes the mass differences into account in a non-perturbative way.

After much trial-and-error, we found that the following ansatz for the weak hadronic current gives the best results for all processes listed above:

$$J_\mu^{Had.} = \cos\theta_c (V_\mu^{-1} + A_\mu^{-1}) + \sin\theta_c (V_\mu^{-2} + A_\mu^{-2}) + h.c. \tag{14}$$

where

$$\begin{aligned} V_\mu^\beta &= \{M, \{M, \hat{V}_\mu^\beta\}\} \\ A_\mu^\beta &= \{M, \{M, \hat{A}_\mu^\beta\}\} \end{aligned} \tag{15}$$

and where $\hat{V}_\mu^\beta$ and $\hat{A}_\mu^\beta$ are SU(3)-octet operators which fulfill eq. (6b). Cartan notation is used for the SU(3) indices [7], and M is the mass operator. In the symmetry "limit", $[M, E_\alpha] = 0$, eq. (14) is equivalent to the usual Cabibbo current, except for constant factors. One may consider eq. (15) to be the phenomenologically-obtained formula which expresses the behavior of the currents away from the symmetry limit.

In an experimental test of the dynamical SU(n) approach, it is important to separate effects which test the various assumptions individually. The ratios of decay rates (i.e., the suppression factors) depend upon both eqs. (15) and (7). However, there are other predictions which are independent of eq. (15) and therefore provide a test of eq. (7). These are:

a.) Ratios of different form factors for a single process. Using eqs. (6c) and (14), we find for the current in $\alpha \longrightarrow \pi^\circ \ell \nu$ decay ($\alpha = \pi^+$ or K^+)

$$\begin{aligned} &\langle \hat{p}'\pi^\circ | J_\mu^{Had.} | \alpha \hat{p} \rangle = \\ &= C_F(\pi^\circ, \{{}^{-1}_{-2}\}, \alpha) \sin\theta_c (m_\alpha + m_\pi)^2 F_+(\hat{t})(\hat{p} + \hat{p}')_\mu \end{aligned} \tag{16}$$

where the velocity transfer $\hat{t}$ is related to the usual momentum transfer t^α by

$$\hat{t} = (\hat{p} - \hat{p}')^2 = (m_\alpha m_\pi)^{-1} [t^\alpha - (m_\alpha - m_\pi)^2] \quad . \tag{17}$$

The factor $(m_K + m_\pi)^2$ originates in the assumption (15). The form factor F_- has been set equal to zero because it is "second class" [7]. Expression (16) may be compared to the conventional form (13), using eq. (17), to yield the form factor ratio which is measurable:

$$\xi(t) = \frac{f_-^K(t)}{f_+^K(t)} = \frac{m_\pi - m_K}{m_\pi + m_K} = -0.57 \quad . \tag{18}$$

This prediction is independent of the factor in front of F_+ in eq. (16), so it is a test of the Werle relation (7).

Prediction (18) is probably the single most important test of the dynamical SU(n) approach, because it depends upon the fundamental assumption (7) rather than the detailed form of the current. If this prediction is not experimentally confirmed, then the whole approach would have to be rejected. For this reason, the experimental situation should be examined carefully. We will just give a summary here [7] : The data is statistically dominated by two experiments, one of which gives a value of $\xi(t)$ near zero, and the other of which gives a value compatible with eq. (18). In short, the experimental situation is presently unclear, but eq. (18) is certainly an experimentally favored value [7].

b.) Ratios of the Cabibbo suppression for leptonic and semi-leptonic processes. Phenomenologically, in the conventional Cabibbo model using eq. (13), one needs two Cabibbo angles:

$$J_\mu^{Had.} = \cos\Theta_V V_\mu^{-1} + \cos\Theta_A A_\mu^{-1} + \dots \tag{19}$$

Experimentally [19],

$$\Gamma(K \to \mu\nu)/\Gamma(\pi \to \mu\nu) \quad \text{gives} \quad \tan\Theta_A = 0.276 \tag{20}$$

and

$$\Gamma(K \to \pi\ell\nu)/\Gamma(\pi \to \pi\ell\nu) \quad \text{gives} \quad \tan\Theta_V = 0.224 \tag{21}$$

with very small errors. The value for Θ_c for baryon decays lies in the same neighborhood.

In the dynamical SU(3) approach, the value determined for the $\tan\Theta$ appearing in eq. (14) is the same as $\tan\Theta_V$ in eq. (20), and is determined from the same data. However, the value determined from the semileptonic data is different from $\tan\Theta_V$. The reason is the following [20]: The definition of the experimentally measured $\tan\Theta_V$ is (see eq. (13))

$$\tan\Theta_V = \frac{f_+^K(t^K=0)}{f_+^\pi(t^\pi=0)} = 0.224 \tag{22}$$

where both form factors are evaluated at zero momentum transfer. In the dynamical group approach, however, both form factors should be evaluated at the same value of the SU(3)-invariant transfer $\hat{t}$. Choosing the value $\hat{t} = 0$, which alsways in the physical region, we have

$$\hat{t}=0 \longleftrightarrow \begin{array}{l} t^\pi \simeq 0 \\ t^K = (m_K - m_\pi)^2 = t^K_{maximum} \end{array} \tag{23}$$

so the suppression is

$$\tan\Theta = \frac{f_+^K(t^K = t^K_{max.})}{f_+^\pi(t^\pi=0)} = 0.276 \tag{24}$$

in agreement with the value determined from the leptonic decay. This result depends also on the form factor parameter λ_+ in

$$f_+(t) = f_+(0)\left(1+\lambda_+\frac{t}{m_\pi^2}\right) \tag{25}$$

so this is not as clean a test of the framework as eq. (18). Nevertheless, it is clear, that this approach allows one to eliminate the use of different Cabibbo suppression factors for these different processes.

Although the equality of the vector and axialvector suppression factors is essentially a consequence of eq. (7), the precise value of the Cabibbo angle depends upon the exact form of the operators appearing in eq. (14) (e.g., on assumption (15)). That is, the Cabibbo suppression is explained not only by the presence of Θ in eq. (14) but also on the transformation property of the currents, as expressed for example by eq. (15). The form (15) seems to be consistent with the data, but the agreement for baryons is not impressive. Thus eq. (15) may well have to be changed, but this will not change the value of $\tan\Theta$ very much because of the small baryon mass differences. One may therefore consider eq. (24) to be a prediction of the model which is not very subject to change.

Acknowledgements

We would like to thank our collaborators and colleagues for innumerable discussions about this subject.

References

[1] E.P. Wigner, Ges. Wiss. Gött., Math.-Phys. Klasse (1932) p. 546. Reprinted in "Group Theory and Solid State Physics", Vol. I, P.H. Meijer, Editor (Gordon and Breach, New York, 1964) pp. 265-278.
V. Bargmann, Annals of Math. 59, 1 (1954).
G. Ludwig, "Grundlagen der Quantum Mechanik", (Springer, ;954) p. 101.

[2] A.O. Barut and A. Böhm, Phys. Rev. 139B. 1107 (1965).

[3] Y. Dothan, M. Gell-Mann, and Y. Ne'eman, Phys. Rev. Lett. 17. 145 (1965)

[4] N. Mukunda, L. O'Raifeartaigh, and E.C.G. Sudarshan, Phys. Rev. Lett. 15, 1041 (1965).

[5] A. Böhm, "Quantum Mechanics",Chap. II (Springer, New York,1978)!

[6] This reation was first suggested by J. Werle, in "On a Symmetry Scheme Described by a Non-Lie-Algebra", ICTP report (Trieste,1965, unpublished). It was used in A. Böhm, Phys. Rev. 175, 1767 (1968), and in connection with spectrum-generating SU(3) in A. Böhm, Phys. Rev. 158, 1408 (1967); D7, 2701 (1973); and A. Böhm and E.C.G. Sudarshan, Phys. Rev. 178, 2264 (1969). The present scheme was formulated in A. Böhm and J. Werle, Nucl. Phys. B106, 165 (1976) and in Ref. [7]. The same idea was suggested also by van Dam and Biedenharn (see Ref. [9]).

[7] A. Böhm, Phys. Rev. D13, 2110 (1976).

[8] The theoretical possibility of a mass-dependent correction factor has been noticed several times before: M. Gourdin, in Symmetries and Quark Models, edited by R. Chand (Gordon and Breach, New York, 1970), R.P. Feynman, Photon-Hadron Interaction (Benjamin,New York, 1972), D.R. Yennie, Phys. Rev. Lett. 34, 239 (1975), G.J. Aubrecht II and M.S.K. Razmi, Phys. Rev. D12, 2120 (1975). An octet-brokem SU(3) model incorporating current mixing has been used in L.M. Brown, P. Singer, Phys. Rev. D15, 3438 (1977) where further references are given. Mass dependent correction factors have also been derived from the Duffin-Kemmer-Petiau formalism:

B.G. Kenney, D.C. Peaslee and M.M. Nieto, Phys. Rev. D13, 757 (1976), and references therein. The DKP formalism has also other results in common with our dynymical group approach;cf. footnote 16 of Ref. [7]. Another way of describing symmetry breaking effects by mass dependent correction factors in the effective coupling constants has been suggested by J. Schwinger, Proceedings of the 7th Hawaii Topical Conference on Particle Physics, Honolulu (1977) and L.F. Urrutia, "Leptonic and Radiative Meson Decays", UCLA preprint (1977) (where further references to the description of symmetry breaking effects are given).

[9] H. van Dam and L.C. Biedenharn, Phys. Rev. D14, 405 (1976).

[10] P. Kielanowski, Warsaw University preprint (1978).

[11] D.R. Yennie, Phys. Rev. Lett. 34, 239 (1975)

[12] A. Böhm and R.B. Teese, University of Texas preprint ORO-3992-317 (to appear in Phys. Rev. D).

[13] A. Böhm and R.B. Teese, Phys. Rev. Lett. 38, 629 (1977) and references therein. See also Ref. [12].

[14] R. Settles, in R. Armenteros et al., "Physics from Friends" Geneva, 1978) p. 127.

[15] A. Böhm, Univ. of Texas preprint ORO 316 (to appear in Phys. Rev. D).

[16] K. Kleinknecht, in XVII International Conference on High Energy Physics, J.R. Smith, Editor (London, 1974).

[17] A. Garcia, Phys. Rev. D3. 2638 (1971);
D9, 177 (1974 ;
D10, 2839 (1974); and
D12, 2692 (1975).

[18] A. Böhm, R.B. Teese, A. Garcia and J.S. Nilsson, Phys. Rev. D15, 689 (1977).

[19] M. Ross, Phys. Lett. 36B, 130 (1971).

[20] A. Böhm, M. Igarashi and J. Werle, Phys. Rev. D15, 2461 (1977).

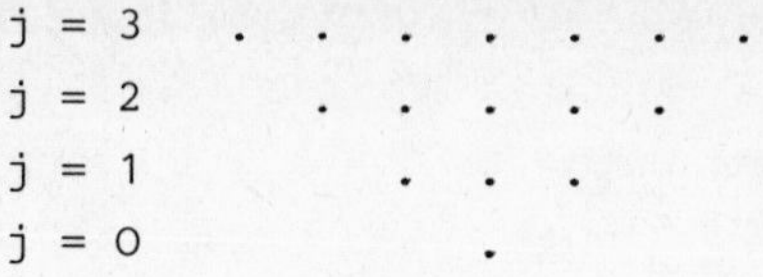

Fig. 1. The weight diagram of the rotator. The ordinate is j and the abscissa is j_3.

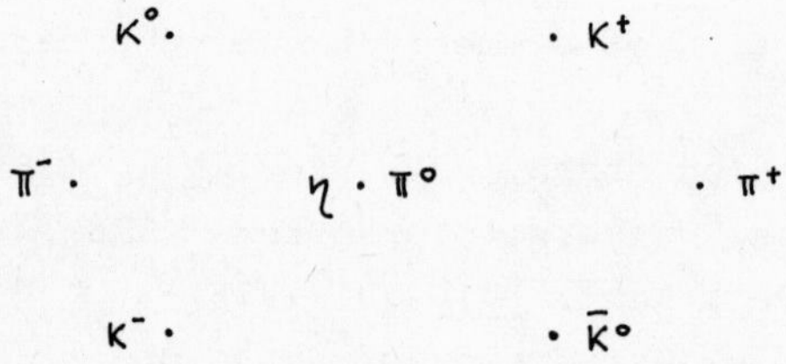

Fig. 2. The weight diagram of the pseudoscalar meson octet. The ordinate is Y and the abscissa is I_3.

The Phase Space for the Yang-Mills Equations

I.E. Segal,
Massachusetts Institute of Technology

0. Introduction

The phase-space point of view regarding a dynamical system is now a well-established one. In its simplest form it requires that the possible states (classical, i. e.) of the system form a symplectic manifold M (i. e. has defined on it a fundamental differential form Ω, which in suitable local coordinates $p_1, p_2, \ldots, q_1, q_2, \ldots$ may be expressed as

$$\Omega = \sum_i dp_i \, dq_i$$

on which temporal evolution acts as a 1-parameter family T_t of displacements preserving the form Ω (i.e., as "automorphisms" of the symplectic manifold (M, Ω)).

In cases of interest there are often symmetries (or essentially equivalently, conservation laws); the phase space and motion may then be said to be "covariant", with respect to a given group G of automorphisms of (M, Ω); the most basic symmetry is temporal invariance, (corresponding to conservation of energy), definable by the group property for the T_t: $T_{t+t'} = T_t T_{t'}$.

In finite-dimensional cases (i.e. when M is a finite dimensional manifold) these ideas, apart from matters of mathematical clarity and generality, go back well over a century, and are widely regarded as virtually axiomatic and beyond dispute. In the infinite-dimensional case, however, the ideas are more recent (e.g. Noether's theorem) and both the mathematics and physics lack the same order of universal formulation and acceptance as in the finite-dimensional case.

Some years ago I initiated a direct approach to the phase-space standpoint for non-linear relativistic systems (Segal, 1960). I think it has been useful in clarifying the relation between the classical and quantum theories for the "same" system, i.e. defined by the same partial differential equation; in treating the exact quantization of linear perturbations in the vicinity of the background (classical) field; and in providing an alternative approach to quantization, quite different from that, involving the investigation of singular functions as in the theory initiated by Dirac, Heisenberg, and Pauli, but nonetheless comparably consistent with fundamental physical constraints.

This work dealt with the quantization of non-linear relativistic partial differential equations for which the situation at a fixed time was formally well understood; and in the first instances, at least, quantization could be reduced to the solution of a Cauchy problem with relatively simple initial data. This considerable reduction of the problem has thus far been lacking in the treatment of the Yang-Mills (1954) equations (a fortiori, the equations of General Relativity), in which case the non-linear fixed-time constraints and the gauge-invariance issues interact non-trivially with the quantization problem.

I propose here to make an initial step in the reduction of the Yang-Mills system (as a prototypical gauge system with non-linear fixed-time constraints) to fixed-time gauge-invariant considerations, by establishing the solution manifold as a covariant phase space which is definable modulo the gauge group in terms of the fixed-time situation. This seems to require a limitation to the original hyperbolic equations, rather than the better-behaved elliptic ("euclidean") variant, for which there is no apparent fundamental form Ω; this physically imposed limitation leads to substantial real-analysis issues, which I can only work modulo in this general presentation.

1. Covariant phase space for unconstrained scalar equations

To illustrate the underlying ideas, consider the relativistic p.d.e.

$$\Box \varphi + p(\varphi) = 0$$

where p is a given polynomial, in Minkowski space. The solution manifold M (for specificity, say, of φ which are in C_0^∞ on space at every fixed time, and defined at least for times t near a fixed time t_0) forms a phase space covariant with respect to the Poincaré group, whose fundamental form Ω is describable as follows.

A tangent vector to M at a point $\varphi_0 \in M$ is essentially a solution η of the first-order variational equation at φ_0 (cf. Segal, 1965)

$$\Box \eta + p'(\varphi_0)\cdot \eta = 0 .$$

If η_1 and η_2 are two arbitrary tangent vectors at φ_0,

$$\Omega_{\varphi_0}(\eta_1,\eta_2) = \int (\eta_1 \frac{\partial \eta_2}{\partial x_0} - \frac{\partial \eta_1}{\partial x_0} \eta_2) \, d_3x \quad .$$

Although Ω appears to depend on the initial time t_o and the Lorentz frame, it is in fact independent of these, and so is Lorentz-invariant. Since it has constant coefficients on a flat space, it is closed, and is plainly non-degenerate.

In the case of the conformally-covariant equation $\Box\varphi + g\cdot\varphi^3 = 0$, Ω is in addition conformally invariant. To check this it suffices to consider the case of scale and conformal inversion invariance. The former is immediate, and for the latter it suffices to consider the action of conformal inversion on a space-like section, or more precisely, on the Cauchy data thereon. A calculus computation, analogous to that involved in showing the Lorentz invariance of Ω, confirms the invariance under inversion.

Gauge-invariance may be illustrated, although only in its "global" form, by the equation $\Box\varphi + g\cdot\varphi\cdot|\varphi|^2 = 0$, where φ is now complex. The gauge- transformations $\varphi \rightarrow e^{i\alpha}\varphi$ (α a real number) leave invariant the solution manifold.Now define Ω as:

$$\Omega(\eta_1,\eta_2) = \mathrm{Im} \int (\eta_1 \frac{\partial \bar{\eta}_2}{\partial x_0} - \frac{\partial \eta_1}{\partial x_0} \bar{\eta}_2) \, d_3x$$

Ω is both conformally and gauge invariant, and remains non-degenerate.

In the case of Minkowski space, the conformal group operates only with singularities, so the covariance described above is only infinitesimal, as regards the full conformal group, although it is global for the Lorentz and scale subgroups. Covariance may be made global for the full conformal space (more precisely, its universal covering space), as in the work of Ørsted (1979).

2. The Yang-Mills equations in evolutionary form

The Yang-Mills equations (YME) introduces three significant complications into the foregoing considerations: (1) non-linear fixed-time constraints; (2) a local gauge group; (3) greater singularity from a p.d.e. point of view, resulting in uncertainties regarding global existence and serious doubts concerning temporal asymptotics. On the other hand, they are exceptionally symmetrical, and it is difficult to approximate them by simple equations which are comparably so.

In order to treat the equations analytically, they are best formulated in function-theoretic terms, and then concern a vector function A on a given 4-dimensional space-time manifold ("cosmos", for short) to a given Lie algebra $\underline{G}$. The main analytical issues are present already when the cosmos is simply Minkowski space; the dimensionality of $\underline{G}$ is irrelevant to the general analytical considerations, but the compactness or non-compactness of the associated Lie group G plays a role. For simplicity, and in as much as it covers all cases usually considered, it is here assumed that G is compact, although parts of the theory remain valid when G is non-compact, or even infinite-dimensional.

In terms of the components A_μ of the "potential" A, where μ = o, 1, 2, 3 and 0 denotes the time components, the equations take the form

$$F_{\mu\nu} = \partial_\mu A_\nu - \partial_\nu A_\mu \quad ,$$

$$\sum_\mu \varepsilon_\mu (\partial_\mu F_{\mu\nu} - [A_\mu , F_{\mu\nu}]) = 0$$

where $\varepsilon_0 = 1$, $\varepsilon_j = -1$, ($j = 1, 2, 3$).

To treat existence and symplectic aspects of these equations it is convenient to write them out component-wise in the "temporal" gauge, in which $A_0 = 0$; however the gauge is not fully specified by this constraint. More specifically, to deal with the existence issue one must select appropriate function spaces both for the unknown functions A_μ and for the admissible gauge transformations. The rigorous possibility of transforming via a gauge transformation a given solution A_μ into one with $A_0 = 0$ depends on an appropriate relation between the function space for the A_μ and for the gauge transformations.

For differential equations purposes, it is desirable (and in some respect necessary) to formulate the group Γ of admissible "local" gauge transformations, formally, the group of smooth maps from M to G, - as a Banach-Sobolev-Lie group. That is, Γ should be a Banach-Lie group whose Lie algebra is constrained analytically to have its components in a designated Sobolev space. Purely for group-theoretic purposes, - independently of the relation to the Yang-Mills equations,- it is necessary to require a number of derivatives at least equal to half the dimension of the space; otherwise analytical control is lacking, and a corresponding Banach-Lie group is not clearly definable.

Furthermore, in line with the present symplectic-dynamical approach, it is natural to consider gauge transformations $U(t,\vec{x})$ as maps $U(t,.)$ from the time t to the group of only space dependent gauge transformations $V(\vec{x})$. The requisite gauge group is then definable in the quite general terms given in the

Theorem 1. Let G denote an arbitrary separable Hilbert-Lie group; and assume that the Hilbert structure in the Lie algebra $\underline{G}$ is invariant, i.e. for all $X \in \underline{G}$, the map $Y \longrightarrow [X,Y]$ is skew-adjoint as a linear operator on $\underline{G}$.

Let $L_{p,r}(S,\underline{G})$ denote the Lie algebra of all maps from the n-dimensional manifold S (assumed to be either $\mathbb{R}^n$ or to be compact) to $\underline{G}$, which together with their first r derivatives, are pth power integrable. If $p > 1$ and $r > n/p$, there exists a unique Banach-Lie group, say $\Gamma_{p,r}(S,\underline{G})$, which consists of mappings from S to $\underline{G}$, and whose Lie algebra is (in the canonical sense) $L_{p,r}(S,\underline{G})$.

It is natural to choose $p = 2$ because of the form of energy for the YME, and because of the poor regularity properties of temporal evolution with other choices of p shown by Littman in even the vastly simpler case of the wave equation. Thus for Minkowski space, it suffices to deal with space-dependent gauge transformations in the group $\Gamma_{2,2}(\mathbb{R}^3)$. The admitted space-time-dependent gauge transformation will then be those represented by C^2 mappings $t \longrightarrow V(t)$ from $\mathbb{R}^1$ to $\Gamma_{2,2}(\mathbb{R}^3)$. With this group, it then follows from non-linear semi-group considerations (Segal, 1963), that one has the

Corollary. A C^3 solution of the YME on Minkowski space is gauge-equivalent to a solution which is in the temporal gauge.

The latter means, more specifically, that $A_o = 0$, i. e. the equations take the following form, setting $\underline{A} = (A_1, A_2, A_3)$, $F_{ij} = \partial_i A_j - \partial_j A_i - [A_i, A_j]$, $\underline{E} = (F_{10}, F_{20}, F_{30})$, $\underline{B} = (F_{32}, F_{21}, F_{12})$:

$$\partial_o \underline{A} = -E \qquad \partial_o \underline{E} = -\nabla \times \underline{B} + \underline{A} \times \underline{B} + \underline{B} \times \underline{A} .$$

to these evolutionary equations must be added the fixed-time constraint

$$\nabla \cdot E = A \cdot E - E \cdot A \quad .$$

3. Regularity aspects of the YME

In order to treat tangent vectors and the like to the solution manifold $\underline{M}$ of the YME, it is necessary to develop solutions in appropriate Sobolev spaces which are valid at least locally in time. It is possible to show the existence of global quasi-solutions for arbitrary "finite-energy" Cauchy data, but the analytical control on such solutions is insufficient for function-space differential geometry.

Accordingly, the point of view is taken of consideration of strong solutions throughout space which at a fixed time have data in a designated Sobolev space; and on which the action of the conformal group is continuous in a neighborhood of the group unit, which neighborhood may depend also on the solution acted on. (Again, to deal properly with non-Lorentzian transformations, it is necessary to replace Minkowski space by a covering of conformal space; otherwise, one has only infinitesimal conformal invariance, and must require an additional derivative for the data.)

As in the case of non-linear wave equations, there is a scale of spaces for the Cauchy data, any member of which is invariant under temporal evolution locally in time. For the wave equation these spaces give the following norm at time t to a solution φ :

$$\|(-\Delta)^{a+1}\varphi\|_2^2 + \|(-\Delta)^a\dot{\varphi}\|_2^2$$

where $\dot{\varphi}$ denotes the time derivative of φ , and a is an arbitrary non-negative integer. Similar spaces are applicable in the case of the YME; cf. Segal 1979. It will suffice here to employ solutions for which at a fixed time the potential A has components in $L_{2,2}$ and the "electric field" E has components in $L_{2,1}$. (It then follows that the "magnetic field" B also has components in $L_{2,1}$). Precise statements of the existence theorems involved in the following are given in loc. cit.

4. The Pre-Symplectic Structure for the YME

As indicated, the point of view taken is that, the set $\underline{M}$ of all solutions of the YME is to be formulated as a manifold whose tangent vectors are identified canonically with the solutions of the first-order variational equations. By a "pre-symplectic" structure will be meant one satisfying all of the conditions for symplecticity except that of non-degeneracy. In a general gauge, a symplectic structure on sufficiently smooth tangent vectors (it will suffice to require their Cauchy data at each time to lie in $L_{2,2} \oplus L_{2,1}$, for the potential $\oplus$ electric field) is definable as follows. Three inner products are here involved: one in $\underline{G}$, one in Minkowski space, and one in $L_2(S)$. The combination of the first two will be denoted as $\langle\langle .,. \rangle\rangle$, so that if e_μ and e'_μ are the components of $\underline{G}$-valued vectors in Minkowski space,

$$\langle\langle e, e' \rangle\rangle = \sum_\mu \varepsilon_\mu \langle e_\mu, e'_\mu \rangle_{\underline{G}} .$$

If $A \in \underline{M}$ and E is the corresponding electric field, it is convenient to denote this point of $\underline{M}$ in terms of the data (A,E) at an arbitrary fixed time t_o, and a corresponding tangent vector likewise as (a,e). If (a',e') is another tangent vector at the same point, the equation

$$\Omega_{A,E}(a \oplus e, a' \oplus e') = \int_S (\langle\langle a(t_o,x), e'(t_o,x) \rangle\rangle - \langle\langle e(t_o,x), a'(t_o,x) \rangle\rangle)\, dx$$

defines an anti-symmetric bilinear form on the Cauchy data space for the first-order variational equation. The properties of Ω are summerized by the

Theorem. Ω <u>is a gauge- and Lorentz-invariant closed differential form on M, and its radical in</u> $L_{2,3} \oplus L_{2,2}$ <u>is spanned by tangent vectors gauge-equivalent to zero; in the case of chronometric space,</u> Ω <u>is invariant also under the finite action (of elements sufficiently near the identity) of the conformal group</u>.

The main novelty in the proof, apart from the greater general complication, is the determination of the radical of Ω , i.e. the set of tangent vectors λ such that $\Omega(\lambda, \lambda') = 0$ for all tangent vectors λ' at the point in question. This is based on a duality-Hilbert space argument involving the computation of the adjoint of the differential system which defines membership in the radical.

5. Discussion

These results relate to two interesting theoretical physical issues.

One is the question of the physical measurability of a perturbation of a non-vanishing electromagnetic field. In the tangent space at the vanishing solution of the YME, the form Ω is non-degenerate, and one is dealing essentially with Maxwell's equations; the direct physical nature of the fields e and b (the linear variation in the magnetic field B) is without question. However, at any non-vanishing solution, this is definitely not the case. In fact, a perturbation which is gauge-equivalent to zero, and so must be regarded as physically non-measurable, may be represented by non-vanishing e and b. More specifically, for any arbitrary test function f in $C_0^\infty(S,\underline{G})$, a tangent vector at (A,E) is defined by the equations

$$a = -f + [A,f] \quad , \quad e = [E,f] \quad , \quad b = [B,f] \quad .$$

Nevertheless, the gauge transformation suitably generated by f carries this vector into zero. Clearly, if E and B are non-vanishing, a choice for F exists which is such that e and b are non-vanishing. Thus only certain linear forms in e and b (namely, those which are gauge-invariant) can correspond to directly physically measurable quantities. The same may be true, although less clearly visible because of the less symmetrical context, for perturbations of Maxwell's equations expressed in terms of a potential. It cannot be taken for granted that what appear as the electric and magnetic field components of the perturbation are indeed such, although in low-energy quantum electrodynamics such gauge-invariance has sometimes been assumed as axiomatic.

The other is the formulation of a Kähler or almost Kähler structure on the Yang-Mills phase space. The foregoing shows that this phase space,- the quotient of the solution manifold by the gauge group,- is, modulo regularity issues involved in the definition of tangent spaces, symplectic. In the case of non linear wave equations, it follows from scattering theory (Segal, 1974) that phase space admits an invariant almost Kähler structure which extends the given symplectic one. There is some reason to believe the same is true of the Yang-Mills phase space, at least near sufficiently well-behaved solutions. However, it will probably require a totally different method of establishment, in view of the greater difficulty of dealing with the global existence and scattering issue, and serious doubts arising from current philosophies regarding the physical role of the equations as to whether these issues can have affirmative resolutions. On the other hand, Glassey and Strauss

(1979) have shown that global solutions to the equations do decay somewhat weakly in finite space regions, if they exist at all, which points not in an opposite but orthogonal direction.

All this illustrates the minimality of our knowledge of the Yang-Mills equations, whose study in their hyperbolic form has just begun. At this juncture, it is not known, for example, if there exist solutions to the Cauchy problem of arbitrarily small spatial support, at least for short times, although this is virtually immediate for nonlinear wave equations.

REFERENCES

Branson, T (1979). Ph.D. Dissertation, M.I.T., forthcoming.

Glassey, R.T., and Strauss, W.A. (1979). J. Funct. Anal., in press.

Ørsted, B. (1979). (1979) J. Funct. Anal., in press.

Segal, I.E. (1960). J. Math. Phys. 1, 468.

Segal, I.E. (1979). J. Funct. Anal., in press.

Segal, I.E., (1965). J. Math. Pur. Appl. 13, 71.

Segal, I.E. (1963). Ann. Math. 78, 339.

Segal, I.E. (1976). Mathematical cosmology and extragalactic astronomy, Academic Press, New York.

Segal, I.E. (1974). Symposia Mathematica XIV, 99.

Yang, C.N., and Mills, R.L. (1954). Phys. Rev. 96, 191

Instantons in Nonlinear σ-Models, Gauge Theories and General Relativity +)

Michael Forger
Institut für theoretische Physik
Freie Universität Berlin
Arnimallee 3, D-1000 Berlin 33, Germany

Abstract. We consider nonlinear σ-models, gauge theories and general relativity as three classes of models of field theory which are of an intrinsically geometric nature as well as (possibly) topologically nontrivial, and explore the role of instantons as the basic tool for new perturbative schemes in these models. In particular, we emphasize the close analogy between nonlinear σ-models and pure gauge theories. We also establish a new, manifest type of analogy by extending them to "nonlinear σ-models with gauge symmetry" and "pure gauge theories in a frame field formulation", respectively.

+) Extended and revised version of a talk given at the Conference on Differential Geometric Methods in Mathematical Physics, Clausthal-Zellerfeld, July 1978. The material presented here also overlaps partially with the author's PhD thesis.

1. The Role of Instantons in Field Theory

In the last ten years, great progress has been made towards the construction of realistic models of quantum field theory describing the three fundamental interactions of elementary particle physics, namely the strong, electromagnetic and weak interactions. On the one hand, quantum electrodynamics (QED), the well-known theory of electromagnetic interactions, has been extended to the Weinberg-Salam model which presently appears to be the best candidate for a unified theory of weak and electromagnetic interactions. On the other hand, quantum chromodynamics (QCD) has emerged as the leading candidate for a theory of strong interactions. The common feature of all these models is that they are examples of a gauge theory with fermions, in which the interaction between the fundamental fermions, i.e. the quarks and the leptons, is mediated by gauge vector bosons. More specifically, in QED the gauge group is U(1), the fundamental fermions are the quarks and the charged leptons[1)], and the gauge vector boson is the photon[2)] while in QCD the gauge group is $SU(3)_{color}$, the fundamental fermions are the quarks[1)], and the gauge vector bosons are the gluons[2)].

In spite of the formal similarity between QED and QCD, there are crucial differences between the two models which seem to be related to the fact that QED is an abelian and QCD is a nonabelian gauge theory. In fact, in any gauge theory, the gauge vector bosons are subject to self-interactions which involve the structure constants of the gauge group as coefficients, and these vanish if and only if the gauge group is abelian. Physically speaking, the photon field does not act as its own source besause it does not carry electric charge, but the gluon field do act as their own sources because they do carry color charge. This fact is generally believed to lead to drastic differences in the behavior of the electromagnetic force and the strong force between two quarks, say as a function of their distance[3)]: The electromagnetic force is large at short distances and decreases by a Coulomb inverse sqare law at long distances; it can be calculated directly from QED by using ordinary renormalized perturbation theory. The strong force, however, is small at short distances and increases at long distances. The first property is known as ultraviolet freedom or asymptotic freedom and means, roughly speaking, that the quarks move around freely inside the hadrons when they are close together, while the second property is known as infrared slavery or confinement and means, roughly speaking, that the quarks are bound together permanently to form hadrons via an attractive force that becomes large when one tries to pull them apart. So far, a method for calculating this type of behavior directly from QCD is yet to be found, although there are models (such as the string model or the MIT-bag model) intended to derive it from simple phenomenological assumptions. For more details on these topics, the reader is referred

1) The quark and some of the leptons (electron and muon) carry electric charge, while other leptons (electron-neutrino and muon-neutrino) don't. The quarks carry color charge, while leptons don't.

2) The number of gauge vector bosons is given by the dimension of the gauge group, so there are 1 photon and 8 gluons.

3) Of course, this formulation uses concepts adapted to classical rather than quantum physics: One really asks for the interaction energy i.e. the expectation value of the Hamiltonian between appropriate quark one-particle states.

to Nambu's review article on quark confinement [30].

As a particular consequence of this picture, we see that ordinary renormalized perturbation theory is an insufficient tool for QCD since it is a good approximation only in the short-distance limit and since QCD is expected to be trivial at short distances (asymptotic freedom) and nontrivial at long distances (confinement) rather than the other way round. In the last few years, however, new techniques of perturbation theory have been developed to cope with such a situation. Let me explain briefly in what respect these new methods differ from the old ones; for details, the reader is referred to Coleman's lectures [6] and the literature quoted there.

One possible basis for all approximation techniques in quantum field theory (or quantum mechanics) is the formalism of functional integrals (or path integrals) first introduced by Feynman [13] and developed further by Fadeev and Popov [12] ; see also [14] , [34]. The mathematical status of (Euclidean) functional integrals is still obscure since they involve integrating over the infinite-dimensional space of all (Euclidean) classical field configurations[4)] with respect to some functional measure, and the existing mathematical theory of Wiener integrals, Gaussian processes etc. essentially just covers the free field case. Still, if one is willing to accept that manipulating functional integrals is (presently) a formal business, they provide an extremely useful tool for systematically deriving various perturbative schemes and for passing from one of them to the other. The standard scheme is the semiclassical approximation or weak-coupling approximation[5)] based on an expansion of the integrand in powers of Planck's constant $\hbar$ or the coupling constant g, respectively, around certain configurations; then the one-loop term (i.e. the first nontrivial term in the expansion which gives the first quantum corrections) involves only Gaussian integrals and can be evaluated explicitly. The natural question is of course around what type of configurations the expansion is to be performed. Stationary phase arguments show that they should at least be solutions to the Euclidean classical field equations, but in order to make the

4) These field configurations satisfy certain boundary conditions and are usually supposed to be continuous, but not necessarily differentiable or smooth; in fact, the smooth configurations typically form a set of measure zero, and so do the configurations of finite action.

5) This amounts to the same thing since the expansion parameter is really a dimensionless constant such as $g^2\hbar$, it is an asymptotic expansion which may start with some negative power.

Gaussian integrals in the one-loop term converge, they should in fact be (local) minima of the Euclidean action rather than just stationary points; in particular, they should have finite Euclidean action. More specifically, given a model of classical field theory whose Euclidean action is positive semidefinite, we will typically be in one of the following situations:

1) the model is of a topologically trivial nature in the sense that the space of smooth field configurations[4)] is connected. There is a unique (absolute) minimum of the Euclidean action, namely the configuration where all the fields vanish identically. Performing the expansion around this single point, one is led to ordinary perturbation theory.

2) The model is of a topologically nontrivial nature in the sense that the space of smooth field configurations[4)] is disconnected, and there is a topological charge parametrizing its connected components. Within each connected component, there is an entire manifold of (absolute) minima of the Euclidean action called instantons, and one can introduce (local) coordinates for this manifold called collective coordinates. Performing the expansion around each of these minima simultaneously and carrying out the Gaussian integrals, the functional integral (in the one-loop approximation) is reduced to a finite-dimensional integral over the collective coordinates (which may still diverge and thus need to be regularized). Although one does not yet know how to deal with higher orders in this type of expansion, even the lowest order already leads to results inaccessible to ordinary perturbation theory.

In sec. 2, we shall briefly describe two entire classes of models of field theory which are both of an intrinsically geometric nature and which fit nicely into this picture: nonlinear σ-models and pure gauge theories. In particular, it turns out that there is a strikingly complete analogy between nonlinear σ-models in 2 space-time dimensions and pure gauge theories in 4 space-time dimensions (see e.g. table 1). This is one of the main reasons for being interested in the former since one can hope to test the power, as well as the limits, of the strategy described above by checking the results against computations based on other techniques of 2-dimensional field theory. (There exist quite a number of such techniques, e.g. completely integrable systems, dual symmetry, infinitely many conserved currents and charges, Bäcklund transformations, factorization of S-matrices, 1/N-expansions, ..., which so far have no analogue in 4-dimensional field theory). It should also be mentioned that in both classes, there are models which fall into category 1) - such as "pure QED", i.e. QED "without fermions", in 4 dimensions - as well as models which fall into category 2) - such as "pure QCD", i.e. QCD "without fermions", in 4 dimensions.

Real QED and real QCD do of course contain fermions, which makes the total Euclidean action indefinite. However, the formal rules for integrating over Euclidean classical Fermi fields are different from those for integrating over Euclidean classical Bose fields, in such a way as to remove the need for definiteness. In fact, the total action being bilinear in the Fermi fields, the functional integrals over the Fermi fields are exactly Gaussian and can be evaluated explicitely. Thus one is left with the functional integrals over the gauge fields, but with a modified effective action which is the sum of the action for the "pure" theory, i.e. the theory "without fermions", and a term involving the renormalized functional determinant of the fermionic Dirac operator in the external gauge field (as a function of the gauge field). Unfortunately, this term is nonlocal and sufficiently complicated to make every single step of the strategy described above a formidable task. For example, it is still a conjecture that this effective action is positive semidefinite. [27].

The boundary conditions to be imposed, both on the field configurations[4] in general and on the instantons in particular, depend on the type of correlation functions one wants to compute. As an example, let us compare correlation functions (e.g. vacuum → vacuum amplitudes) for temperature $T = 0$ and for temperature $T > 0$: For $T = 0$, there are no a priori boundary conditions on the field configurations, and the requirement that the instantons have finite Euclidean action is a fall-off condition at infinity in all directions. If in addition, the action is conformally invariant, this essentially amounts to the requirement that they can be defined over the sphere S^n - the one-point compactification of flat Euclidean space $\mathbb{R}^n$ - because (the inverse of) stereographic projection maps flat Euclidean space conformally onto the sphere minus a single point [6]. For $T > 0$, the field configurations have to be periodic in Euclidean time with period $\sim 1/T$, and the requirement that the instantons have finite Euclidean action is a fall-off condition at infinity in the spatial directions. Unfortunately, there seems to be no simple property of the action which would guarantee that this essentially amounts to the

6) Usually, the number n of space-time dimensions is either 2 or 4.

requirement that they can be defined over some compactification of $S^1 \times \mathbb{R}^{n-1}$ such as $S^1 \times S^{n-1}$, say. 6)

Recently, these new techniques have also been applied to quantum gravity (QG). The additional problem there is that the Euclidean action even of the "pure" theory, i.e. the theory "without matter", is indefinite; in fact, varying a metric just within its conformal equivalence class, one can make its action as negative as on wants. Thus if one performs the semiclassical = weak-coupling approximation 5) around a configuration which is a solution to the Euclidean vacuum Einstein equations, there will always be divergent Gaussian integrals in the one-loop term corresponding to the negative eigenvalues of the relevant differential operator. Gibbons, Hawking and Perry have shown [19], [20] that formally, these can be made convergent by Wick rotating the contours of integration for the coefficients belonging to the negative eigenvalues to lie along the imaginary rather than real axis. Although the meaning of this prescription in terms of properties of the functional measure of QG is somewhat obscure, it serves as a motivation for the definition of gravitational instantons as topologically nontrivial solutions to the Euclidean vacuum Einstein equations, with or without cosmological term, depending on the boundary conditions 4):

a) Asymptotically Euclidean (AE) and Asymptotically locally Euclidean (ALE) boundary conditions are relevant for certain temperature $T = 0$ correlation functions (e.g. vacuum → vacuum amplitudes) and require that outside some compact region, the manifold is $\mathbb{R}^+ \times S^3$ and $\mathbb{R}^+ \times S^3/\Gamma$, respectively, and the metric approaches the standard flat Euclidean metric. Here, the $\mathbb{R}^+$-factor is a 4-dimensional Euclidean radial variable, and Γ is a discrete transformation group of S^3 making S^3/Γ into a lens space. It turns out that for AE boundary conditions, there are no instantons [19,] [20], while for ALE boundary conditions, instantons have been constructed recently by Eguchi, Hanson, Gibbons and Hawking [10], [18].

b) Asymptotically flat (AF) boundary conditions are relevant for certain temperature $T > 0$ correlation functions (e.g. vaccuum → vacuum amplitudes) and require periodicity in Euclidean time with period $\sim 1/T$ as well as that outside some spatially compact region, the manifold is $S^1 \times \mathbb{R}^+ \times S^2$ and the metric approaches the standard metric there. Here, the S^1-factor is Euclidean time and the $\mathbb{R}^+$-factor is a 3-dimensional spatial radial variable. See [17], [25].

c) Compact boundary conditions have been discussed by Hawking [24], and instantons have also been constructed by Gibbons, Pope [21] and

Back, Forger, Freund [4].

2. Nonlinear σ-Models, Pure Gauge Theories, and Instantons

In this section, we briefly review nonlinear σ-models and pure gauge theories as two classes of models of classical field theory which are both of an intrinsically geometric nature. The exposition is intended to bring out the close analogies between them, with special emphasis on the concept of instantons in 2-dimensional σ-models and 4-dimensional gauge theories. These common features are collected in table 2.1, and further analogies, leading to a partially unified formulation ot the two classes of models, are discussed in Sec. 2.3.

Models of classical field theory (in n dimensions) are defined by specifying the possible field configurations ϕ, i.e. the types of fields or potentials involved, and by writing down the Lagrangian $L = L(\phi)$ and its integral, the action $S = S(\phi)$. The action being a functional on the space of field configurations, one sets up a variational problem by requiring the classically allowed configurations to be its stationary points; then a standard argument from the calculus of variations shows that these are precisely the solutions to the corresponding Euler-Lagrange equations, which are therefore identified with the field equations of the model. Usually one works over flat Minkowski space $X = \mathbb{R}^n$ or over some suitable spatial compactification such as $X = \mathbb{R} \times S^{n-1}$ (which amounts to imposing certain boundary conditions at spatial infinity), but if a gravitational background field is present, it may become necessary to work over some more general Lorentz manifold X. In any case, the field equations are expected to constitute a hyperbolic system of partial differential equations with a well-posed Cauchy problem whose solution yields the time-evolution of the configuration from given initial data on some Cauchy Hypersurface. For reasons discussed in sec. 1, one is also interested in the Euclidean version of the model, where one usually works over flat Euclidean space $X = \mathbb{R}^n$ or over some suitable compactification such as $X = S^n$ (which amounts to imposing certain boundary conditions at infinity), but if a gravitational background field is present which admits a well-behaved Euclidean section, it may become necessary to work over some more general Riemann manifold X.

Thus let X be a connected, oriented n-dimensional pseudo-Riemannian manifold with metric g and corresponding volume form η, possibly with boundary ∂X; in terms of local coordinates x^μ on X:

g on TX: $g_{\mu\nu} = g(\partial_\mu, \partial_\nu)$, g on T^*X: $g^{\mu\nu} = g(dx^\mu, dx^\nu)$

$$\eta \text{ on } TX: \quad \eta = \frac{1}{n!}\eta_{\mu_1\dots\mu_n} dx^{\mu_1}\wedge\dots\wedge dx^{\mu_n} =$$

$$= |\det g|^{1/2}\, dx^1\wedge\dots\wedge dx^n \in \Gamma(\Lambda^n T^*X) \tag{2.1}$$

$$\eta \text{ on } T^*X: \quad \eta = \frac{1}{n!}\eta^{\mu_1\dots\mu_n}\partial_{\mu_1}\wedge\dots\wedge\partial_{\mu_n} =$$

$$= |\det g|^{-1/2}\,\partial_1\wedge\dots\wedge\partial_n \in \Gamma(\Lambda^n TX)$$

We distinguish two cases:

a) The Euclidean Case: g is a Riemann metric, i.e. of type ++...+.
b) The Minkowski Case: g is a Lorentz metric, i.e. of type +-...-. in addition, X is assumed to be time-oriented.

We shall repeatedly make use of the Cartan-Hodge calculus for bundle-valued differential forms on X: Let V be a real vector bundle over X carrying a nondegenerate (symmetric or antisymmetric) bilinear form $\omega \in \Gamma(\otimes^2 V^*)$, and let D be a linear connection in V preserving ω. Writing $\Omega^p(X)$ and $\Omega^p(X,V)$ for the spaces of ordinary real-valued and of V-valued p-forms on X, respectively, we then have the following operations [7]:

a) An exterior product $\wedge_\omega : \Omega^p(X,V) \times \Omega^q(X,V) \longrightarrow \Omega^{p+q}(X)$ induced by ω :

$$(s\otimes\alpha)\wedge_\omega(t\otimes\beta) = \omega(s,t)\,\alpha\wedge\beta \quad \text{for} \quad s,t\in\Gamma(V),\ \alpha\in\Omega^p(X),\ \beta\in\Omega^q(X) \tag{2.2}$$

b) A bilinear pairing $\omega(.,.) : \Omega^p(X,V) \times \Omega^q(X,V) \longrightarrow C^\infty(X)$ induced by ω and g:

$$\omega(s\otimes\alpha, t\otimes\beta) = \omega(s,t)\,(\alpha,\beta)_g \quad \text{for} \quad s,t\in\Gamma(V),\ \alpha\in\Omega^p(X),\ \beta\in\Omega^q(X) \tag{2.3}$$

It is zero except when p = q .

7) s is the index of g, i.e. s = 0 in the Euclidean case and s = n-1 in the Minkowski case. In c), we assume ω to be symmetric.

c) A star operator $*: \Omega^p(X,V) \xrightarrow{\cong} \Omega^{n-p}(X,V)$ satisying

(2.4) $\alpha \wedge_\omega {*\beta} = (-1)^s \; \omega(\alpha,\beta)\,\eta \;, \quad **\alpha = (-1)^{p(n-p)+s}\alpha$ for $\alpha,\beta \in \Omega^p(X,V)$.

It is just the tensor product of the ordinary star operator $* : \Omega^p(X) \longrightarrow \Omega^{n-p}(X)$ with the identity on the V-part.

d) Three differential operators, namely the covariant exterior derivative $d^V : \Omega^p(X,V) \longrightarrow \Omega^{p+1}(X,V)$ induced by D, the covariant exterior coderivative $\delta^V = -(-1)^{n(p+1)+s} * d^V * :$ $\Omega^p(X,V) \longrightarrow \Omega^{p-1}(X,V)$ which is its formal adjoint since

$$(2.5) \qquad \omega(d^V\alpha,\beta) - \omega(\alpha,\delta^V\beta) = d(\alpha \wedge_\omega {*\beta})$$

for $\alpha \in \Omega^p(X,V)$, $\beta \in \Omega^{p+1}(X,V)$

and the covariant Laplace-de Rham operator

$$\Delta^V = (d^V\delta^V + \delta^V d^V) : \Omega^p(X,V) \longrightarrow \Omega^p(X,V)$$

(Euclidean case)

or d'Alembert - de Rham operator

$$\Box^V = (d^V\delta^V + \delta^V d^V) : \Omega^p(X,V) \longrightarrow \Omega^p(X,V) \quad .$$

(Minkowski case)

For more details, see e.g. [8], [15], [23], [26], [29], [32].

2.1 Nonlinear σ-Models

Fix a Riemannian manifold M with metric h. (in all cases of practical interest, M will be a homogeneous space, and the metric h will be invariant under the relevant group; cf. the examples at the end of sec. 2.3.) The space of (smooth) field configurations is the space of (smooth) maps $\sigma: X \longrightarrow M$. It decomposes into connected components parametrized by the set $[X,M]$ of homotopy classes of maps from X to M, so the model is of a topologically nontrivial nature if and only if this set is nontrivial. In particular, for $X = S^n$, $[X,M] = \pi_n(M)$. The Lagrangian is

$$(2.6) \qquad L(\sigma) = \frac{1}{2}\, h\,(d\sigma, d\sigma) = \frac{1}{2}\, g^{\mu\nu} \cdot h(\partial_\mu\sigma, \partial_\nu\sigma),$$

and the action is its integral

$$(2.7) \qquad S(\sigma) = \frac{1}{2} \int_X g^{\mu\nu}\; h(\partial_\mu\sigma, \partial_\nu\sigma)\; d\,\mathrm{vol} =$$

$$= \frac{1}{2} \int_X d\sigma \wedge_h {}^* d\sigma \quad .$$

the field equations are

(2.8) $$\delta^\sigma d\sigma = 0 \quad ,$$

or in terms of local coordinates x^μ on X

(2.9) $$g^{\mu\nu} (D_\mu \partial_\nu \sigma - \Gamma^\kappa_{\mu\nu} \partial_\kappa \sigma) = 0,$$

where the $\Gamma^\kappa_{\mu\nu}$ are the Christoffel symbols of the Levi-Cività connection in X with respect to the x^μ. (Compare (2.13) below.) We also have the identity

(2.10) $$d^\sigma d\sigma = 0,$$

or in terms of local coordinates x^μ on X

(2.11) $$D_\mu \partial_\nu \sigma - D_\nu \partial_\mu \sigma = 0 \, .$$

To explain the notation, observe that given any field configuration σ, we can define a real vector bundle V over X carrying a Riemannian fibre metric h, and a linear connection D in V preserving h, as follows: V is the pull-back σ^*TM of the tangent bundle TM of M to X via σ, h on V is the pull-back of h on TM, and D in V is the pull-back of the Levi-Cività connection in M. Thus the sections of V are precisely the vector fields on M along σ, and we write d^σ and δ^σ for the corresponding covariant exterior derivative and coderivative, respectively.

Finally, the tangent map $T\sigma : TX \longrightarrow TM$ to σ can be viewed as a distinguished V-valued 1-form $d\sigma \in \Omega^1(X,V)$ on X to which the operations sketched in a) - d) before can be applied. In particular, we see that for any field configuration σ, the lhs of (2.8) is a vector field $\delta^\sigma d\sigma \in \Gamma(V)$ on M along σ; its negative is called the <u>tension field</u> of σ. In the Euclidean case, its vanishing means, due to (2.10), that $d\sigma$ is a harmonic V-valued 1-form on X, and then σ itself is said to be <u>harmonic</u>.

If the u^α are local coordinates for M, the map σ and its tension field $\delta^\sigma d\sigma$ are locally given by ordinary functions σ^α and $(\delta^\sigma d\sigma)^\alpha$ on X, where

2.12) $$(\delta^\sigma d\sigma)(x) = (\delta^\sigma d\sigma)^\alpha (x) \frac{\partial}{\partial u^\alpha}\Big|_{\sigma(x)} \quad ,$$

and in terms of local coordinates x^μ on X, we have

(2.13) $$(\delta^\sigma d\sigma)^\alpha = - g^{\mu\nu} (\nabla_\mu \nabla_\nu \sigma^\alpha + \Gamma^\alpha_{\beta\gamma} \nabla_\mu \sigma^\beta \nabla_\nu \sigma^\gamma)$$

$$= - g^{\mu\nu} (\partial_\mu \partial_\nu \sigma^\alpha - \Gamma^\kappa_{\mu\nu} \partial_\kappa \sigma^\alpha + \Gamma^\alpha_{\beta\gamma} \partial_\mu \sigma^\beta \partial_\nu \sigma^\gamma)$$

Here, ∇ is the Levi-Cività connection in X, and the $\Gamma^\kappa_{\mu\nu}$ and $\Gamma^\alpha_{\beta\gamma}$ are the Christoffel symbols of the Levi-Cività connection in X and M with respect to the local coordinates x^μ and u^α, respectively. On the other hand, assume that M is isometrically embedded into some Euclidean space E. Thus the trivial vector bundle M x E over M splits into the orthogonal direct sum of the tangent bundle TM and the normal bundle NM of M, and if $(.)_T$ denotes taking the tangential component, then we have

$$\text{(2.14)} \qquad \begin{aligned} \delta^\sigma d\sigma &= - (\Delta \sigma)_T & \text{(Euclidean case)} \\ \delta^\sigma d\sigma &= - (\Box \sigma)_T & \text{(Minkowski case)} \end{aligned}$$

where Δ is the Laplacian and $\Box$ is the d'Alembertian on $C^\infty(X,E)$. From both formulations, it is clear that (2.8) is a system of 2nd order nonlinear partial differential equations for the field σ, which justifies the name "non-linear σ-model". However, the nature of the nonlinearity is quite different from that encountered in many other models of field theory as it does not appear directly in the Lagrangian (2.6); rather, it is due to the curvature of the manifold M. Actually, (2.13) shows that we can make the nonlinear term vanish (by using local coordinates u^α with $\Gamma^\alpha_{\beta\gamma} = 0$) if and only if M is flat, while (2.14) expresses the nonlinearity concisely in terms of an extrinsic curvature defined via an embedding. Physically speaking, it is the constraint of being confined to $M \subset E$ which generates the interaction.

There is an extensive mathematical literature on the subjects discussed above and other aspects of harmonic maps; we have mainly used [8], [9].

Concerning instantons in nonlinear σ-models, the case of two dimensions has two distinguished features: 1) the action (2.7) is conformally invariant. (Thus in particular, mapping flat Euclidean space $\mathbb{R}^2$ conformally onto S^2 minus a single point via (the inverse of) stereographic projection, the action taken with respect to S^2 coincides with the action taken with respect to $\mathbb{R}^2$, so that S^2 being compact, both are finite for any field configuration σ defined over S^2).

2) The distinguished 1-form $d\sigma \in \Omega^1(X,V)$ sits in the middle dimension, so that we have its dual $^*d\sigma \in \Omega^1(X,V)$ as another distinguished 1-form at our disposal. Therefore, let us assume for the rest of sec. 2.1 that we are in the Euclidean case and in two

dimensions, i.e. X is a connected, oriented 2-dimensional Riemann manifold with metric g and corresponding volume form η ; for simplicity, we also assume X to be compact and without boundary.

We also want to be more specific about the manifold m by supposing it to be a Hodge manifold [26]. In other words, M is a Kähler manifold - i.e. a complex manifold with a Hermitean metric $\langle .,. \rangle$ whose real part is the Riemannian metric h used before and whose imaginary part is a symplectic form ω defining a real cohomology class $[\omega] \in H^2(M,\mathbb{R})$ - , and this cohomology class is integral, i.e. $[\omega] \in H^2(M,\mathbb{Z})$. For simplicity, we also assume M to be compact; then by a theorem of Kodaira [26], we are actually requiring M to be a compact complex algebraic Kähler manifold. In particular, given any field configuration σ , pulling back yields a complex vector bundle V over X carrying a Hermitean fibre metric $\langle .,. \rangle$ with real part h and imaginary part ω , and a linear connection D in V preserving all these as well as the complex structure (multiplication by i in the fibres).

Under these circumstances, we can define the topological density

$$\frac{1}{2}\,\omega(d\sigma, {}^*d\sigma) = -\frac{1}{2}\,\eta^{\mu\nu}\,\omega(\partial_\mu\sigma, \partial_\nu\sigma) \tag{2.15}$$

(compare (2.6)), and the topological charge is its integral

$$\begin{aligned} q(\sigma) &= -\frac{1}{2}\int_X \eta^{\mu\nu}\,\omega(\partial_\mu\sigma, \partial_\nu\sigma)\,d\,\mathrm{vol} = \\ &= \frac{1}{2}\int_X d\sigma \wedge_\omega d\sigma \end{aligned} \tag{2.16}$$

(compare (2.7)), q is a topological invariant also called the <u>instanton number</u>; it takes only integral values: In fact, under the isomorphism

$$\begin{array}{ccc} H^2(X,\mathbb{R}) & \xrightarrow{\ \simeq\ } & \mathbb{R} \\ \cup & & \cup \\ H^2(X,\mathbb{Z}) & \longrightarrow & \mathbb{Z} \end{array} \tag{2.17}$$

given by integration, $q(\sigma)$ corresponds to the pull-back $\sigma^*[\omega] \in H^2(X,\mathbb{Z})$ of $[\omega] \in H^2(M,\mathbb{Z})$ via σ , which depends only on the homotopy class $[\sigma] \in [X,M]$ of σ.
Moreover, we have the important inequality

$$S(\sigma) \geqslant |q(\sigma)| \quad , \tag{2.18}$$

and the equation holds if and only if

$$(2.19)\qquad \begin{aligned} {}^*d\sigma &= +\, i\, d\sigma \qquad \text{if } q(\sigma) > 0 \\ {}^*d\sigma &= -\, i\, d\sigma \qquad \text{if } q(\sigma) < 0 \end{aligned}$$

These statements follow from the estimate

$$\begin{aligned} 0 \quad &\leqslant \quad \langle d\sigma \pm i\,{}^*d\sigma ,\, d\sigma \pm i\,{}^*d\sigma \rangle \\ &= h(d\sigma, d\sigma) \pm i(\langle d\sigma, {}^*d\sigma\rangle - \overline{\langle d\sigma, {}^*d\sigma\rangle}) + h({}^*d\sigma, {}^*d\sigma) \\ &= 2\left\{ h(d\sigma, d\sigma) \mp \omega(d\sigma, {}^*d\sigma)\right\} \end{aligned}$$

since $*$ is an isometry.

The analysis in this paragraph goes back to a discussion with Prof. F. Hirzebruch; in the meantime, it has also been carried through by Perelomov [33]. See also Golo and Perelomov [22].

The <u>instanton equations</u> (2.18) are a system of 1st order (linear) partial differential equations for the field σ which due to the identity (2.10) imply the field equations (2.8). They are directly related to complex analysis since they are nothing but Cauchy-Riemann equations for σ : In fact, as a connected, oriented 2-dimensional Riemann manifold, X is automatically a Kähler manifold with complex structure (multiplication by i in the tangent spaces) given by the ordinary star operator , and (2.18) amounts to the statement that σ is holomorphic (if $q(\sigma) > 0$) or antiholomorphic (if $q(\sigma) < 0$). For more details on this aspect, we refer the reader to [9], [29],.

2.2 Pure Gauge Theories

Fix a compact connected Lie group G with an Ad-invariant inner product (.,.) on its Lie algebra $\mathfrak{g}$. The space of (smooth) field configurations is the space of gauge equivalence classes [P,A] of pairs (P,A), where P is a (smooth) principle G-bundle over X and A is a (smooth) connection form on P, and where (P,A) and (P',A') are gauge equivalent if and only if there exists a gauge transformation between them, i.e. an isomorphism $f: P \longrightarrow P'$ of principle G-bundles over X such that $f^*A' = A$. This space decomposes into connected components parametrized by the set $k_G(X)$ of isomorphism classes of principal G-bundles over X, so the model is of a topologically nontrivial nature if and only if this set is nontrivial. In particular, for $X = S^n$, $k_G(X) = \pi_{n-1}(G)$ [28],[35]. The Lagrangian is [7])

$$(2.20) \qquad L = (-1)^s \tfrac{1}{4}(F,F) = (-1)^s \tfrac{1}{4} g^{\mu\kappa} g^{\nu\lambda} (F_{\mu\nu}, F_{\kappa\lambda}) \quad ,$$

and the action is its integral

$$(2.21) \qquad S = (-1)^s \frac{1}{4} \int_X g^{\mu\kappa} g^{\nu\lambda} (F_{\mu\nu}, F_{\kappa\lambda}) \, d\,vol = $$
$$= (-1)^s \frac{1}{4} \int_X F \wedge_{(.,.)} {}^* F \quad .$$

The field equations are the pure Yang-Mills equations

$$(2.22) \qquad \delta^A F = 0 \quad ,$$

or in terms of local coordinates x^μ on X

$$(2.23) \qquad g^{\kappa\lambda} (D_\kappa F_{\lambda\mu} - \Gamma^\nu_{\kappa\lambda} F_{\nu\mu} - \Gamma^\nu_{\kappa\mu} F_{\lambda\nu}) = 0 \quad ,$$

where the $\Gamma^\kappa_{\mu\nu}$ are the Christoffel symbols of the Levi-Cività connection in X with respect to the x^μ. (Compare (2.29) below.) We also have the Bianchi identity

$$(2.24) \qquad d^A F = 0 \quad ,$$

or in terms of local coordinates x^μ on X

$$(2.25) \qquad D_\kappa F_{\lambda\mu} + D_\lambda F_{\mu\kappa} + D_\mu F_{\kappa\lambda} = 0 \quad .$$

To explain the notation, observe that given any field configuration [P,A] , we can define a real vector bundle V over X carrying a Riemannian fibre metric (.,.), and a linear connection D in V preserving (.,.), as follows: V is the Lie algebra bundle $P\times_G \mathfrak{g}$ associated to P and the adjoint representation Ad of G on $\mathfrak{g}$, (.,.) on V is induced from (.,.) on $\mathfrak{g}$, and D in V is induced from A on P. Thus the sections of V are precisely the infinitesimal gauge transformations, i.e. the infinitesimal automorphisms of P, and we write d^A and δ^A for the corresponding covariant exterior derivative and coderivative, respectively. Finally, the curvature from curv A = $dA + \frac{1}{2}[A,A]$ of A can be viewed as a distinguished V-valued 2-form $F \in \Omega^2(X,V)$ on X to which the operations sketched in a) - d) before can be applied. In particular, we see that for any field configuration [P,A] , the lhs of (2.22) is a V-valued 1-form $\delta^A F \in \Omega^1(X,V)$ on X; its negative is called the <u>current form</u> of [P,A] . In the Euclidean case, its vanishing means, due to (2.24), that F is a harmonic V-valued 2-form on X, and then [P,A] itself is said to be <u>harmonic</u> .

If we choose a local trivialisation of P, the connection form A, its curvature form F and its current form $\delta^A F$ are locally given by

$\mathfrak{g}$-valued forms on X which we also denote by A,F and $\delta^A F$, respectively, and if we choose generators T_a for $\mathfrak{g}$ defining structure constants f^a_{bc} by

$$[T_b, T_c] = f^a_{bc}\, T_a \quad , \tag{2.26}$$

they are locally given by ordinary forms A^a, F^a and $(\delta^A F)^a$ on X, respectively, such that

$$A = A^a T_a \quad , \quad F = F^a T_a \quad , \quad \delta^A F = (\delta^A F)^a T_a \quad . \tag{2.27}$$

In terms of local coordinates x^μ on X, we have

$$\begin{aligned} &A = A_\mu dx^\mu \;, \quad F = \tfrac{1}{2} F_{\mu\nu}\, dx^\mu \wedge dx^\nu = \sum_{\mu<\nu} F_{\mu\nu}\, dx^\mu \wedge dx^\nu \;, \\ &\delta^A F = (\delta^A F)_\mu\, dx^\mu \\ &A^a = A^a_\mu dx^\mu \;, \quad F^a = \tfrac{1}{2} F^A_{\mu\nu}\, dx^\mu \wedge dx^\nu = \sum_{\mu<\nu} F^a_{\mu\nu}\, dx^\mu \wedge dx^\nu, \\ &(\delta^A F)^a = (\delta^A F)^a_\mu\, dx^\mu \end{aligned} \tag{2.28}$$

and

$$\begin{aligned} (\delta^A F)_\mu &= - g^{\kappa\lambda} (\nabla_\kappa F_{\lambda\mu} + [A_\kappa, F_{\lambda\mu}]) \\ &= - g^{\kappa\lambda} (\partial_\kappa F_{\lambda\mu} - \Gamma^\nu_{\kappa\lambda} F_{\nu\mu} - \Gamma^\nu_{\kappa\mu} F_{\lambda\mu} + [A_\kappa, F_{\lambda\mu}]) \;, \end{aligned} \tag{2.29}$$

or

$$\begin{aligned} (\delta^A F)^a_\mu &= - g^{\kappa\lambda} (\nabla_\kappa F^a_{\lambda\mu} + f^a_{bc} A^b_\kappa F^c_{\lambda\mu}) \\ &= - g^{\kappa\lambda} (\partial_\kappa F^a_{\lambda\mu} - \Gamma^\nu_{\kappa\lambda} F^a_{\nu\mu} - \Gamma^\nu_{\kappa\mu} F^a_{\lambda\nu} + f^a_{bc} A^b_\kappa F^c_{\lambda\mu}) \;, \end{aligned} \tag{2.30}$$

Here, ∇ is the Levi-Civitá connection in X, and the $\Gamma^\kappa_{\mu\nu}$ are its Christoffel symbols with respect to the local coordinates x^μ. From the definition

$$F_{\mu\nu} = \partial_\mu A_\nu - \partial_\nu A_\mu + [A_\mu, A_\nu] \tag{2.31}$$

or

$$F^a_{\mu\nu} = \partial_\mu A^a_\nu - \partial_\nu A^a_\mu + f^a_{bc} A^b_\mu A^c_\nu \tag{2.32}$$

it is clear that (2.22) is a system of 2nd order nonlinear partial differential equations for the field A. The nonlinearity is due to the noncommutativity of the Lie group G. Actually, (2.29) - (2.32) show that we can make the nonlinear term vanish (since the structure constants f^a_{bc} vanish) if and only if G is abelian.

Concerning instantons in pure gauge theories, the case of four dimensions has two distinguished features : 1) The action (2.21) is conformally invariant. (Thus in particular, mapping flat Euclidean space $\mathbb{R}^4$ conformally onto S^4 minus a single point via (the inverse of) stereographic projection, the action taken with respect to S^4 coincides with the action taken with respect to $\mathbb{R}^4$, so that S^4 being compact, both are finite for any field configuration $[P,A]$ defined over S^4.)

2) The distinguished 2-form $F \in \Omega^2(X,V)$ sits in the middle dimension, so that we have its dual $^*F \in \Omega^2(X,V)$ as another distinguished 2-form at our disposal. Therefore, let us assume for the rest of sec. 2.2 that we are in the Euclidean case and in four dimensions, i.e. X is a connected, oriented 4-dimensional Riemann manifold with metric g and corresponding volume form η ; for simplicity, we also assume X to be compact and without boundary.

Under these circumstances, we can define the topological density

$$(2.33) \qquad N(F,{}^*F) \;=\; \frac{N}{4}\,\eta^{\mu\nu\kappa\lambda}(F_{\mu\nu},F_{\kappa\lambda})$$

(compare (2.20)), and the topological charge is its integral

$$(2.34) \qquad q \;=\; \frac{N}{4}\int_X \eta^{\mu\nu\kappa\lambda}(F_{\mu\nu},F_{\kappa\lambda})\,d\,vol \;=\; N\int_X F\wedge_{(.,.)}F$$

(compare (2.21)). $N > 0$ is a normalization constant that we could have absorbed into the definition of the inner product (.,.) on $\mathfrak{g}$. q is a topological invariant also called the <u>instanton number</u> ; it takes only integral values if N is chosen appropriately: In fact, under the isomorphism

$$(2.35) \qquad \begin{array}{ccc} H^4(X,\mathbb{R}) & \longrightarrow & \mathbb{R} \\ \cup & \simeq & \cup \\ H^4(X,\mathbb{Z}) & \longrightarrow & \mathbb{Z} \end{array}$$

given by integration, q corresponds to some characteristic class which is integral if N is chosen appropriately and which depends only on the isomorphism class $[P] \in k_G(X)$ of (P,A). For example, if $G = SU(n)$, $\mathfrak{g} = su(n)$, and $(X,Y) = -\,\mathrm{trace}\,XY$ for $X,Y \in su(n)$, we choose $N = 1/8\pi^2$ and obtain

$$(2.36) \qquad S \;=\; -\frac{1}{4}\int_X \mathrm{trace}\;F\wedge{}^*F \quad , \quad q \;=\; -\frac{1}{8\pi^2}\int_X \mathrm{trace}\;F\wedge F \quad ,$$

so q corresponds to the second Chern class. Moreover, we have the important inequality

$$(2.37) \qquad S \;\geqslant\; \frac{4}{N}|q| \quad ,$$

and the equality holds if and only if

$$(2,38) \qquad {}^{*}F = +F \quad \text{if } q > 0$$
$$\qquad {}^{*}F = -F \quad \text{if } q < 0 .$$

These statements follow from the estimate

$$0 \leq (F \mp {}^{*}F, F \mp {}^{*}F) = (F,F) \mp 2(F,{}^{*}F) + ({}^{*}F,{}^{*}F)$$
$$= 2\left\{(F,F) \mp (F,{}^{*}F)\right\}$$

since $*$ is an isometry.

The analysis is this paragraph goes back to Belavin, Polyakow, Schwarz and Tyupkin [5] .

The instanton equations (2.38) are a system of 1st order nonlinear partial differential equations for the field A which due to the identity (2.24) imply the field equations (2.22). They are intimately related to complex analysis via the concept of twistor spaces; for more details on this aspect, we refer the reader to [1] , [2] , [3], [16], [36].

No	Property	2-dim nonlinear σ-models	4-dim pure gauge theo.
1	Field Configurations (Dynamical Variable)	σ: map $\sigma: X \to M$	[P,A]: P principal G-bundle, A connectionf.
2	Field Strength	$d\sigma \in \Omega^1(X,V)$, $V = \sigma^{*}TM$	$F \in \Omega^2(X,V), V = P \times_G \mathfrak{g}$
3	Action	(2.7)	(2.21)
4	Action conformally invariant	yes	yes
5	Field equations (system of 2nd order partial diff. eq.s)	(2.8), (2.9); see (2.13)	(2.22), (2.23) see (2.29), (2.30) Yang-Mills Equations
6	Identities	(2.10), (2.11)	(2.24),(2.25) Bianchi Identities
7	Topological Charge	(2.16)	(2.34)
8	Estimate	(2.18)	(2.37)
9	Instanton Equations (system of 1st order partial diff. eq.s)	(2.19) Cauchy-Riemann equations $\Longleftrightarrow$ (2.18) has = $\Rightarrow$ (2.8)	(2.38) Self-Duality equations $\Leftrightarrow$(2.37) has = $\Rightarrow$(2.22)
10	Relation to Complex Analysis	yes, via holomorphic or antiholomorphic maps from X to M	yes,via holomorphic bundles over twistor spaces fibered over X

Table 2.1

2.3 A Unified Formulation for σ-Models and Gauge Theories

The strikingly complete analogy between nonlinear σ-models and pure gauge theories suggests that they might just be two different aspects of some other class of models which in some sense incorporates them both and which should also be of an intrinsically geometric nature. In the following, we want to indicate briefly how such a partial unification can be achieved.

Fix a Riemannian manifold M with metric h and a compact connected Lie group G with an Ad-invariant inner product (.,.) on its Lie algebra $\mathfrak{g}$. In addition, fix a principle G-bundle Q over M with connection : We write $\rho : Q \longrightarrow M$ for the bundle projection, Ver Q for the vertical bundle, τ for the canonical isomorphism

$$(2.39)\quad \tau : \begin{array}{ccc} Q \times \mathfrak{g} & \longrightarrow & \text{Ver } Q \\ (q,X) & \longmapsto & \frac{d}{dt}\, q\cdot \exp tX \Big|_{t=0} \end{array} ,$$

Hor Q for the corresponding horizontal bundle and H_* resp. V_* for the corresponding horizontal resp. vertical projection, related as follows [23]:

$$TQ \;=\; \text{Ver } Q \;\oplus\; \text{Hor } Q$$

$$(2.40)\quad \begin{array}{l} H_* : TQ \longrightarrow TQ,\; H_*^2 = H_*,\; \ker H_* = \text{Ver } Q,\; \text{im } H_* = \text{Hor } Q \\ V_* : TQ \longrightarrow TQ,\; V_*^2 = V_*,\; \ker V_* = \text{Hor } Q,\; \text{im } V_* = \text{Ver } Q \end{array}$$

$$H_* + V_* \;=\; id_{TQ} .$$

Moreover, we write α for the corresponding connection form on Q given as the composition

$$(2.41)\qquad \alpha = \tau^{-1} \circ V_* \;:\; TQ \longrightarrow Q \times \mathfrak{g} ,$$

and we introduce a G-invariant Riemannian metric on Q, determined uniquely by the following three conditions

$$\text{Ver } Q \;\perp\; \text{Hor } Q$$

$$(2.42)\qquad \tau^{-1} \;:\; \text{Ver } Q \longrightarrow Q \times \mathfrak{g} \text{ is an isometry}$$

$$T\rho\Big|_{\text{Hor } Q} :\; \text{Hor } Q \longrightarrow TM \text{ is an isometry .}$$

(in all cases of practical interest, Q and M will both be homogeneous spaces, and the connection as well as the metrics will also be invariant under the relevant global symmetry group; cf. the examples at the end of sec. 2.3.) The space of (smooth) field configurations is the space of gauge equivalence classes $[P,\phi]$ of pairs (P,ϕ), where P is a (smooth) principal G-bundle over X and ϕ is a (smooth) homomorphism $\phi : P \longrightarrow Q$ of principal G-bundles, and where (P,ϕ) and (P',ϕ') are gauge equivalent if and only if there exists a gauge transformation between them, i.e. an isomorphism $f: P \longrightarrow P'$ of principal G-bundles over X such that $\phi' \circ f = \phi$. This space decomposes into connected components parametrized by the disjoint union

(2.43) $$\dot{\bigcup_{[\sigma]\in[X,M]}} \pi_{[\sigma]}$$

of groups $\pi_{[\sigma]}$, $[\sigma]\in[X,M]$, which are determined uniquely up to an isomorphism by the requirement that given any map $\sigma: X \longrightarrow M$, $\pi_{[\sigma]}$ is isomorphic to the group $\pi_0(\mathrm{Aut}(\sigma^*Q))$ of connected components - i.e. the 0th homotopy group - of the group $\mathrm{Aut}(\sigma^*Q)$ of gauge transformations - i.e. of automorphisms - in the pull-back σ^*Q of Q to X via σ [23].

It is obvious that the space of field configurations as defined above admits a projection

(2.44) $$[P,\phi] \longrightarrow \sigma$$

to the space of field configurations for the nonlinear σ-model by taking σ to be the base map $\sigma: X \longrightarrow M$ induced by $\phi : P \longrightarrow Q$, as well as a projection

(2.45) $$[P,\phi] \longrightarrow [P,A]$$

to the space of field configurations for the pure gauge theory by taking the connection form A on P to be the pull-back $A = \phi^*\alpha$ of the connection form α on Q via $\phi : P \longrightarrow Q$. In terms of a commutative diagram,

(2.46) $$\begin{array}{ccc} (P,A) & \xrightarrow{\ \phi\ } & (Q,\alpha) \\ \Big\downarrow \pi & & \rho \Big\downarrow \\ X & \xrightarrow{\ \sigma\ } & M \end{array}$$

where we write $\pi: P \longrightarrow X$ for the bundle projection, and taking tangent maps, we obtain the commutative diagram

(2.47) $$\begin{array}{ccc} TP & \xrightarrow{\ T\phi\ } & TQ \\ \Big\downarrow T\pi & & T\rho \Big\downarrow \\ TX & \xrightarrow{\ T\sigma\ } & TM \end{array}$$

Now in the case of nonlinear σ-models, the pull-back π^*V of the vector bundle V over X to P via π is

(2.48) $$\pi^* V = \pi^*(\sigma^* TM) \cong \phi^*(\rho^* TM) \cong \phi^*(\text{Hor } Q) ,$$

and the horizontal part $D\phi = H_* \circ T\phi : TP \longrightarrow \text{Hor } Q$ and the tangent map $T\phi : TP \longrightarrow TQ$ to ϕ can be viewed as a distinguished equivariant, horizontal, $(\pi^* V)$-valued 1-form $D\phi \in \Omega^1_{EH}(P, \pi^* V)$ on P which under the isomorphism $\Omega^1_{EH}(P, \pi^* V) \cong \Omega^1(X,V)$ corresponds to $d\sigma \in \Omega^1(X,V)$. On the other hand, in the case of pure gauge theories, the pull-back $\pi^* V$ of the vector bundle V over X to P via π is

(2.49) $$\pi^* V = \pi^*(P \times_G \mathfrak{g}) \cong P \times \mathfrak{g} \cong \phi^*(Q \times \mathfrak{g}) \cong \phi^*(\text{Ver } Q),$$

and the horizontal part $D\phi = H_* \circ T\phi : TP \longrightarrow \text{Hor } Q$ of the tangent map $T\phi : TP \longrightarrow TQ$ to ϕ, together with the Lie bracket $[.,.]$ for vector fields on Q and the connection form α on Q, gives rise to a distinguished equivariant, horizontal, $(\pi^* V)$-valued 2-form $-\alpha[D\phi, D\phi] \in \Omega^2_{EH}(P, \pi^* V)$ $(= \Omega^2_{EH}(P, \mathfrak{g}))$ on P which under the isomorphism $\Omega^2_{EH}(P, \pi^* V) \cong \Omega^2(X,V)$ corresponds to $F \in \Omega^2(X,V)$.

The term "projection" for (2.44) and (2.45) suggests a surjectivity statement: This is trivially true for (2.44) because given σ, choose P to be the pull-back $\sigma^* Q$ of Q to X via σ and ϕ to be the natural lift of σ [23]. A couple of deep theorems [28],[35], [31] show that it is also true for (2.45) if $\rho : Q \longrightarrow M$ and α are chosen appropriately: they have to be <u>universal</u> in dimensions $\leq$ dim X. (This concept is briefly explained below.) In both cases, it is clear that all the quantities appearing both in sec. 2.1 and sec. 2.2 may be reexpressed entirely in terms of the configuration $[P,\phi]$, and we call the resulting models <u>nonlinear σ-models with gauge symmetry</u> and <u>pure gauge theories in a frame field formulation</u>, respectively. (the last term is due to the fact that in some sense, the relation between A and ϕ is similar to the relation between the Levi-Cività connection (Christoffel symbols) and the metric (or orthonormal frame fields) in general relativity.

We conclude by giving examples which in a way are typical and fundamental: Let N and k be positive integers, $N > k$, and Let $\mathfrak{M}(N,k)$ be the space of all complex $(N \times k)$-matrices (N rows and k columns), endowed with its natural positive definite Hermitean form $(z_1, z_2) \longrightarrow \text{trace } z_1^* z_2$, where $z^* \in \mathfrak{M}(k,N)$ denotes the Hermitean adjoint of $z \in \mathfrak{M}(N,k)$. The real submanifold

(2.50) $$V(N,k) = \{ z \in \mathfrak{M}(N,k) \ / \ z^* z = 1_k \}$$

of $\mathfrak{M}(N,k)$ is naturally identified with the Stiefel manifold $Q = V(N,k)$ of orthonormal k-frames in $\mathbb{C}^N$. On the one hand, the "big" unitary group U(N) and its subgroup SU(N) act transitively on V(N,k) by matrix multiplication from the left, and computing stability subgroups of the distinguished element

$$(2.51) \qquad \begin{matrix} k \updownarrow \\ N-k \updownarrow \end{matrix} \begin{pmatrix} 1_k \\ 0 \end{pmatrix} \in \quad V(N,k),$$

one obtains the following identifications of V(N,k) as a homogeneous space:

$$(2.52) \qquad V(n,k) \cong U(N)/U(N-k) \cong SU(N)/SU(N-k) .$$

On the other hand, the "small" unitary group U(k) acts on V(N,k) by matrix multiplication from the right, and the quotient V(N,k)/U(k) is naturally identified with the Grassmann manifold $M = G(N,k)$ of k-planes (k-dimensional subspaces) in $\mathbb{C}^N$. Using (2.50) - (2.52), one obtains the following identifications of G(N,k) as a homogeneous space:

$$(2.53) \qquad G(N,k) \cong U(N)/U(N-k) \times U(k) \cong SU(N)/S(U(N-k) \times U(K)).$$

The projection $\rho : V(N,k) \longrightarrow G(N,k)$ just takes an orthonormal k-frame to the k-plane it generates, and it defines a principal U(k)-bundle called the Stiefel bundle, on which U(N) acts transitively from the left. Moreover, the u(k)-valued 1-form $\alpha = z^* dz$ defines a connection on it, called the Stiefel connection, which is U(N)-invariant. Finally, the inclusion $V(N,k) \subset \mathfrak{M}(N,k)$ implies that V(N,k) admits a normal bundle NV(N,k) as well as its tangent bundle TV(N,k), defining an orthogonal decomposition

$$(2.54) \qquad V(N,k) \times \mathfrak{M}(N,k) = TV(N,k) \oplus NV(N,k)$$

of the trivial vector bundle $V(N,k) \times \mathfrak{M}(N,k)$, and the connection can also be described in terms of the orthogonal decomposition

$$(2.55) \qquad TV(N,k) = \text{Ver } V(N,k) \oplus \text{Hor } V(N,k)$$

of the tangent bundle TV(N,k) of V(N,k). Explicitly, for $z \in V(N,k)$

$$(2.56) \qquad \begin{aligned} T_z V(N,k) &= \{ a \in \mathfrak{M}(N,k) \,/\, z^* a + a^* z = 0 \} \\ N_z V(N,k) &= \{ a \in \mathfrak{M}(N,k) \,/ a = zb \text{ with } b \in \mathfrak{M}(k,k) \text{ hermitean} \} \end{aligned}$$

$$\mathrm{Ver}_z V(N,k) = \{ a \in \mathfrak{M}(N,k) \ / a = zb \text{ with } b \in \mathfrak{M}(k,k) \text{ antihermetian} \}$$

$$\mathrm{Hor}_z V(N,k) = \{ a \in \mathfrak{M}(N,k) \ / \quad z^* a = 0 \text{ and } a^* z = 0 \}$$

and the corresponding orthogonal projections are

$$(2.57) \quad \begin{array}{ccc} \mathfrak{M}(N,k) & \longrightarrow & T_z V(N,k) \\ a & \longrightarrow & a - \frac{1}{2} z(z^* a + a^* z) \end{array} \qquad \begin{array}{ccc} \mathfrak{M}(N,k) & \longrightarrow & N_z V(N,k) \\ a & \longrightarrow & \frac{1}{2} z(z^* a + a^* z) \end{array}$$

$$V_* : \begin{array}{ccc} T_z V(N,k) & \longrightarrow & \mathrm{Ver}_z V(N,k) \\ a & \longrightarrow & zz^* a \end{array} \quad , H_* : \begin{array}{ccc} T_z V(N,k) & \longrightarrow & \mathrm{Hor}_z V(N,k) \\ a & \longrightarrow & a - zz^* a \end{array}$$

(Observe that $V(N,k)$ being a real submanifold of $\mathfrak{M}(N,k)$, the notion of orthogonality refers to the positive definite, symmetric, real bilinear form $(z_1, z_2) \longrightarrow \mathrm{Re\ trace}\ z_1^* z_2 = \frac{1}{2} \mathrm{trace}\ (z_1^* z_2 + z_2^* z_1)$ on $\mathfrak{M}(N,k)$.) As indicated above, the particular role of this construction is due to its universality: More specifically the Stiefel bundle is <u>universal in dimension $\leq n$</u> if $N \geq n/2 + k$ [28], [35], which means that given a manifold X of dimension $\leq n$ and a principal $U(k)$-bundle P over X, there exists a <u>classifying map</u> $\sigma : X \longrightarrow G(N,k)$ such that P is isomorphic to the pull-back $\sigma^* V(N,k)$ of $V(N,k)$ to X via σ. In fact, it is well known [26], [33] that this prescription establishes a one-to-one correspondence between the set $k_{U(k)}(X)$ of isomorphism classes of principal $U(k)$-bundles over X and the set $[X, G(N,k)]$ of homotopy classes of maps from X to $G(N,k)$. Moreover, the Stiefel connection is also <u>universal in dimensions $\leq n$</u> if $N \geq (n-1)(2n+1)k^3$ [31], which means that given a manifold X of dimensions $\leq n$, a principal $U(k)$-bundle P over X and a connection form A on P, there exists a <u>classifying homomorphism</u> $\phi : P \longrightarrow V(N,k)$ of principal $U(k)$-bundles such that A is the pull-back $A = \phi^* \alpha$ of α via ϕ.

This example can be generalized in several directions: On the one hand, one may cover the cases where the gauge group G is the orthogonal group $O(k)$ or the symplectic group $Sp(k)$ rather than the unitary group $U(k)$ by working over the field of real numbers or the algebra of quaternions rather than the field of complex numbers, obtaining real or quaternionic rather than complex matrices, Stiefel and Grassmann manifolds, etc. On the other hand, if G is some closed subgroup of $U(k)$, $O(k)$ or $Sp(k)$ [8], one can still use the Stiefel manifold $Q = V(N,k)$ but has to replace the Grassmann manifold $G(N,k)$ by the quotient $M = V(N,k)/G$: For example, let $G = SU(k)$. The quotient $V(N,k)/SU(k)$ is naturally identified with the <u>oriented Grassmann manifold</u> $M = SG(N,k)$ of oriented k-planes (oriented k-dimensional subspaces) in $\mathbb{C}^N$. Using (2.50) - (2.52), one obtains the following identifications of $SG(N,k)$ as a homogeneous space:

$$(2.52) \qquad SG(N,k) \cong U(N) \Big/ U(N-k) \times SU(k) \cong SU(N) \Big/ SU(N-k) \times SU(k).$$

The projection $\varsigma : V(N,k) \longrightarrow SG(N,k)$ just takes an orthonormal k-frame to the k-plane it generates, endowed with the induced orientation, and it defines a principal SU(k)-bundle, called the <u>oriented Stiefel bundle</u>, on which U(N) acts transitively from the left. Moreover, the su(k)-valued 1-form $\alpha = z^*dz - \frac{1}{k}\,\mathrm{trace}\,(z^*dz)$ defines a connection on it, called the <u>oriented Stiefel connection</u> which is U(N)-invariant. For general G, the procedure is similar. In all these cases, the universality statements are preserved. Moreover, the limit of large N (for k fixed) which plays a role there is precisely the limit of the 1/N-expansion - one of the most powerful tools of (2-dimensional) quantum field theory which has been used by D'Adda, Di Vecchia and Luescher to derive statements about confinement in 2-dimensional nonlinear σ-models with a U(1) gauge symmetry [7]. (Observe that

$$(2.59) \qquad V(N,1) = S^{2N-1} \quad , \quad G(N,1) = \mathbb{C}P^{N-1} \quad ,$$

which is why - following Eichenherr [11] who was the first to consider them - they term them $\mathbb{C}P^{N-1}$ nonlinear σ-models.) It seems hard to believe that this is merely an accident, and we conjecture that our approach leads to a new type of 1/N-expansion for gauge theories which may be free of the problems arising in the 1/k-expansion that has been investigated so far.

<u>Acknowledgements.</u>
It is a pleasure to acknowledge fruitful discussions with G.W. Gibbons, F. Hirzebruch, M. Luescher, B. Schroer and L. Stuller.

8) Recall that any compact connected Lie group G admits a finite-dimensional, faithful, unitary representation and may therefore be considered as a closed subgroup of U(k) for suitable k.

I also want to thank L. Stuller for a critical reading of part of the manuscript.

REFERENCES

1) M.F. Atiyah, R.S. Ward: Commun. Math. Phys. 55, 117 (1977)
2) M.F. Atiyah, N.J. Hitchin, I.M. Singer: Proc. Roy. Soc. London 362, 425 (1978)
3) M. Atiyah, V.G. Drinfeld, N.J. Hitchin, Yu.I. Manin: Phys. Lett. 65A , 185 (1978)
4) A. Back, M. Forger, P.G.O. Freund: Phys. Lett. 77B, 181 (1978)
5) A.A. Belavin, A.M. Polyakov, A.S. Schwarz, Y.S. Tyupkin: Phys. Lett. 59B, 85 (1975)
6) S. Coleman: "The Uses of Instantons", in: Proc. 1977 Summer School Subnuclear Phys., Erice
7) A. D'Adda, P. Di Vecchia, M. Lüscher: Nucl. Phys. B146, 63 (1978)
8) J. Eells, J.H. Sampson: Am. J. Math 86, 109 (1964)
9) J. Eells, L. Lemaire: Bull. London Math. Soc. 10, 1 (1978)
10) T. Eguchi, A. Hanson: Phys. Lett. 74B, 249 (1978)
11) H. Eichenherr: Nucl. Phys. B146, 215 (1978)
12) L.D. Fadeev, V.N. Popov: Phys. Lett. 25B, 29 (1967)
13) R.P. Feynman: Phys. Rev. 80, 440 (1950)
14) R.P. Feynman, A.R. Hibbs: "Quantum Mechanics and Path Integrals", Mc Graw-Hill, New York (1965)
15) H. Flanders: "Differential Forms", Academic Press, New York (1963)
16) M. Forger: " Gauge Theories, Instantons and Algebraic Geometry", in: Proc. Diff. Geom. Math. Phys., Clausthal-Zellerfeld (1977); to appear in Rep. Math. Phys.
17) G.W. Gibbons, S.W. Hawking: Phys. Rev. D15. 2752 (1977)
18) G.W. Gibbons, S.W. Hawking: Phys. Lett. 78B, 430 (1978)
19) G.W. Gibbons, S.W. Hawking, M.J. Perry: Nucl. Phys. B138, 141 (1978)
20) G.W. Gibbons, M.J. Perry: Nucl. Phys. B146, 90 (1978)
21) G.W. Gibbons, C.N. Pope: Commun. Math. Phys. 61, 239 (1978)
22) V.L. Golo, A. M. Perelomov: Lett. Math. Phys. 2, 477 (1978)
23) W. Greub, S. Halperin, R. Vanstone: "Connections, Curvature and Cohomology", Vol. I (1972) and Vol. II (1973), Academic Press, New York
24) S.W. Hawking: "Spacetime Foam", DAMTP preprint, Cambridge, UK (1978)
25) S.W. Hawking: Phys. Rev. D18, 1747 (1978)

26) F. Hirzebruch: "Topological Methods in Algebraic Geometry", 3rd edition, Springer, New York (1966)

27) H. Hogreve, R. Schrader, R. Seiler: Nucl. Phys. B142, 525 (1978)

28) D. Husemoller: "Fibre Bundles", 2nd edition, Springer, Berlin (1966)

29) A. Lichnerowicz: "Applications harmoniques et variétés kähleriennes", in: Symp. Math. Bologna III, Academic Press, New York (1970)

30) Y. Nambu: "The Confinement of Quarks ", in : Scientific American 235 , 48 (Nov. 1976)

31) M.S. Narasimhan, S. Ramanan: Am. J. Math. 83, 563 (1961)

32) E. Nelson: "Tensor Analysis", Princeton University Press, Princeton (1967)

33) A.M. Perelomov: Commun. Math. Phys. 63, 237 (1978)

34) V.N. Popov: "Functional Integrals in Quantum Field Theory", CERN preprint TH 2424 (Dec. 1977)

35) N. Steenrod: "The Topology of Fibre Bundles", Princeton University Press, Princeton (1951)

36) R. Stora: "Yang-Mills Instantons, Geometrical Aspects", in: Proc. 1977 Summer School Mathematical Phys., Erice, Lecture Notes in Physics, Vol. 73, Springer, Berlin (1978)

Gauge-Theoretical Foundation of Color Geometrodynamics

Eckehard W. Mielke

Institut für Reine und Angewandte Kernphysik
der Christian-Albrecht-Universität Kiel,
Olshausenstraße 40-60, 2300 Kiel 1
Federal Republic of Germany

Abstract:
Salam's SL(6,$\mathbb{C}$) gauge theory of strong interactions is generalized to one having GL(2f,$\mathbb{C}$) $\otimes$ GL(2c,$\mathbb{C}$) or the affine extension thereof as structure group. The concept of fibre bundles and Lie-algebra-valued differential forms are employed in order to exhibit the geometrical structure of this gauge-model. Its dynamics is founded on a gauge-invariant Einstein-Dirac-type Lagrangian. The Heisenberg-Pauli-Weyl non-linear spinor equation generalized to a curved space-time of hadronic dimensions and Einstein-type field equations for the strong f-metric are then derived from variational principles. It is shown that the nonlinear terms are induced into the Dirac equation by Cartan's geometrical notion of torsion. It may be speculated that in this geometrical model extended particles are represented by f x c quarks which are (partially) confined within geon-like objects.

I. Introduction:

"The formulation of Dirac's theory of the electron in the frame of general relativity has to its credit one feature which should be appreciated even by the atomic physicist who feels safe in ignoring the role of gravitation in the building-up of the elementary particles:

) Work supported by the Deutsche Forschungsgemeinschaft, Bonn

Its explanation of the quantum mechanical principle of "gauge invariance" that connects Dirac's ψ with the electromagnetic potentials".

This view put forth 1950 by Herman Weyl is revived in color-geometrodynamics (CGMD): Matter is represented by f x c fundamental spinor fields $\psi^{(q_f, q_c)}$ (distinguished by f flavor and c color degrees of freedom) which are coupled to a Lagrangian invariantly constructed from the gauge potentials of strong interactions. However, unlike quantum-chromodynamics (Gell-Mann et al., 1978) which assumes $U(f) \otimes U(c)$ as "gauge group", according to Weyl, the tensor forces of strong gravity (Isham et al., 1971) should play an equivalently important role for a description of strong interactions. Following this idea, the group $GL(2f,\mathbb{C}) \otimes GL(2c,\mathbb{C})$ is taken as the gauge group of CGMD whereas its dynamics determined by a gauge-invariant generalization of the Einstein-Hilbert action together with a Dirac Lagrangian generalized to a curved space-time of hadronic dimension. The latter is characterized by the modified Planck length

$$\ell^* = (8\pi\hbar G_s/c^3)^{1/2} = \sqrt{8\pi}\ \hbar/c M^* \tag{1.1}$$

or the Planck mass $M^* \sim 1$ GeV of strong gravity.

As is well-known from general relativity with spin and torsion (Hehl et al., 1976) Cartan's notion of torsion (Cartan, 1922-23) of the underlying space-time induces nonlinear spinor terms into the Dirac equation. In the generalization considered here the resulting Heisenberg-Pauli-Weyl spinor equation (Weyl, 1950) gives rise to a nonlinear coupling also among the different fundamental spinor fields, similarly as in Heisenberg's unified field theory of elementary particles (Heisenberg, 1966; 1974).

In this paper the semi-classical spinor equation as well as the Einstein-type field equations of CGMD will be derived in mathematical detail by employing gauge-covariant differential forms.

Thereby it is possible to bring out the underlying gauge-theoretical structure of CGMD in a more concise form compared to other similar approaches (Trautman 1972-73 ; Hehl et al., 1976) which are limited to the Poincaré group $SL(2,\mathbb{C}) \rtimes \mathbb{R}^4$ as gauge group.

II. Differential Forms

As is well-known, the fundamental geometric structure of gauge theories can be concisely brought out by the use of differential forms. In order to establish a notation, their main properties will

be collected (following, e.g. Flanders, 1963; Wheeler, 1962, Kobayashi and Nomizu, 1963):
On a differential manifold M^n of dimension n, the bundle $D^{(p)}(M^n)$ of differential forms

$$A^{(p)} = \frac{1}{p!} A_{\alpha_1 \ldots \alpha_p} dx^{\alpha_1} \wedge \ldots \wedge dx^{\alpha_p} \tag{2.1}$$

is defined as a skew-symmetric covariant tensor field of degree p. (KN, p. 33). More generally, a $\mathfrak{g}$ -valued p-form takes values in the Lie algebra $\mathfrak{g}$ of G, i.e. with respect to a basis $E_1, \ldots, E_r$ for $\mathfrak{g}$ its components may be written as

$$A_{\alpha_1 \ldots \alpha_p} = A^j_{\alpha_1 \ldots \alpha_p} E_j \; ; \; j = 1, \ldots, \dim \mathfrak{g} \, . \tag{2.2}$$

Then, the exterior (or wedge) product is defined by

$$A^{(p)} \wedge B^{(q)} = \frac{1}{p!\,q!} A_{\alpha_1 \ldots \alpha_p} B_{\beta_1 \ldots \beta_q} \times \\ \times \, dx^{\alpha_1} \wedge \ldots \wedge dx^{\alpha_p} \wedge dx^{\beta_1} \wedge \ldots \wedge dx^{\beta_q} \tag{2.3}$$

whereas the commutator of $\mathfrak{g}$-valued forms may be denoted by

$$[A^{(p)}, B^{(q)}] = A^{(p)} \wedge B^{(q)} - (-1)^{pq} B^{(q)} \wedge A^{(p)} \tag{2.4}$$

The collection of all forms

$$D(M^n) = \bigwedge_{p=0}^{\infty} D^p(M^n) \tag{2.5}$$

constitutes the exterior algebra over $\mathbb{R}$.
The exterior derivative d acts on forms as total derivative obeying the supplementary rules

$$d(A^{(p)} \wedge B^{(q)}) = dA^{(p)} \wedge B^{(q)} + (-1)^p A^{(p)} \wedge dB^{(q)} \tag{2.6}$$

and

$$dd\,A^{(p)} = 0 \, . \tag{2.7}$$

On M^n regarded as a (pseudo-) Riemannian manifold with the metric tensor $f_{\mu\nu}$ and signature s the dual of a p-form is the n-p form

$$^*A^{(n-p)} = \frac{1}{(n-p)!\,p!} \sqrt{|f|}\; \varepsilon_{\alpha_1 \ldots \alpha_p \beta_1 \ldots \beta_{n-p}} A^{\alpha_1 \ldots \alpha_p} \times \\ \times \, dx^{\beta_1} \wedge \ldots \wedge dx^{\beta_{n-p}} \tag{2.8}$$

which (for p ⩽ n) results from the application of the Hodge star operator *. Here the determinant of the metric is abbreviated with $f = \det f_{\mu\nu}$, whereas $\varepsilon_{\alpha_1 \ldots \alpha_p}$ denotes the completely antisymmetric Levi-Civita tensor (MTW, p. 87). Accordingly, the volume form on M^n may be expressed as

$$d\mu(M^n) = {}^*1 = \sqrt{|f|}\, dx^1 \wedge \ldots \wedge dx^n \tag{2.9}$$

Up to a sign, the double dual gives back the original form

$$**A^{(p)} = (-1)^{p(n-p)+(n-s)/2} A^{(p)} \quad . \tag{2.10}$$

The star operator induces an inner product on p-forms:

$$\begin{aligned} Tr(A^{(p)} \wedge {}^* B^{(p)}) &= Tr(B^{(p)} \wedge {}^* A^{(p)}) = \\ &= \frac{(-1)^{(n-s)/2}}{p!} A^{j\alpha_1 \ldots \alpha_p} B^{k}{}_{\alpha_1 \ldots \alpha_p} Tr(E_j E_k) \; {}^*1 \quad . \end{aligned} \tag{2.11}$$

Furthermore, the divergence δ may be generalized to p-forms. The p-1 form

$$\delta A^{(p)} = (-1)^{np+n+1+(n-s)/2} \; {}^*d^*A^{(p)} \tag{2.12}$$

is the result.

III. Fibre Bundle Geometry of Gauge Theories

The generalization of the SL(6,$\mathbb{C}$) gauge theory of strong interactions (Isham et al., 1973) to one with additional flavor and (hidden) color degrees of freedom is formally straightforward.

To this end a principal fibre bundle $P(M^4, G, \pi)$ (Kobayashi & Nomizu, 1963) over a pseudo-Riemannian curved space-time M^4 with signature $s = -2$, i.e.

$$\mathrm{diag}(\eta_{\alpha\beta}) = (\underbrace{1}_{(n+s)/2}, \underbrace{-1,-1,-1}_{(n-s)/2}) \tag{3.1}$$

will be introduced.

a) Structure group

the Lie group

$$\begin{aligned} G &= GL(2f, \mathbb{C}) \otimes GL(2c, \mathbb{C}) \\ &\supset {}^*U(f)_L \otimes U(f)_R \otimes {}^*U(c)_L \otimes U(c)_R \end{aligned} \tag{3.2}$$

will be considered as structure group of $P(M^4, G, \pi)$, where f and c denote the number of flavored and colored internal degrees of freedom. The unitary subgroups act on the involved fermion fields with left (L) or right (R) helicities, only. In order to have the option on a broken f-g gauge theory (Isham et al., 1974) (3.2) has to be tensored with the extended Lorentz group $GL(2,\mathbb{C})$ which would account for conventional gravity.
By choosing the affine groups

$$A(2k,\mathbb{C}) = GL(2k,\mathbb{C}) \otimes \mathbb{C}^{2k} \qquad (3.3)$$

instead, an affine extension (Lord, 1978) of the $SL(6,\mathbb{C})$ gauge theory would have been obtained which would be much more closely modeled after the Poincaré gauge theory (Hehl et al., 1976) of gravity. The former is the semi-direct product of the general linear group and the k-dimensional vector group over the field of complex numbers. It will be remarked lateron, in which sense both approaches are related.

Following Isham et al. (1973), the infinitesimal generators of $GL(2k,\mathbb{C})$ are realized in the so-called Dirac basis by

$$\{E_j = \sigma^{\alpha\beta}\lambda_i, \lambda_i, i\gamma^5\lambda_i \mid j = 1,\dots,8k^2;\ i = 1,\dots,k^2\} . \qquad (3.4)$$

With respect to the Dirac matrices γ^α satisfying

$$\gamma^\alpha\gamma^\beta + \gamma^\beta\gamma^\alpha = 2\eta^{\alpha\beta} \qquad (3.5)$$

the generators of the covering group $SL(2,\mathbb{C})$ of the Lorentz group $SO(1,3)$ will be denoted by

$$\sigma^{\alpha\beta} = \frac{i}{2}[\gamma^\alpha,\gamma^\beta] \qquad (3.6)$$

(The conventions of Bjorken & Drell (1964) are used throughout). Furthermore the generalized Gell-Mann matrices λ_i normalized to

$$\mathrm{Tr}(\lambda_i\lambda_j) = 2\delta_{ij} \qquad (3.7)$$

have been employed in the representation (3.4). They are the k^2 vector operators of $U(k)$ and fulfill the following commutation- and anti-commutation relations (Gell-Mann & Ne'eman , p. 180)

$$[\lambda_i, \lambda_j] = 2i f_{ij}{}^{l}\lambda_l \quad , \qquad (3.8)$$

$$\{\lambda_i , \lambda_j\} = 2 d_{ij}^{\ \ l} \lambda_l \tag{3.9}$$

b) The Bundle of Affine Frames

In order to introduce spinor fields, the bundle $L(M^4)$ of linear frames (KN, p.55) has to be considered. With respect to the affine structure group (3.3), $L(m^4)$ may be regarded as a subbundle of the bundle $A(M^4)$ of affine frames (KN, p. 126). To $L(M^4)$ corresponds the matrix-valued canonical 1-form

$$L = L_\mu \, dx^\mu \tag{3.10}$$

(Although not necessary, because of familiarity all local expressions refer to holonomic coordinate charts x^μ). L may also be regarded as a spinor based version of Cartan's $\binom{1}{1}$ unit tensor $d\mathcal{P} = e_\mu dx^\mu$ (MTW, p. 376), as

$$\begin{aligned} L_\mu = {} & \tfrac{1}{2}\left\{ L^{(f)\ \ j}_{\mu\alpha} \gamma^\alpha + i \, {}^*L^{(f)\ \ j}_{\mu\alpha} \gamma^\alpha \gamma^5 \right\} \lambda^{(f)}_j \\ & + \tfrac{1}{2}\left\{ L^{(c)\ \ j}_{\mu\alpha} \gamma^\alpha + i \, {}^*L^{(c)\ \ j}_{\mu\alpha} \gamma^\alpha \gamma^5 \right\} \lambda^{(c)}_j \end{aligned} \tag{3.11}$$

is a convenient representation. Dual forms may be constructed from L according to the rule

$$^*(\underbrace{L \wedge \ldots \wedge L}_{p}) = -i L^5 \underbrace{L \wedge \ldots \wedge L}_{4-p} \quad , \tag{3.12}$$

where

$$\begin{aligned} L^5 &= \frac{i}{4!} |f|^{-1/2} \varepsilon^{\alpha_1 \ldots \alpha_4} L_{\alpha_1} \wedge \ldots \wedge L_{\alpha_4} \ , \\ L^5 L^5 &= 1 \end{aligned} \tag{3.13}$$

is a matrix-valued 0-form.

Since the curved space-time manifold M^4 is assumed to be paracompact, the G bundle $L(M^4)$ (set of all $8(f^2+c^2)$ -bein fields in space-time) admits the gauge-invariant fibre metric (KN, p. 116)

$$f_{\mu\nu} = f^{(f)}_{\mu\nu} \oplus f^{(c)}_{\mu\nu} = \frac{1}{4} \mathrm{Tr}(L_\mu L_\nu) \quad . \tag{3.14}$$

As the bundle $L(M^4)$ corresponds to the canonical 1-form L, an affine connection $\tilde{B}$ can be introduced which is equivalent to a linear connection (KN, p. 129). The $\mathfrak{g}$-valued connection 1-form corresponding to a linear connection is denoted by

$$B = B_\mu dx^\mu \quad . \tag{3.15}$$

The gauge potentials

$$B_\mu = \frac{1}{4} B^{(f)\,j}_{\mu\alpha\beta} \sigma^{\alpha\beta} \lambda^{(f)}_j \oplus \frac{1}{4} B^{(c)\,j}_{\mu\alpha\beta} \sigma^{\alpha\beta} \lambda^{(c)}_j \oplus A_\mu \tag{3.16}$$

may, as usually, be expanded in terms of the $8(f^2+c^2)$ infinitesimal Hermitian generators of the non-compact group G. Because of later importance the contributions from the unitary subgroups will be listed separately by

$$\begin{aligned} A_\mu = & \frac{1}{2} \{ A^{(f)j}_\mu + {}^*A^{(f)j}_\mu \gamma^5 \} \lambda^{(f)}_j \\ & \oplus \frac{1}{2} \{ A^{(c)j}_\mu + {}^*A^{(c)j}_\mu \gamma^5 \} \lambda^{(c)}_j \quad . \end{aligned} \tag{3.17}$$

c) Gauge transformations

Consider now a diffeomorphism $\Omega : P \rightarrow P$ of the principal fibre bundle $P(M^4, G, \pi)$ such that (1) $\Omega(gp) = g\,\Omega(p)$, $g \in G$, $p \in P$, (2) Ω preserves each fibre $F_m = \pi^{-1}(m)$, i.e. acts trivially on the base space $M^4 \ni m$. An element of the corresponding infinite dimensional group $\mathcal{G}$ of gauge transformations may be realized (Isham et al., 1973) by

$$\begin{aligned} \Omega = \exp i \{ & \frac{1}{2} [\frac{1}{2} \omega^{(f)j}_{\alpha\beta} \sigma^{\alpha\beta} + \omega^{(f)j} + {}^*\omega^{(f)j} \gamma^5] \lambda^{(f)}_j \\ \oplus & \frac{1}{2} [\frac{1}{2} \omega^{(c)j}_{\alpha\beta} \sigma^{\alpha\beta} + \omega^{(c)j} + {}^*\omega^{(c)j} \gamma^5] \lambda^{(c)}_j \} , \end{aligned} \tag{3.18}$$

where the ω's are real functions on M^4. The action on the frame bundle $L(M^4)$ is locally that of conjugation:

$$\Omega^{-1}(L) = \Omega^{-1} L\, \Omega \quad , \quad \Omega \in \mathcal{G} \quad . \tag{3.19}$$

Objects which transform like (3.19) are called gauge-covariant forms. In the local cross-section

$$\psi = \{ \psi^{(q_f, q_c)} \mid q_f = 1 \ldots f \; ; \; q_c = 1 \ldots c \} \tag{3.20}$$

of the bundle of f x c Dirac spinors associated with $L(M^4)$, $\mathcal{G}$ acts as

$$\Omega^{-1}(\psi) = \Omega^{-1} \psi \quad . \tag{3.21}$$

For the Dirac adjoint defined by

$$\bar{\psi} = \psi^{+} \gamma^{0} \otimes \lambda_{0} \quad , \tag{3.22}$$

the gauge transformation reads:

$$\Omega^{-1}(\bar{\psi}) = \bar{\psi}\Omega \quad . \tag{3.23}$$

Only the connection form (3.15) transforms inhomogeneously (KN, p. 66) according to

$$\Omega^{-1}(B) = \Omega^{-1} B \Omega + i \Omega^{-1} d\Omega \quad . \tag{3.24}$$

This allows to define the gauge-covariant differentiation of spinors by

$$D\psi = d\psi + i B\psi \tag{3.25}$$

and that of $\mathfrak{g}$ -valued, gauge-covariant p-forms by

$$DA^{(p)} = dA^{(p)} + i\,[B, A^{(p)}] \tag{3.26}$$

i.e., the exterior covariant derivative (KN, p. 77).

In order to link the internal gauge symmetry to the curved space-time M^4 the "metric condition"

$$\nabla_{\mu} L^{\alpha} = D_{\mu} L^{\alpha} + \Lambda^{\alpha}{}_{\beta\mu} L^{\beta} = 0 \tag{3.27}$$

may be imposed on the covariant derivative ∇ with respect to an reduced subbundle of $P(M^4, G, \pi)$ (KN, p. 118). Then, geometric objects can be defined which are invariant not only with respect to the local gauge group $\mathcal{G}$ but also with respect to the diffeomorphism group $\mathcal{D}$ of general coordinate transformations (Isham et al., 1973)

d) Structure equations and Bianchi identities

In terms of the gauge-covariant derivative D a torsion 2-form

$$T = \frac{1}{2} L_{\alpha} T^{\alpha}{}_{\mu\nu}\, dx^{\mu} \wedge dx^{\nu} \tag{3.28}$$

can be defined by

$$T = DL = dL + i\,[B, L] \tag{3.29}$$

i.e. via the 1st structure equation of E. Cartan.

The curvature 2-form ("curvature operator", MTW, p. 365)

$$C = \frac{1}{2} C_{\mu\nu} dx^\mu \wedge dx^\nu = \frac{i}{4} L_\alpha \wedge L_\beta R^{\alpha\beta}{}_{\mu\nu} dx^\mu \wedge dx^\nu \qquad (3.30)$$

is then given by the <u>2nd structure equation of E. Cartan:</u>

$$C = dB + i B \wedge B \quad . \qquad (3.31)$$

The local version of (3.31) yields the familiar relation for the <u>gauge field strength</u>:

$$C_{\mu\nu} = \partial_\mu B_\nu - \partial_\nu B_\mu + i [B_\mu, B_\nu] \quad . \qquad (3.32)$$

Note that the curvature form $\tilde{C}$ corresponding to an <u>affine</u> connection $\tilde{B}$ (KN, p. 128) is given by <u>one</u> structure equation $\tilde{C} = d\tilde{B} + i\tilde{B} \wedge \tilde{B}$ which, because of $\gamma^* \tilde{B} = L + B$, comprises (3.29) and (3.31). Differentiating the torsion form and inserting the structure equations yields

$$\begin{aligned} dT &= ddL + i\, dB \wedge L - i\, B \wedge dL + i\, dL \wedge B - i\, L \wedge dB \\ &= i[C,L] + [B \wedge B, L] - i[B,T] - [B,[B,L]] \quad . \end{aligned} \qquad (3.33)$$

This result may be expressed in a gauge-covariant manner by the <u>1st Bianchi identity:</u>

$$DT = i[C,L] \quad . \qquad (3.34)$$

A similar derivation for the curvature form

$$dC = ddB + i\, C \wedge B - i\, B \wedge C + B \wedge B \wedge B - B \wedge B \wedge B \qquad (3.35)$$

yields the <u>2nd Bianchi identity</u>:

$$DC = 0 \quad . \qquad (3.36)$$

IV. The Geometrodynamical Lagrangian

The <u>geometrodynamical Lagrangian 4-form</u> which couples the fundamental spinor fields of matter to the hypothetical <u>strong</u> gravity consist out of three pieces (Salam, 1973):

$$L_{GMD} = \hbar c\, \ell^{*-2} \{ L_W(f) - 2\Lambda {}^*1 + \ell^{*2} L_0(\psi) \} \qquad (4.1)$$

having ℓ^* given by (1.1) as sole coupling constant. The first two parts govern the vacuum dynamics of the strong f-metric (3.14), whereas the third Lagrangian accounts for the dynamics of the constituent spinor fields (3.20) of particles.

It is known in general relativity that the Einstein-Hilbert action with cosmological term up to a complete divergence is the only Lagrangian which gives rise to . second order Euler-Lagrange equations for the metric (Rund & Lovelock), 1972, Theorem 5.3).

a) Weyl's Lagrangian

Therefore, its G-gauge-invariant generalization should be provided by the Lagrangian 4-form

$$\mathcal{L}_W = i\,\mathrm{Tr}\left\{ C \wedge {}^*(L\wedge L)\right\} \tag{4.2}$$

proposed already by Weyl (1929). To begin with, note that (4.2) is equivalent to

$$\mathcal{L}_W = \mathrm{Tr}\,(C\wedge L^5 L\wedge L) \tag{4.3}$$

because of (3.12). Using the local expansion (3.30) of the curvature operator C and the obvious relation

$$dx^\mu \wedge dx^\nu \wedge dx^\varkappa \wedge dx^\lambda = \varepsilon^{\mu\nu\varkappa\lambda}\, d^4x \tag{4.4}$$

$$\mathcal{L}_W = \frac{1}{4}\,\mathrm{Tr}\left(L_\alpha \wedge L_\beta\, R^{\alpha\beta}{}_{\mu\nu} \wedge L_\varkappa \wedge L_\lambda L^5\right)\varepsilon^{\mu\nu\varkappa\lambda}\, d^4x \tag{4.5}$$

can be obtained. From the definition (3.13) of L^5 follows

$$L_\alpha \wedge L_\beta \wedge L_\varkappa \wedge L_\lambda = i\,\varepsilon_{\alpha\beta\varkappa\lambda}\sqrt{|f|}\, L^5 \tag{4.6}$$

The insertion yields

$$\mathcal{L}_W = -\frac{1}{4}\,\mathrm{Tr}\,({}^*R^* L^5 L^5)\, {}^*1 \quad . \tag{4.7}$$

Since the contracted double dual (MTW, p. 325) of the Riemann tensor is via

$${}^*R^* = -R \tag{4.8}$$

related to the scalar curvature, the equivalence of Weyl's Lagrangian (4.2) to the Einstein-Hilbert Lagrangian

$$\mathcal{L}_W = R\sqrt{|f|}\, d^4x \tag{4.9}$$

has been proven.

Although conventional renormalization procedures in quantum field theory would rather favor (Fairchild, 1977) the Yang-Mills-type (Yang & Mills, 1954) Lagrangian

$$\mathcal{L}_{Y\text{-}M} = Tr\,(C \wedge {}^*C) \tag{4.10}$$

this alternative approach will not be pursued in the following.

Instead, the torsion content of Weyl's Lagrangian will be exhibited for later purposes.

To this end, the 2nd structure equation (3.31) will be substituted in (4.2):

$$\mathcal{L}_W = i\,Tr\{dB \wedge {}^*(L \wedge L)\} - Tr\{B \wedge B \wedge {}^*(L \wedge L)\}\ . \tag{4.11}$$

A subtraction of the total divergence

$$\begin{aligned} \mathcal{L}_\infty &= i\,Tr\{d\,(B \wedge {}^*(L \wedge L))\} \\ &= i\,Tr\{dB \wedge {}^*(L \wedge L)\} - \delta(L \wedge L) \wedge {}^*B\} \end{aligned} \tag{4.12}$$

and a rearrangement of terms leads to

$$\begin{aligned} \mathcal{L}_W - \mathcal{L}_\infty = &\ \frac{i}{8}\,Tr\{\delta(L \wedge L) \wedge L \wedge {}^*[B,L]\} \\ &+ \frac{1}{2}\,Tr\{[B,L] \wedge {}^*[B,L]\}\ . \end{aligned} \tag{4.13}$$

The torsion 2-form T may now be inserted via the 1st structure equation (3.29):

$$\begin{aligned} \mathcal{L}_W - \mathcal{L}_\infty = \frac{1}{2}\,Tr\Big\{ &\frac{1}{4}\,\delta(L \wedge L) \wedge L \wedge {}^*(T - dL) \\ &- (T - dL) \wedge {}^*(T - dL)\Big\}\ . \end{aligned} \tag{4.14}$$

For vanishing torsion this expression reduces to

$$\mathcal{L}_W - \mathcal{L}_\infty = -\frac{1}{2}\,Tr\{dL \wedge {}^*dL + \frac{1}{8}\,d(L \wedge L) \wedge d^*(L \wedge L)\}\ . \tag{4.15}$$

b) Dirac's Lagrangian

The appropriate gauge-invariant Lagrangian 4-form for the Dirac fields ψ is

$$L_D = -\tfrac{1}{2} i \, \bar{\psi} L \wedge {}^*D\psi + \tfrac{i}{2} \overline{D\psi} \wedge {}^*L\psi - \left(\tfrac{\mu c}{\hbar}\right) \bar{\psi}\psi \, {}^*1 \tag{4.16}$$

(see, e.g., Isham, 1978).

As the adjoint connection is commonly required to satisfy

$$\bar{B} = \gamma^0 \otimes \lambda_0 B \gamma^0 \otimes \lambda_0 = B \tag{4.17}$$

(4.16) may be written as

$$\begin{aligned} L_D = & -\tfrac{i}{2} \bar{\psi} L \wedge {}^*d\psi + \tfrac{i}{2} d\bar{\psi} \wedge {}^*L\psi \\ & - \left(\tfrac{\mu c}{\hbar}\right) \bar{\psi}\psi \, {}^*1 + \tfrac{1}{2} \bar{\psi} \, [B, {}^*L] \, \psi \; . \end{aligned} \tag{4.18}$$

Expressing L with the aid of (3.12) and then inserting the structure equation (3.29) reveals that in general the Dirac Lagrangian may also depend on Cartan's torsion:

$$\begin{aligned} L_D = & -\tfrac{i}{2} L \wedge {}^*d\psi + \tfrac{i}{2} d\bar{\psi} \wedge {}^*L\psi - \left(\tfrac{\mu c}{\hbar}\right) \bar{\psi}\psi \, {}^*1 \\ & - \tfrac{1}{2} \bar{\psi} \, (T - dL) \wedge L \wedge L^S L \psi \; . \end{aligned} \tag{4.19}$$

Our basic model defined by (4.1) will be referred to as (classical color geometrodynamics (CGMD), since it is known that a complete Rainich geometrization of the fermion fields is in principle possible for G = Gl(2,ℂ) (Kuchař, 1965). The main reason being, that the spin-unitary-spin current can be related to the torsion of the underlying space-time, as will be shown in section V. b).

V. Field Equations

Since a survey of the general theory of invariant variational principles as applied to the theory of relativity has already been presented by Rund & Lovelock (1972), the following analysis may focus on the particular cases at hand.

a) Gravitational Field Equations

In the derivation of the Einstein-type field equations variational principles similar to those which Weyl introduced 1929 with respect to local tetrad fields are employed.

A more elegant but equivalent procedure is to vary for the 1-form

L corresponding to the linear frame bundle, and at the same time, to insert (denoted by in (...)) the local expression of its independent components. By applying this procedure to (4.3)

$$\frac{\delta \mathcal{L}_W}{\delta L} \text{ in } (L_\mu \wedge dx^\nu) = -\frac{1}{4} \mathrm{Tr}\, ({}^*R^*{}_{\alpha\beta}{}^{\alpha\nu} L^\beta L^5)\, {}^*1 \tag{5.1}$$

is obtained. The contracted double dual Riemann tensor is via

$${}^*R^*{}_{\alpha\mu}{}^{\alpha\nu} = G_\mu{}^\nu = R_\mu{}^\nu - \frac{1}{2}\delta_\mu{}^\nu R \tag{5.2}$$

related (MTW, p. 325) to the (nonsymmetric) Einstein tensor $G_\mu{}^\nu$. Varying the Lagrangian (4.16) according to the same prescription yields the canonical stress-energy tensor for the Dirac fields:

$$\frac{\delta \mathcal{L}_D}{\delta L} \text{ in } (L_\mu \wedge dx^\nu) = \frac{i}{2}\{\overline{D^\nu\psi} \wedge L_\mu \psi - \bar\psi L_\mu \wedge D^\nu \psi\}\, {}^*1 \;. \tag{5.3}$$

Finally, from the geometrodynamical Lagrangian (4.1) the Einstein-type field equations with "cosmological" term follow

$$G_{\mu\nu} + \Lambda f_{\mu\nu} = \frac{i}{2} e^{*2}\{\overline{D_\nu\psi} \wedge L_\mu \psi - \bar\psi L_\mu \wedge D_\nu \psi\} \;. \tag{5.4}$$

b) Cartan's torsional equation

From the Einstein Cartan theory of general relativity it is known (Hehl & Datta, 1971) that the proper torsion is dual to an axial spin current. In conformity with this result the Ansatz

$$T - dL = L \wedge L \wedge L \wedge {}^*S = -i\, S \wedge L^5 L \tag{5.5}$$

may be inserted in (4.14) and (4.19). Variation of $\mathcal{L}_{GMD}$ for the scalar-valued (may be placed before the trace) 1-form S yields

$$\begin{aligned} \frac{\delta \mathcal{L}_{GMD}}{\delta S} = \frac{1}{2}\hbar e^{*-2}\Big[&\mathrm{Tr}\{-\tfrac{i}{4} S (L \wedge L) \wedge L \wedge L^5 L\} \\ &+ \mathrm{Tr}\{(L \wedge L \wedge L) \wedge {}^*(L \wedge L \wedge L)\} \wedge {}^*S\Big] \\ &+ \tfrac{i}{2}\hbar \bar\psi L^5 L \wedge L \wedge L^5 L \psi = 0 \;. \end{aligned} \tag{5.6}$$

After evaluating the trace and remembering (3.13) and (3.12) the dual of (5.6) reads:

$$S = \frac{3!}{16} e^{*2} \bar{\psi} L^5 L \psi - \frac{3!}{4^3} \mathrm{Tr} \{ {}^*(\delta(L \wedge L) \wedge {}^*(L \wedge L)) \} \quad . \tag{5.7}$$

Therefore, Cartan's torsion form

$$\begin{aligned} T = & - \frac{3!}{16} i e^{*2} \bar{\psi} L^5 L \psi \wedge L^5 L + dL \\ & + \frac{3!}{4^3} i \mathrm{Tr} \{ L \wedge L \wedge d^*(L \wedge L) \} \wedge L^5 L \end{aligned} \tag{5.8}$$

depends not only on the axial vector of the canonical spin-unitary spin current but, in a curved space-time, also on the object dL of anholonomity (Hehl et al., 1976).

c) Heisenberg-Pauli-Weyl nonlinear spinor equation

The expression (5.8) for the torsion may be resubstituted into Dirac's Lagrangian (4.19) with the result

$$\begin{aligned} \mathcal{L}_D = & - \frac{i}{2} \bar{\psi} L \wedge {}^* d\psi + \frac{i}{2} d\bar{\psi} \wedge {}^* L \psi - \left(\frac{\mu c}{\hbar} \right) \bar{\psi} \psi {}^* 1 \\ & - i \bar{\psi} L \wedge L \wedge L \psi \wedge \Big[\frac{3}{4^3} \mathrm{Tr} \{ {}^*(\delta(L \wedge L) \wedge {}^*(L \wedge L) \} \\ & \qquad - \frac{3}{16} e^{*2} \bar{\psi} L^5 L \psi \Big] \quad . \end{aligned} \tag{5.9}$$

Then, the variation for $\delta \mathcal{L}_D / \delta \bar{\psi}$ yields:

$$\begin{aligned} & -i L \wedge {}^* d\psi - \frac{i}{2} \delta L \psi - \frac{3}{4^3} \mathrm{Tr} \{ \delta(L \wedge L) \wedge {}^*(L \wedge L) \} \wedge L^5 L \psi \\ & + \frac{3}{8} e^{*2} \bar{\psi} L^5 L \psi \wedge {}^* L^5 L \psi - \left(\frac{\mu c}{\hbar} \right) \psi {}^* 1 = 0 \quad . \end{aligned} \tag{5.10}$$

Its local expression

$$\begin{aligned} & \Big\{ i L^\mu \partial_\mu + \frac{i}{2} \partial_\mu L^\mu + \frac{3}{4^3} \mathrm{Tr} \left([L_\mu, L_\nu] \partial_\varsigma [L^\varsigma, L^\nu] \right) L^5 L^\mu \\ & - \frac{3}{8} e^{*2} \bar{\psi} L^5 L_\mu \psi L^5 L^\mu - \left(\frac{\mu c}{\hbar} \right) \Big\} \psi = 0 \end{aligned} \tag{5.11}$$

will be referred to as the G-gauge-invariant Heisenberg-Pauli-Weyl spinor equation. In a SL(2,ℂ) gauge theory of gravitation, Weyl derived already 1950 a similar equation with a self-interaction of the axial-vector type. The first three terms of (5.11) generalize the Dirac operator to the curved space-time of strong gravity, whereas the fourth term due to torsion generates a more general self-coupling of the spinor fields compared to the equation proposed 1958 by Heisenberg and

Pauli (Heisenberg, 1966). The latter was originally devised to be invariant only with respect to the group U(2) of isotopic spin, which incorporates only "iso-torsion".

VI. Concluding remarks

It may be suspected (Mielke, 1977) that in CGMD the constituent fields $\psi^{(q_f, q_c)}$ are bound together via the nonlinear self-interaction of the fundamental spinor equation (5.12). Moreover, via the Einstein-type field equations (5.4) the semi-classical spinor fields ψ may produce a strongly curved hadronic background with respect to the f-(gluon) metric. Therefore, it may be speculated that in CGMD these quark-type fields ψ are (partially) confined (Salam and Strathdee, 1978) in geon-type objects (Wheeler, 1962). This expectation is backed up by the self-attracting feature of the tensor gluons of CGMD, a property which it does not share with Yang-Mills gauge theories (Coleman & Smarr), 1977). Since the color may become transcendent (Wheeler, 1971) far from the center of the black geon (-soliton, Salam & Strathdee, 1976) it could be rightly regarded as a classical prototype on an extended particle.

As a first preliminary step in the search for such soliton-type solutions, radially localized solutions of the Heisenberg-Klein-Gordon equation

$$\left(\Box - \frac{3\varepsilon \mu c}{8\hbar} \ell^2 |\varphi|^2 + \frac{9}{256} \ell^4 |\varphi|^4 + \left(\frac{\mu c}{\hbar}\right)^2 \right) \varphi = 0 \qquad (6.1)$$

for a complex scalar field have been computed numerically in a flat and exterior Schwarzschild background (Deppert & Mielke, 1979). The HKG equation (6.1), which is (but not identically) related to the "squared" form of (5.12), provides a simplified model admitting stable (Anderson, 1971) solutions.

Acknowledgement

I would like to express my sincere gratitude to Professor F.H. Hehl and Professor J.A. Wheeler for their continuous support and encouragement. Furthermore, I thank Professor H.D. Doebner for the invitation to give a talk on CGMD at the Clausthal meeting.

References

Anderson, D.L.T. (1971). J. Math. Phys. 12, 945.

Bjorken, J.D., and Drell, S.D. (1964) Relativistic Quantum Mechanics (Mc-Graw-Hill, San Francisco).

Cartan, E. (1922). Acad. Sci. Paris, Comptes Rend. 174, 593.

Cartan, E. (1923). Ann. École Norm. Sup. 40, 325.

Coleman, S. and Smarr, L. (1977). Commun. math Phys. 56, 1.

Deppert, W. and Mielke, E.W. (1979) "Localized Solutions of the non-linear Heisenberg-Klein-Gordon equation. In flat and exterior Schwarzschild space-time" Phys. Rev. D 20, 1303.

Fairchild, Jr., E.E. (1977) Phys. Rev. D 16 , 2438.

Flanders, H. (1963). Differential Forms with Applications to Physical Sciences (Academic Press, New York).

Gell-Mann, M. and Ne'eman, Y. (1964). The Eightfold Way (Benjamin New York).

Gell-Mann, Ramond, P., and Slansky, R. (1978). Rev. Mod. Phys. 50, 721.

Hehl, F.W., and Datta, B.K. (1971). J. Math. Phys. 12, 1334.

Hehl, F.W., von der Heyde, P., Kerlick, G.D., and Nester, J.M. (1976). Rev. Mod. Phys. 48, 393-

Heisenberg, W. (1966). Introduction to the Unified Field Theory of Elementary Particles (Wiley, London).

Heisenberg, W. (1974). Naturwissenschaften 61, 1.

Isham, C.J., Salam, A., and Strathdee, J.
(1971). Phys. Rev. D 3, 867;
(1973). Phys. Rev. D 8, 2600;
(1974). Phys. Rev. D 9. 1702.

Isham, C.J. (1978). Proc. R. Soc. London. A. 364 , 591.

Kobayashi, S., and Nomizu, K. (1963). Foundations of Differential Geometry, Vol. I. (Interscience, New York) (quoted as KN).

Muchar, K. (1965). Acta. Phys. Polon. 28, 695.

Lord, E.A., (1978). Phys. Lett. 65A. 1.

Mielke, E.W. (1977). Phys. Rev. Lett. 39, 530.

Misner, C.W., Thorne, K.S., and Wheeler, J.A. (1973). Gravitation (Freeman and Co., San Francisco), (quoted as MTW).

Rund, H., and Lovelock, D. (1972). Über. Deutsch. Math.-Verein, 74, 1.

Salam, A. (1973). In: Fundamental Interactions in Physics, edited by B. Kursunoglu et al. (Plenum, New York), p. 55.

Salam, A., and Strathdee, J. (1976). Phys. Lett. 61B, 375.

Salam, A., and Strathdee, J. (1978). Phys. Rev. D 18, 4596.
Trautman, A. (1972-73). Bull. Acad. Pol. Sci. Ser. Sci. Math. Astron. Phys. 20, 185, 503, 895; 21, 345.
Weyl, H. (1929), Z. Physik 56, 330.
Weyl, H. (1950). Phys. Rev. 77, 699.
Wheeler, J.A. (1962). Geometrodynamics (Academic Press, New York).
Wheeler, J.A. (1971). In Cortona Symposium on "The Astrophysical Aspects of the Weak Interactions", edited by L. Radicati (Academia Nazinale Dei Lincei, Roma) p. 133.
Yang, C.N., and Mills, R.L. (1954). Phys. Rev. 96, 191.

Non-Associative Algebras and Exceptional Gauge Groups[+]

L.C. Biedenharn[++)]

Institut für Theoretische Physik der Johann Wolfgang Goethe Universität, Frankfurt/Main, Germany (BRD)

L.P. Horwitz[+++)]

Tel Aviv University, Ramat Aviv, Israel

1. Introduction

One of the central problems of theoretical physics, if not t h e central problem, is the problem of understanding the internal space - and hence the internal structure - of hadrons. In other words, the problem posed by the concept of a quark.

There is an enormous amount of empirical information now known on this problem, and there is, in fact, essentially a consensus as to the overall framework for the answer: quantum gauge field theory realized as a fiber bundle over Minkowski space (as the base space) and a non-Abelian gauge group G as the symmetry group acting on the fiber. The mathematics of fiber bundles and differential geometry is thus the appropriate discipline and the subject of this conference.

The arbitrary element in this consensus view is the choice of the symmetry group, the gauge group G. The recent impressive successes of the standard model of weak and EM interactions clearly requires G to contain the SU(2) x U(1) group of this model and it is reasonable

+) Research supported in part by the National Science Foundation

++) Alexander von Humboldt Foundation Senior U.S. Scientist Award, on leave from Duke University, Durham, North Carolina, 27706, USA.

+++) Research supported by the Binational Science Foundation, Jer. Israel to include the color group ($SU3^c$) so that a minimal requirement is that: $G \supset SU3^c \times SU2 \times U(1)$. Cosmological evidence limits the rank of G from above (via an upper limit on the number of neutrinos). Unification would require that G be a simple group (aside from discrete symmetries).

Is there any principle, or principles, by which one could hope to delimit the symmetry group G?

Without any claim to completeness, I would like to mention here a remarkable fact emphasized by Faulkner and Ferrar[(1)] in their recent review. They point out that: all notions of exceptionality in algebra and in geometry are manifestations of one underlying structure, that is,

non-classical Lie- algebras
non-associative alternative algebras
non-special Jordan algebras
non-de Sarguesian projective planes

are related , in one way or another, to the octonions (Cayley numbers).

This strongly suggests that a similar phenomenon should exist in motivating the choice of the symmetry group G. By no means is this a new idea, and the fundamental work of P. Jordan [(2)] (who initiated the inquiry) and of I. Segal [(3)] who(with Sherman [(4)]) codified it, long predates hadronic physics. The current revival of interest in this approach was initiated primarily by Gürsey [(5)], although there were important contributions by Pais, Gamba, Goldstine and others.

Let us indicate why this approach is intuitively hopeful:

(a) Gürsey has noted that if one specializes one of the seven non-scalar Cayley units to play the role of the imaginary unit in standard quantum mechanics (call this unit e_7) then one automatically achieves a rationale for $SU3^{color}$.

In particular, the five exceptional Lie groups exhibit a color-flavor structure [(5)]:

G2:	$SU3^c$	F4:	$SU3 \times SU3^c$	E7:	$SU6 \times SU3^c$
		E6:	$SU3 \times SU3 \times SU3^c$	E8:	$E6 \times SU3^c$

(b) It is believed that non-associativity may (somehow!) be connected with the problem of confinement. It is quite suggestive that the distance function for non- de Sarguesian geometries contains a part directly due to non-associativity (called $R_{\alpha\beta}$ in Ref. (6)). The interpretation of this fact is difficult, however, since $R_{\alpha\beta}$ is not invariant under the group.

(c) Most importantly there are but a finite number of possible models, which can be tested against known results. In particular, G2, F4 are eliminated (flavor group too small); E6 seems viable, but E7 recently seems to be ruled out.[7,8]

The purpose of the present discussion is to accept the validity of the general approach sketched above and to review the progress in implementing it.

2. Brief Survey of the Jordan Algebra Approach

In order to give the necessary background for the discussion of the progress in applying non-associative algebras in quantum mechanics let us summarize briefly the research program that began with P. Jordan in 1932. In effect, Jordan attempted to capture the essence of the Hermitian matrix algebra (that characterized quantum mechanics) by eliminating all reference to the underlying wave function concept, by focussing attention only on the algebraic properties of observables, and by eliminating the explicit use of the imaginary unit i. This latter via his "formally real axiom": $a^2 + b^2 = 0 \Rightarrow a=b=0$.

For Hermitian matrices the operations of: Multiplication by real scalars, $x \rightarrow \alpha x$; addition, $(x+y)$; and formation of powers, x^n were all taken over, but the only allowed product is the symmetric one: $xy + yx$ (since the lack of i forbids commutators).

The axioms for a Jordan algebra were taken to be: (1) $x \circ y = y \circ x$ (commutativity) and (2) $(x^2 \circ y) \circ x = x^2 \circ (y \circ x)$ (non-associativity). (The role of this second axiom is exactly the same as the Jacobi axiom in Lie algebras; it ensures that one has an integration process - the Jordan analog to the Baker-Campbell-Hausdorff identity.)

It is remarkable that this technique - which, by contrast to the Dirac q-number approach, swaps commutativity for non-commutativity and non-associativity for associativity - is essentially identical to standard quantum mechanics. The one exception, M_3^8, 3x3 Hermitian matrices over octonions [with $x \circ y = 1/2(xy + yx)$] is the first known example of a quantum mechanics for which there is no Hilbert space, and no wave function.

Let us contrast the Dirac and Jordan approaches.

	Standard (Dirac) Formulation:	Jordan Formulation[2,6]
State	Hilbert space vector bra-vectors: $\langle\phi\vert$ ket-vectors: $\vert\psi\rangle$, $\langle\psi,\psi\rangle=1$.	rank 1 idempotent $P_\psi \circ P_\psi = P_\psi$, $trP_\psi = 1$
Probability:	$P(\phi,\psi)=\vert\langle\phi\vert\psi\rangle\vert^2$	$P(\phi,\psi)=trP_\phi\circ P_\psi$
Phase:	Yes, $\vert\psi\rangle \cong e^{i\alpha}\vert\psi\rangle$	Concept undefined
Probability Amplitude:	$\langle\phi\vert\psi\rangle \in \mathbb{C}$	Concept undefined
Basis:	$\{\vert i\rangle\}$, $\langle i\vert j\rangle=\delta_{ij}$	$\{P_i\}$, $trP_i\circ P_j=\delta_{ij}$ $P_i\circ P_j = \begin{cases} P_k & i=j=k \\ 0 & \text{otherwise}\end{cases}$
Super-position:	$\sum_i a_i\vert\phi_i\rangle$ is a state $a_1 \in \mathbb{C}$ (unnormalized)	$tr\, P_a\circ P_b \times P_c = 0$ is condition that states a,b,c are in linear super-position (X defined in Sec.5)
"Resolution of unit":	$\langle\phi\vert\psi\rangle=\sum_i\langle\phi\vert i\rangle\langle i\vert\psi\rangle$ $\{\vert i\rangle\}$ = basis	$tr\, P_\phi\circ P_\psi = \sum_i tr\{P_\phi,P_i,P_\psi\}$ ($\{xyz\}$ defined in Sec. 4) $\{P_i\}$ = basis
Symmetry:	Invariance group of Probability (Wigner-Artin)	Invariance group of Probability (Automorphism group of Jordan algebra)
Compatible Observables:	$[A,B] = 0$	$(A,X,B) = 0\ \forall X$ (A,X,B) = associator $= (A\circ X)\circ B - A\circ(X\circ B)$
Algebraic Properties:	Non-commutative Associative	Commutative Non-associative

3. Further Remarks on the Jordan Program

The most important limitation in the Jordan program is the assumption of finite dimensionality. This limitation was recognized very early, and in his axioms for quantum mechanics, Segal [3] assumed, besides a norm, a real vector space structure on observables in which squares could be formed. He thus adopted the Jordan product, $a\circ b = 1/2\,[(a+b)^2 - a^2 - b^2]$. This was sufficient for spectral theory, but Sherman [4] pointed out that one needs stronger assumptions - the additional Jordan postulate - to rule out uninteresting cases. Thus the observables are again taken to be a Jordan algebra.

The nicest algebras with infinite dimension are, from a representation theory viewpoint, the Banach star algebras (B^* algebras), which, by the famous Gel'fand-Naimark theorem, always have a faithful representation as a C^* algebra of operators on a complex Hilbert space. So we are back to standard quantum mechanis.

A significant advance has been made by Størmer and his collaborators [9,10] when they defined a Jordan-Banach algebra (JB-algebra). This is a Jordan algebra over $\mathbb{R}$ with identity which is a Banach space with respect to a norm satisfying for all a,b:

$$\text{(JB-1)} \quad \|a \circ b\| \leqslant \|a\| \, \|b\| \quad ,$$
$$\text{(JB-2)} \quad \|a^2\| = \|a\|^2 \quad ,$$
$$\text{(JB-3)} \quad \|a^2\| \leqslant \|a^2 + b^2\| \quad ,$$

The exceptional algebra $\mathcal{M}_3^8$ can be contained in JB-algebras as a Jordan ideal.[11]

A second major limitation of the Jordan algebra approach is the unsolved problem defining a suitable tensor product on the exceptional algebra $\mathcal{M}_3^8$. Very recently Hanche-Olsen [12] has shown that JB-algebras with tensor products are, in fact, C^* algebras.

Finite dimensionality is not necessarily objectionable, and may be desirable, for modelling charge space, as Gürsey advocates. In fact, the algebraic properties of finite dimensional charge spaces may be a critical feature in resolving the quark puzzle.[13] The problem with this approach lies in combining the algebraic structure with the fiber bundle structure. The vector space structure can be readily combined, but the additional algebraic structure has so far proved incompatible. Thus all current uses of exceptional algebraic charge spaces combines only the vector space structure into the fiber bundle. Applications of E6 and E7 as exceptional gauge groups thus use only the group theoretic structure und relinquish the algebraic structure.

A direct attempt to use octonians as the scalars in a Hilbert space structure was carried out by Goldstine and Horwitz.[14] A spectral theorem for self-adjoint operators was shown. If, however, Fourier series expansions were to be obtained, it is necessary to consider the closure of linear manifolds under the multiplicative action of the Cayley algebra. It can be shown [14] that every vector generates a linear manifold of at most 128 dimensions. This construction leads to a module structure for the"Hilbert space" . Horwitz and Biedenharn [15] have shown that the propositional calculus associated with this type of space constitutes a complete, weakly modular, orthocomplemented atomic lattice, and hence a quantum mechanics, but with superselection rules. This "matrix Hilbert space" approach has been applied just recently [16] to incorporate the proposal of Gürsey, and of Günaydin, [17] to implement color SU3 by taking the octonionic unit e_7 as the imaginary unit in quantum mechanics. It is premature to assess as to whether this approach can lead to progress in resolving the finite dimensionality limitation of non-associative algebraic structures but it does appear helpful.

The situation is much clearer as to finding finite dimensional non-associative algebraic models for quantum mechanical charge spaces other than M_3^8, and we survey the progress here in the succeeding sections.

4. Developments in the Mathematics of Jordan Algebras [1,18]

Although the Jordan program began in physics, most of the interest, and developments, in Jordan algebras have been in mathematics. This progress has led to considerable change in the basic viewpoints, and to the forging of new concepts that promise to be of genuine importance in physics. We will discuss here two developments of this type:

(a) the concept of a quadratic Jordan algebra (McCrimmon [19]), and the related concept of inner ideals; and

(b) the concept of structural group and Jordan pairs.

We will illustrate each of these developments by a physical application. Let us turn to the concept of quadratic Jordan algebra. Here the idea is to model everything on the product $U_x(y) = xyx$, which is quadratic in x, rather than on the bilinear operation: $x \circ y = 1/2(xy + yx)$. The three axioms given by MacCrimmon are:

(Q1) $U_1 = \text{identity}$,

(Q2) $U_x V_{y,x} = V_{x,y} U_x$,

(Q3) $U_{U_x(y)} = U_x U_y U_x$,

where: $V_{x,y}(z) = (U_{x+z} - U_x - U_y)y$.

As an example, one verifies that for an associative algebra the product $U_x(y) = xyx$ yields a quadratic Jordan algebra A^+. As a second example, the exceptional Jordan algebra M_3^8 has the quadratic product: $U_x(y) = 2(x \circ y) \circ x - (x \circ x) \circ y$. The additional product takes the form: $1/2 V_{x,y}(z) = \{xyz\} = (x \circ y) \circ z + x \circ (y \circ z) - (x \circ z) \circ y$.

It is hardly obvious that such a complicated structure is really a step forward! We can indicate that it is by noting these points:

(1) Nothing is lost - quadratic Jordan algebras are categorically equivalent to the linear Jordan algebras whenever the latter is defined (i.e., characteristic not 2).

(2) The quadratic algebra allows composition with the "generalized determinant", the norm form N(y). Thus: $N(U_x(y)) = (N(x))^2 N(y)$. There is nothing analogous to this in the linear case.

(3) There is a structure theory for the quadratic algebras which is closely analogous to that for the associative algebras.

Let us explain the significance of this last point. For a physicist the Jordan approach is unhandy largely because it banishes the concept of wave function (more precisely, bra and ket vectors) with only the density matrix remaining. In mathematical language what has happened is this: the concept of a bra (or ket) vector is the concept of a left ideal, a subset $\mathcal{N}$ of the associative algebra A such that: $n \cdot A \subset \mathcal{N}$ if $n \in \mathcal{N}$. (There is one analogous result for right ideals.) The wave function concept is enormously useful, and possibly even essential, (for example, in enforcing Hamiltonian constraints in relativistic dynamics).

In a non-associative algebra there is no such concept. What replaces it comes from the quadratic algebra: the concept of an inner ideal. An inner ideal $\mathcal{M}$ is a subset of a quadratic algebra J such that: $U_x(J) \subset \mathcal{M}$ if $m \in \mathcal{M}$.

To appreciate further the physical importance of these ideas let us note that the propositional calculus approach to quantum mechanics (Piron [19]) shows that quantum mechanics can be viewed as a projective geometry. The relation between (left/right) ideals and projective geometry is given by the theorem cited in Ref. (1), (cf. 8.9). The projective geometry of the space of n-tuples with entries in a field Φ is isomorphic to the geometry of left ideals in Φ_n (the n x n matrices over Φ) with incidence given by containment. Thus we see how nicely the standard (associative algebra) approach to quantum mechanics is expressible in various equivalent formulations.

If we now remark that inner ideals play an entirely analogous role for non-associative algebras [1,17] we see that the concept of quadratic Jordan algebra, and of inner ideal, are indeed of basic interest in theoretical physics.

We illustrate the application of these concepts in section 4.

(b) Let us turn now to the second conceptual development: the concept of a structural group (Koecher [20]) and Jordan pairs

(Loos [21]). The automorphisms of a given physical structure are a well-known approach to the intrinsic properties of the structure. For an algebra, one studies the automorphisms which preserve the algebraic laws. Accordingly, such transformations always map the unit element into itself.

How does one change the unit element? If u has an inverse, let us replace the product xy in an associative algebra by: $xy \leftrightarrow x\,u^{-1}y$. The new unit element and its inverse are easily computed: $1^{[u]} = u$, $x^{-1[u]} = u\,x^{-1}\,u$.

For associative algebras this new algebra $A^{[u]}$ is, in fact, isomorphic to A but remarkably for non-associative algebras this shift of the unit can produce a different algebra. Such a new algebra is called an isotope $\mathcal{J}^{(u)}$ of the original algebra.

The desire to study not only the Jordan algebra $\mathcal{J}$ but all its isotopes as a single entity leads to the two concepts of structural group and of Jordan pair. The structural group, Str($\mathcal{J}$), is the group of isomorphic mappings of a Jordan algebra $\mathcal{J}$ and its isotopes onto itself: $\mathcal{J}^{(u)} \xrightarrow[T]{} \mathcal{J}^{(T_u)}$. The automorphism group Aut($\mathcal{J}$) is the subject of such mappings fixing the unit element.

To indicate the usefulness of this concept let us remark that $\mathcal{M}_3^8$, the exceptional 27 dimensional Jordan algebra, has as automorphism group the exceptional Lie group F4, which has a faithful representation of lowest dimension 26, the set of the traceless elements of $\mathcal{M}_3^8$. If we allow the unit element to change under the mapping, we obtain the group E6 having a 27 dimensional irrep. (E 6 is the "reduced" structural group, "reduced" by removing the operation of multiplicative scale changes.)

We will discuss in section 5 how these concepts are used in a physical context.

5. Example: Inner Ideals in Formulating the Geometry of $\mathcal{M}_3^8$

The Jordan algebra of the Hermitian 3 x 3 matrices over octonions, $\mathcal{M}_3^8$, is the algebra of matrices of the form:

$$x = \begin{pmatrix} \alpha & a & b \\ \bar{a} & \beta & c \\ \bar{b} & \bar{c} & \gamma \end{pmatrix}, \quad \begin{matrix} \alpha,\beta,\gamma \in \mathbb{R} & \text{with } \bar{a} \text{ being the octonion} \\ a,b,c \in \mathbb{O} & \text{conjugation} \end{matrix}$$

the product $x \circ y \equiv 1/2(xy + yx)$, the product xy being the matrix product. [The octonions have the form

$a = \sum_{i=0}^{7} \alpha_i e_i$, with the unit e_o and the rules:

$e_i e_j = \varepsilon_{ijk} e_k$, $e_i^2 = - e_o$. The conjugate is given by

$\bar{a} = \alpha_o e_o - \sum_1^7 \alpha_i e_i$. The most convenient way to display the product ε_{ijk} is with a diagram:

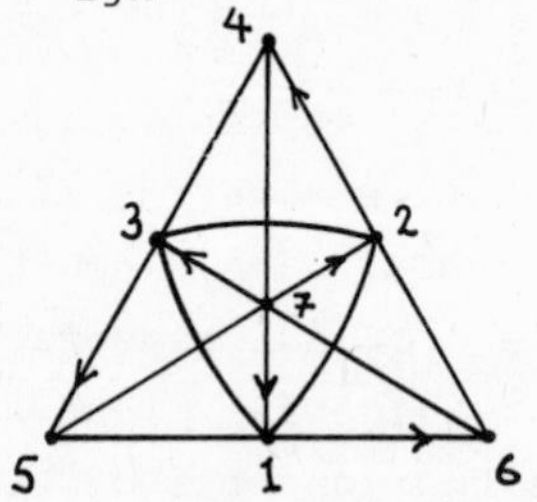

where, for example, $e_4 e_7 = e_1$ since these units lie along a straight line ("circle") for 123) in the positive direction].

In a short note Jordan showed (in 1949) that this algebra described a projective plane, a result found independently by Freudenthal. It follows from the general results of Piron that one has a quantum mechanics for this structure. The discussion of the M_3^8 quantum mechanical system has been developed quite thoroughly and elegantly in papers by Gürsey (5,6), Günaydin (23), and by Günaydin, Piron and Ruegg (23). Our purpose here is to show how the more general concept of inner ideals can be useful in this special case.

The points of this projective geometry are the quantum mechanical pure states. Were there to be a Hilbert space a pure state would have the form: $|\psi\rangle = \begin{pmatrix} a \\ b \\ c \end{pmatrix}$, where $a,b,c \in \mathbb{O}$. This will not work, however, since the density matrix: $\rho_\psi = |\psi\rangle\langle\psi|$, requires associativity to be idempotent: $\rho_\psi \rho_\psi = (|\psi\rangle\langle\psi|)(|\psi\rangle\langle\psi|) \neq |\psi\rangle(\langle\psi|\psi\rangle)\langle\psi|$ in general.

But this technique can be made to work if one notes that in a projective geometry realized in Hilbert space we have the equivalence relation:

$|\psi\rangle \cong d|\psi\rangle$, where $d \neq 0$ belongs to the scalars ($\mathbb{R}$, $\mathbb{C}$, or $\mathbb{Q}$). Applying this idea to $|\psi\rangle$, assuming all three octonions a,b,c not to be zero, we multiply by $(a\bar{a})^{-1}\bar{a}$ to find:

$$|\psi\rangle \cong \begin{pmatrix} 1 \\ a' \\ b' \end{pmatrix} .$$

Since any two octonions associate (this includes associating with

Incidence is defined for inner ideals by containment. The inner ideals for $\mathcal{M}_3^8$ are all of the form:

Points: $x \in \mathcal{J}$ with $x \times x = 0$, tr $x = 1$

Lines: $x \times \mathcal{J}$ with x a point.

Two points x,y lie on a line z if $z = \alpha \cdot x \times y$, and dually. A point x lies on the line y if tr(x,y) = 0. These results are standard.

The importance of the inner ideal concept lies in its generality and in the close analogy between inner ideals (for nonassociative algebras) and left (right) ideals (for associative algebras).

6. The Jordan Pair Concept in Formulating Quantum Mechanics for E6

This example is rather more interesting than the example of section 4 since the results are new, and have only recently been obtained.[28]

It will be recalled that the list of all possible formally real Jordan algebras, and hence possible quantum mechanical spaces, as found by JNW was complete. To obtain a new quantum mechanical space requires relaxing one of the hypotheses of Jordan-Von Neumann-Wigner[2b], and we shall relax the requirement of formal reality, by considering the "complex octonionic" plane.

From the point of view of abstract group theory, the existence of a "complex octonionic plane" is strongly indicated by results discussed in the review by Freudenthal [30]. To see this let us interpret $\mathcal{M}_3^8$ group-theoretically, and then use this interpretation as our model. $\mathcal{M}_3^8$ has the dimension $16 = 2 \cdot 8$, which is interpreted (cf. sec. 4) as the two octonions determining a generic point. Group theoretically $\mathcal{M}_3^8$ may be viewed as the symmetric homogeneous space F4/spin(9) which has dimension 52 - 36 = 16. For the remaining exceptional groups there are the following homogneous symmetric spaces: E6/Spin(10)$\times \mathbb{R}/\mathbb{Z}$ (dimension 32); E7/Spin(12)xSU(2); (dimension 64); and E8/Sp(16); (dimension 128). These spaces have been claimed as models of projective planes over complex octonions, quaternionic octonions, and bi-octonions, respectively. There is no question as to the existence of such spaces, realized as associative structures over $\mathbb{C}$. The problem, however, is to realize these planes as algebraic structures, involving the Cayley numbers, and equipped with an explicit quantum mechanical interpretation. In a sense this group-theoretic hint is even misleading since, as we will show, the proper procedure to obtain the "complex octonions" is not to consider an E6 homogeneous space but rather a homogeneous space the conjugates as well) for this equivalent $|\psi\rangle$ we get: $\rho_\psi^2 = \rho_\psi$, $\mathrm{tr}\,\rho_\psi = 1$ (normalizing $\langle\psi|\psi\rangle = 1$). (For those $|\psi\rangle$ for which one or two of the three octonions is zero, the same associativity already is true.)

We may define a quantum mechanical probability for any two states via:

$$P(\phi,\psi) = \mathrm{tr}\,\rho_\psi \circ \rho_\phi = \mathrm{tr}(\rho_\phi, \rho_\psi).$$

These results were explicitly given by Jordan.

An idempotent having trace 2 is associated with a line. Since the unit element is idempotent with trace 3 the geometry is that of a plane.

The systematic development of the Jordan algebra M_3^8 is greatly helped by the Freudenthal product:

$$x \times y = x\circ y - 1/2x\ \mathrm{tr}\ y - 1/2y\ \mathrm{tr}x - 1/2\left[\mathrm{tr}\ x\circ y - (\mathrm{tr}x)(\mathrm{tr}y)\right] .$$

This product was introduced in order to define the cubic invariant (analog to the determinant) given by: $N(x) = \mathrm{tr}(x\circ(x \times x))$. Our reason for introducing the Freudenthal product is that it plays a role in the quadratic Jordan algebra. [Let us remark that the Freudenthal product does not yield a Jordan algebra (it is not power associative), nor does the related Wedge product [25], (replace 1/2 by 1/3 above) the algebra of traceless elements ("D" matrices)[27].]

Let us now show how the concept of an inner ideal applies to this case. Recalling that for a quadratic Jordan algebra we use the two products:

(a) $\{x\ y\ z\} = (x\circ y)\circ x) + x\circ(y\circ z) - (x\circ z)\circ y$

(b) $U_x(y) = \{x\ y\ z\}$,

we now make use of an identity (deduced from MacCrimmon's axioms) given by Faulkner [28]:

(c) $U_x(y) = \mathrm{tr}(x,y)x - 1/2(x \times x) \times y.$ (*)

An inner ideal is to be the set of elements $\mathcal{N}$ for which: $U_n(\mathcal{J}) \subset \mathcal{N}$, for every $n \in \mathcal{N}$. It follows from (c) that for n to belong to an inner ideal the term $(n \times n) \times y$ must belong to the ideal for all y in the algebra. This can only be true if $n \times n = 0$. This condition for an inner ideal - which one finds directly for the quadratic algebra - is equivalent to the idempotency condition which is less immediate in the linear Jordan approach.

for E7.

Let us first motivate the concept of a Jordan pair. Just as the Jordan algebra itself can be seen as a device for multiplying two symmetric matrices so as to get a third symmetric matrix, so can the Jordan pair structure be seen as a way to multiply rectangular matrices. Let the rxs matrix ($r > s$) be noted $M_{r,s}$. Then to get an algebra of $M_{r,s}$ and $M_{s,r}$ matrices we use:

(1) $M_{r,s}\, M_{s,r}\, M_{r,s} \longrightarrow M_{r,s}$

(2) $M_{s,r}\, M_{r,s}\, M_{s,r} \longrightarrow M_{s,r}$.

Thus we have two spaces $V = M_{r,s}$ and $\tilde{V} = M_{s,r}$ equipped with two quadratic algebraic operations:

(a) $U_v(v') \in V$,

(b) $U_{v'}(v) \in \tilde{V},\ v \in V,\ v' \in \tilde{V}.$

The axioms for a Jordan pair structure were given by Loos: [22]

(JP1) $\{x,y,U_x^{\sigma}(z)\} = U_x^{\sigma}(\{xyz\})$,

(JP2) $\{U_x^{\sigma}(y),y,z\} = \{x,U_y^{-\sigma}(x),z\}$,

(JP3) $U^{\sigma}_{U_x^{\sigma}(y)} = U_x^{\sigma}\, U_y^{-\sigma}\, U_x^{\sigma}$,

where $\sigma = +,-$ denotes the two quadratic products and x,y,z belong to the appropriate two spaces.

To proceed further, we make use of a construction of the E7 Lie algebra given by Koecher [31]. This construction realizes E7 as if it were the symmetric homogeneous space of dimension $2\cdot 27$ having the stability group E6x$\mathbb{R}$/Z (dimension 79). The Koecher construction, however, defines E6x$\mathbb{R}$/Z as the structural group of $\mathcal{M}_3^8$ with the homogeneous space having the form of a Jordan pair $(V,\tilde{V})$, carrying the irrep. lables $(27,\overline{27})$ of E6. (Koecher's construction prededed the concept of Jordan pairs, and he interpreted the space (as a Hermitian bounded symmetric domain) very differently from the way we shall do so below.)

The Lie algebra of E7 splits in this way:

Lie Alg(E7) = $V \dot{+} h \dot{+} \tilde{V}$

27 under E6	E6x$\mathbb{R}$/Z struct. algebra	$\overline{27}$ under E6

$[h,h] \subset h;\ [h,V] \subset V;\ [h,\tilde{V}] \subset \tilde{V};\ [V,V] = [\tilde{V},\tilde{V}] = 0;\ [V,\tilde{V}] = h.$

Let us now indicate the Jordan pair structure explicitly. To do so we first note that:

$$U_V(\tilde{V}) \in V, \quad U_{\tilde{V}}(V) \in \tilde{V} \quad ,$$

as can be read off Eq. (✱, sec. 5) or by noting that Freudenthal product obeys the group theoretic rule: 27 x 27 = $\overline{27}$, and similarly $\overline{27}$ x $\overline{27}$ = 27.

Thus two copies of the M_3^8 algebras, if we adjoin the imaginary unit i, serve as carrier spaces for the 27 and $\overline{27}$ irreps of E6, and function as the Jordan pair.

How can we determine the quantum mechanical structure for this system? To do so, we take over the concept of inner ideal, and idempotent, to the Jordan pair structure [22]. An idempotent for the Jordan pair $(V,\tilde{V})$ is a pair (x,y) such that:

$$U_x(y) = x \text{ and } U_y(x) = y \quad ;$$

with an analogous definition for inner ideals. Clearly the inner ideals carry over just as before, and we obtain the result that: a point in the geometry defined by the Jordan pair algebra is a pair (x,y) so that $x \times x = 0$, $y \times y = 0$, $\mathrm{tr}\, x = \mathrm{tr}\, y = 1$.
It is equally clear that the rank of the space is three so that we have a projective planar geometry over the "complex octonionic" plane (a point being determined in general by four octonions).

The quantum mechanical probability function (determining the "distance" between two points) is given by:

$$P(p,q) = 1/2 \, \mathrm{tr}(x \circ y' + y \circ x')$$

with

$$p = (x,y); \; q = (x',y').$$

This result makes sense group-theoretically since $x,x' \in 27$ and $y,y' \in \overline{27}$ in E6 and the product $(27,\overline{27}) \longrightarrow$ scalar is invariant under E6.

Despite the brevity of this sketch of the construction, one may conclude that there is indeed a quantum mechanics for the "complex octonionic" plane and that the desired structure is that of a Jordan pair algebra. There is much further to be said, and a detailed account is in preparation.

Let us note before concluding that Gürsey has constructed an E6 invariant quantum mechanics previously, [8] using very different considerations. His results appear to be equivalent to the one sketched above, although there are differences in detail (for example, his idempotents do not appear to be allowed elements of underlying M_3^8 algebra). It is our view that the mathematical concepts presented in section 3 will prove to be very useful in physics for they seem to be, from many points of view, an ideal way to proceed.

References

(1) J.R. Faulkner and J.C. Ferrar, Bull. London Math. Soc. 9, 1-35 (1977.

(2) (a) P.Jordan, Nachr. Ges. Wiss. Göttingen, 209 (1933).
(b) P.Jordan, J.v. Neumann, and E.P. Wigner, Ann. Math. 35, 29 (1934).

(3) I.E. Segal, Ann. Math. 48, 930-948 (1947).

(4) S. Sherman, Ann. Math. 64. 593-601 (1956).

(5) F. Gürsey, in Kyoto International Symposium on Mathematical Physics, ed. by H. Araki, p. 189 (Springer, N.Y.,1975).

(6) F. Gürsey, invited paper at the Conference on Non-Associative Algebras, Univ. of Virginia, Charlottesville, Va., March 1977 (unpublished)

(7) C.W. Kim, invited paper at the second Johns Hopkins Workshop on "Current Problems in High Energy Particle Theory", Johns Hopkins University, Baltimore, Md., April 1978, ed. by G. Domokos and S. Kövesi-Domokos, (Baltimore, Md., 1978).

(8) F. Gürsey, invited paper, loc. cit. in Ref. 7.

(9) E. Størmer, Trans. Am. Math. Soc. 120, 438-447 (1965); Acta Math. 115, 165-184 (1966); Trans. Am. Soc. 130, 153-166 (1968).

(10) e. Alfven, E. Schultz and E. Størmer, to appear in Advances in Mathematics.

(11) E. Størmer, Acta Physicy Austriaca, Suppl. XVI, 1-14 (1976).

(12) Harald Hanche-Olsen, preprint ISBN-82-553-0341-3 (University of Oslo), April ;978.

(13) G. Domokos and S. Kövesi-Domokos, J. Math. Phys. 19, 1477 (1978).

(14) H.H. Goldstine and L.P. Horwitz, Proc. Nat. Acad. Sci. 48, 1134 (1962); Math. Ann. 154, 1 (1964); ibid. 164, 291 (1966).

(15) L.P. Horwitz and L.C. Biedenharn, Hel. Phys. Acta 38, 385 (1965).

(16) L.P. Horwitz and L.C. Biedenharn to appear in J. Math. Phys. (a preliminary report was given at the Second Johns Hopkins Workshop, see citation in Ref. 7.)

(17) M. Günaydin, J. Math. Phys. 17, 1875 (1976).

(18) K. MacCrimmon, Bull. Am. Maht. Soc., 84, 612-627 (1977).

(19) K. MacCrimmon, Proc.Nat. Acad. Sci., 56, 1072-1079 (1966).

(20) C. Piron, "Foundations of Quantum Physics" (Benjamin, New York, 1976).

(21) M. Koecher, "On Lie Algebras Defined by Jordan Algebras", Aarhus Univ. Lect. Notes (Aarhus, Denmark) 1967.

(22) O. Loos, "Jordan Pairs", Lecture Notes in Mathematics, Vol. 460, Springer Verlag (New York, 1975).

(23) M. Günaydin, Invited paper at the Second Johns Hopkins Workshop as cited in Ref. 7.

(24) M. Günaydin, C. Piron and H. Ruegg, Univ. of Geneva preprint UGVA-DPT 1977/ 12-154 (to be published in Comm. Math. Phys.).

(25) L.C. Biedenharn, J. Math. Phys. 4, 436 (1963).

(26) L. Michel and L.A. Radicati, Ann. Inst. Henri Poincaré, XVIII, =, 185-214 (1973).

(27) M. Gell-Mann, Phys. Rev. 125, 1097 (1962).

(28) J. Faulkner, Mem. Amer. Math. Soc., No. 104, (1970).

(29) This section incorporates results completed after the conference.

(30) H. Freudenthal, Advances in Math., I, 145 (1965).

(31) M. Koecher, "An Elementary Approach to Bounded Symmetric Domains", Rice Univercity Lecture Notes, (Rice University, Houston, Texas, 1969).

Atiyah-Singer Index Theorem

and

Quantum Field Theory

by

H. Römer

Fakultät für Physik, Universität Freiburg

0. Introduction

Recently, the relevance of global features in quantum field theory has been much emphasized. These properties, which in perturbation theory at most reflect themselves in germinal form , are accessible by global differential geometric and topological methods and are of crucial importance even for the qualitative behaviour of the quantum system.

Much work has been done on the topological and, for gauge fields, also on the differential geometric classification [1] of the stationary configurations of the classical action, which are supposed to contain interesting information about the affiliated quantum field theory.

Here we shall concentrate on a different question, namely on the spectrum of fluctuations about a stationary point of a Euclidean action. This problem arises immediately, already to lowest order in Planck's constant $\hbar$, if one tries to quantize a classical field theory. In mathematical terms it corresponds to the determination of the eigenvalues of an elliptic differential operator. It is known that this eigenvalue distribution reflects fundamental topological and also metrical features of the underlying system. For instance the area, boundary length and number of holes of a swinging membrane can be obtained from the spectrum of its oscillations, it is, indeed, largely possible to "hear the shape of a drum" [2]. The metrical and topological properties of the spectrum of quantum fluctuations are of immediate physical relevance. A beautiful example is provided by the Casimir effect [3], an attraction between two uncharged conducting plates, which is due to the topological modification of the vacuum fluctuations as compared to the vacuum without plates. Closely related [4][5] to this is the trace anomaly, a quantum effect for the energy momentum tensor which depends on both topological and metric invariants, namely the Euler characteristic and certain integral curvatures.

The Atiyah-Singer theorem [6] states that a very special spectral feature, the difference between the number of zero modes of an elliptic operator and its adjoint, the so-called index of the operator has a purely topological meaning and is related to the (differential) topology of the underlying base (space-time) manifold and the winding numbers of the bundles which appear in the problem.

On the other hand, the index can be shown [7] to be related to the anomaly of an associated current, so that current anomalies turn out to have topological significance.

The connecting link between the metrical and topological properties of a system and the spectrum of an elliptic operator D is the asymptotic expansion of the heat transport kernel of the operators D^*D and $D D^*$ [8], which, on the physical side is closely related to certain renormalization schemes like analytic renormalization point splitting [9][5] [19], ζ-function renormalization [4][8][10] and Schwinger's proper time formalism[5].

In this work we shall mainly deal with physical applications of the Atiyah-Singer index theorem. The difficulty in the presentation lies in the fact that most of the mathematical notions employed like manifolds [11], bundles [11] [12], characteristic classes [13] [6], cohomology etc., although intuitive and directly interpretable in physical terms are not so well-known to physicists, and that a reasonably complete description of these notions would lead to an unbalance of the mathematical and physical part of this work.
We try to cope with this difficulty by simply referring to the available good literature for the notions of manifolds [11] and fibre bundles [11][12] and by explaining the contents of the Atiyah-Singer theorem [6] in as non-technical a way as possible, trying to exhibit the essential ideas without striving for completeness and full rigor and not even attempting to describe a proof of the theorem.

The organisation of this work is as follows:
In chapter one we state and explain the contents of the Atiyah-

Singer theorem.
Chapter two contains a description of the heat transport formalism [8], which is vital for understanding the connection between spectral and metrical topological properties and links up with renormalization theory.
In chapter three, as an immediate application of the index theorem, the dimension of the space of non gauge equivalent fluctuations about a self dual instanton configuration is evaluated [14 - 18].
This leads to a determination of the number of parameters of instanton solutions of Yang-Mills' equations for gauge fields an arbitrary compact four dimensional orientable Euclidean space-time manifolds with arbitrary simple gauge group.
Chapter four contains a description of the relationship [7] between current anomalies and index theorem for a large class of anomalous currents, thus establishing the topological significance of current anomalies.
This general insight is applied in chapter five to evaluate the gravitational part of the axial anomaly as well as the Yang Mills part for fermion fields of spin 1/2 [19] and also spin 3/2 [10] (supergravity). In addition new anomalous currents are constructed [19], whose anomalies are related to the Euler characteristic and the signature of Euclidean space time.
Finally chapter six deals with noncompact space time manifolds.
The generalized Atiyah-Singer theorem for this case [20] contains a peculiar non-local boundary term, which helps to resolve a puzzle [21][22] about apparently fractional winding numbers. The index in gravitational background fields is evaluated for spin 1/2 [22] and spin 3/2 [23]. The index theorem is a powerful tool for calculating the additional boundary contribution. The case of spin 3/2 turns out to be particularly interesting because only there the index is nonvanishing, making chirality losses by vacuum tunnelling possible.

For the reader's convenience some essentials about de Rham cohomology, characteristic classes and formal splitting methods are collected in two short appendices.

1. The Atiyah-Singer Index Theorem

In this section we shall briefly explain the content and the meaning of the fundamental index theorem of Atiyah and Singer [6], a theorem on elliptic operators between complex vector bundles over a compact manifold. For all notions of manifolds and vector bundles we refer to the abundant literature [11)12)] on these subjects.

Let E and F be complex vector bundles over a manifold M. The sets of sections of E and F will be denoted by $\Gamma(E)$ and $\Gamma(F)$ respectively. For physical situations M can be interpreted e.g. as a Euclidean space-time manifold and the sections in $\Gamma(E)$ and $\Gamma(F)$ as suitable vector fields (spinors, isospinors etc...)

A differential operator D from E to F (written $E \xrightarrow{D} F$) is by definition a linear mapping

$$D: \quad \Gamma(E) \longrightarrow \Gamma(F) ,$$

which in terms of local coordinates in M and the fibres of E and F assumes the form

$$(Dv(x))_i = \sum_{\substack{|\alpha| \leq m \\ j}} a^{\alpha}_{ij}(x) \frac{\partial^{|\alpha|}}{\partial x^{\alpha}} v_j(x) \tag{1.1}$$

with $1 \leq i \leq \dim F, \quad 1 \leq j \leq \dim E$ (1.1a)

the fibre dimensions of E and F.

The summation on the right hand side of eq. (1.1) runs over the multiindex

$$\begin{aligned} \alpha &= (\alpha_1, \ldots, \alpha_n); \quad n = \dim M \\ |\alpha| &= \alpha_1 + \alpha_2 + \ldots + \alpha_n \end{aligned} \tag{1.1b}$$

The finite natural number m is called the <u>order</u> of the differential operator D. It does not depend on the coordinates chosen. To the operator D we associate a <u>symbol</u> σ_D by taking the "leading part" of the operator and substituting the derivatives $\frac{\partial^{|\alpha|}}{\partial x^{\alpha}}$ by monomials $\xi^{\alpha} := \xi_1^{\alpha_1} \cdot \ldots \cdot \xi_n^{\alpha_n}$, $\xi_i \in \mathbb{R}$:

$$\sigma_D(x,\xi)_{ij} = \sum_{|\alpha| = m} a^{\alpha}_{ij}(x)\, \xi^{\alpha} \tag{1.2}$$

The elements $\sigma_D(x,\xi)_{ij}$ define a dim F x dim E matrix. (It is easy to show that the symbol defines a vector bundle homomorphism $\sigma_D : \pi^* E \longrightarrow \pi^* F$, where π is the projection $\pi : T^*M \longrightarrow M$ of the cotangent bundle over M onto M).

The operator D is called <u>elliptic</u>, if the matrix $\sigma_D(x,\xi)$ is <u>invertible</u> for $\xi \neq 0$.

For instance, the Laplace operator on $\mathbb{R}^3$, $\Delta = \frac{\partial^2}{\partial x_1^2} + \frac{\partial^2}{\partial x_2^2} + \frac{\partial^2}{\partial x_3^2}$, has a symbol $\sigma_\Delta(x,\xi) = \xi_1^2 + \xi_2^2 + \xi_3^2$ (one by one matrix) and is elliptic, whereas the d'Alembert operator on $\mathbb{R}^4$, $\Box = \frac{\partial^2}{\partial x_0^2} - \frac{\partial^2}{\partial x_1^2} - \frac{\partial^2}{\partial x_2^2} - \frac{\partial^2}{\partial x_3^2}$, with symbol $\sigma_\Box = \xi_0^2 - \xi_1^2 - \xi_2^2 - \xi_3^2$ is not elliptic.

We now define the kernel of D as the space of "zero modes" of D:

$$\ker D = \{ v \in \Gamma(E) \mid Dv = 0 \} \tag{1.3}$$

the image of D

$$\operatorname{im} D = \{ w \in \Gamma(F) \mid \exists v \in \Gamma(E) \quad w = Dv \} \tag{1.4}$$

the set of sections in $\Gamma(F)$ of the form Dv with $v \in \Gamma(E)$ and the cokernel of D:

$$\operatorname{coker} D = \Gamma(F)/\operatorname{im} D \tag{1.5}$$

the quotient space of $\Gamma(F)$ modulo im D.
For compact M it is a fact of functional analysis [6)24)] that ker D and coker D are finite dimensional.

Hence, the (analytic) index of the elliptic operator can be defined:

$$\operatorname{ind} D = \dim \ker D - \dim \operatorname{coker} D \quad . \tag{1.6}$$

Equivalently one can consider the adjoint D^* of D with respect to any hermitean metric in the fibres and define

$$\operatorname{ind} D = \dim \ker D - \dim \ker D^* \quad . \tag{1.6a}$$

The amazing statement of the Atiyah-Singer index theorem [6)] is now, that this apparently analytic quantity, defined by the number of solutions of certain linear partial differential equations actually turns out to have a topological meaning. It can also be obtained by evaluating a certain characteristic class [13)] which depends on the operator D and on the tangent bundle TM of M.

It is not possible at this place to give a full description of the concept of characteristic classes of (vector) bundles. Some essentials will be given in appendix II. Here we only mention that for a vector bundle E over a manifold M a characteristic class $\chi(E)$ is a cohomology class on M, which gives us information about the degree of nontriviality of the bundle E. For trivial E $\chi(E)$ is trivial, a nontrivial $\chi(E)$ implies nontriviality of E. The characteristic classes we shall need later on are furthermore representable [25] as closed differential forms which are polynomials in the curvature quantities of the bundle E, obtained by introducing some linear connection on E, in concrete cases polynomials in the Yang -Mills field strength and the Riemannian curvature tensor.

In the cases we are going to consider the Atiyah Singer theorem assumes the following form:

Let E and F be complex vector bundles over a compact n-dimensional manifold M (without boundary). Let $E \xrightarrow{D} F$ be an elliptic operator then

$$\mathrm{ind}\, D = (-1)^{\frac{n}{2}(n+1)} \frac{\mathrm{ch}\, E - \mathrm{ch}\, F}{e(TM)}\, \mathrm{td}\,(TM \otimes \mathbb{C})\,[M] \quad . \qquad (1.7)$$

To arrive at this simple form we have assumed that D has, what is called a universal interpretation [6] which roughly corresponds to the physical statement that the operator can be defined by itself without reference to its concrete realization on the manifold M and the bundles E and F. The Dirac operator is an example of this kind of universality.

On the right hand side of eq. (1.7) ch E, ch F, e(TM), $\mathrm{td}(TM \otimes \mathbb{C})$ are the Chern characters of E and F, the Euler class of the tangent bundle TM and the Todd class of the complexified tangent bundle $TM \otimes \mathbb{C}$. They are all to be understood as closed differential forms of even degree which in turn are polynomials in the Yang Mills field strengths or the Riemannian curvature. The division by e(TM) can really be performed, and the result is a closed differential form on M, not homogeneous in its degree. The symbol [M] indicates that one has to extract the part of degree $n = \dim M$ and integrate it over the compact manifold M to obtain ind D.

Altogether ind D is obtained as the integral over a well defined polynomial in the curvature quantities of the bundles which appear in the problem. More precise definitions and details will be given in appendix II, where it will also be shown how the polynomial in the curvature quantities can readily be obtained from the theory of the characteristic classes. In the next section we shall also indicate a direct but laborious way of calculating the polynomial for a given operator.

Some remarks may be appropriate:

a) Trivially, for self-adjoint operators the index vanishes.

b) From (1.7) it is evident that in the cases considered the index does not depend on the operator but only on the bundles involved and, of course on the nontrivial fact that there is an elliptic operator connecting the two bundles E and F. (This requires for instance dim E = dim F.)

c) There is a useful apparent generalization of the A.S. theorem, which on close inspection turns out to be equivalent to the original formulation, but which is nevertheless worthwhile to formulate:

Take a finite sequence of complex bundles over a compact manifold and of differential operators

$$0 \xrightarrow{D_{-1}} E_0 \xrightarrow{D_0} E_1 \xrightarrow{D_1} E_2 \xrightarrow{D_2} \ldots\ldots \xrightarrow{D_{n-1}} E_n \xrightarrow{D_n} 0 \quad . \tag{1.8}$$

This sequence is called an elliptic complex, if $D_i D_{i-1} = 0$ and if the sequence of symbols σ_{D_i} is exact, i.e.

$$\mathrm{im}\, \sigma_{D_{i-1}} = \ker \sigma_{D_i} \qquad \text{for all } i \quad . \tag{1.9}$$

In this case the spaces

$$H_i = \ker D_i / \mathrm{im}\, D_{i-1} \tag{1.10}$$

have finite dimensions, and, denoting the sequence (1.9) by D, one defines

$$\mathrm{ind}\, D = \sum_{i=0}^{n-1} (-1)^i \dim H_i \quad . \tag{1.11}$$

Then, under the same universality condition as above, the Atiyah-Singer theorem says

$$\mathrm{ind}\, D = (-1)^{\frac{n}{2}(n+1)} \frac{\sum_{i=0}^{n} (-1)^i \mathrm{ch}\, E_i}{e(TM)} \, td(TM \otimes \mathbb{C})[M] \quad . \tag{1.12}$$

Evidently the original theorem (1.7) is subsumed as the special case of a two step complex.

There are also generalizations of the index theorem to manifolds with boundary, to which we shall return in the last chapter.

We conclude this paragraph by mentioning another very important generalization of the index theorem which will be employed later on, the G index theorem [6].

Take again two complex vector bundles E and F over a compact manifold M. In addition we assume that a group G acts on M and on the bundles E and F in such a way that for every $m \in M$ the fibre of E over m is mapped linearly onto the fibre of E over gm, and analogously for F. Then G also acts on the sections of E (and F) in the following way:

$$s \longrightarrow gs \qquad (s,\ gs \in \Gamma(E))$$

with

$$(gs)(m) = g(s(g^{-1}m)) \quad . \tag{1.13}$$

We now consider an elliptic operator $E \xrightarrow{D} F$, which commutes with the action of G:

$$gD = Dg \qquad \text{for all } g \in G \tag{1.14}$$

Then ker D and coker D are stable under G and thus finite dimensional representation spaces of the group G. For every $g \in G$ one can form the traces of g on these representation spaces and define

$$\mathrm{ind}_g D = \mathrm{tr}\, g\big|_{\ker D} - \mathrm{tr}\, g\big|_{\mathrm{coker}\, D} \tag{1.15}$$

Then the G index theorem says that this quantity has topological meaning and is the value of a characteristic class. One has

$$\mathrm{ind}_g D = \frac{i^*(\mathrm{ch}_g E - \mathrm{ch}_g F)\cdot td\,(TM^g)}{\mathrm{ch}_g(\Lambda_{-1} N^g \otimes \mathbb{C})\, e\,(TM^g)}\,[M^g] \tag{1.16}$$

where M^g is the set of fixed points of g, TM^g its tangent bundle, $N^g \otimes \mathbb{C}$ the complexified normal bundle of TM^g in TM, $\Lambda_{-1}(N^g \otimes \mathbb{C})$ the alternating sum of the external powers of $N^g \otimes \mathbb{C}$ and ch_g the so-called equivariant Chern character, a generalization of the usual Chern character, which will be defined in the appendix II.

We see that the G-index is character valued rather than integer valued and that the usual index theorem is recovered for $g = e$, the identity element of the group G.

2. Index Theorem and Heat transport [8)]

Consider a complex vector bundle E over a compact manifold M and a non-negative elliptic operator $\Delta : \Gamma(E) \longrightarrow \Gamma(E)$ on E. For this operator one constructs the heat transport operator

$$h = e^{-t\Delta} \tag{2.1}$$

The kernel of this operator has the spectral decomposition in terms of the eigenfunctions of Δ :

$$h(t,x,y) = \sum_{\lambda} e^{-\lambda t} \varphi_\lambda(x) \overline{\varphi_\lambda(y)} \tag{2.2}$$

where the summation runs over the eigenvalues of Δ , multiple eigenvalues being counted several times.

(For all of our constructions one has, strictly speaking to complete $\Gamma(E)$ to a Hilbert space with an appropriate Sobolev norm. Details can be found in ref. [24]).

The matrix valued kernel function obeys the heat transport equation

$$\left(\frac{\partial}{\partial t} + \Delta_x\right) h(t,x,y) = 0 \tag{2.3a}$$

with initial value

$$h(0,x,y) = \delta(x,y) \quad . \tag{2.3b}$$

For $t \longrightarrow \infty$ it simply tends to the projector onto the space of zero modes of Δ .

It can be shown [8) 24)] that for small positive t the function h(t,x,x) has an asymptotic expansion

$$h(t,x,x) \underset{t \to 0_+}{\sim} \sum_r t^r \mu_r(x) \tag{2.4}$$

with only finitely many negative powers of t. Furthermore the coefficient functions $\mu_r(x)$ can recursively be determined as polynomials in the coefficients of the operator Δ and their derivatives. For differential operators with a geometric meaning only curvature tensors and their derivatives occur. In concrete cases this recursive determination may be very laborious.

The related quantity $h(t) \stackrel{\text{Def}}{=} \operatorname{tr} \int_M dx\, h(t,x,x)$ equals

$$h(t) = \sum_{\lambda} e^{-\lambda t} \tag{2.5a}$$

and has an asymptotic expansion

$$h(t) \underset{t \to 0_+}{\sim} t^r a_r \tag{2.5b}$$

with

$$a_r = tr \int_M \mu_+(x)\,dx \quad . \tag{2.5c}$$

Sometimes it is advantageous to consider the operator Δ^{-s}, whose kernel $\zeta(s,x,y)$ is related to h(t,x,y) by a Mellin transform

$$\zeta(s,x,y) = \frac{1}{\Gamma(s)} \int_0^\infty dt\, t^{s-1} \left(h(t,x,y) - P_o(x,y) \right) \tag{2.6}$$

where $P_o(x,y) = \sum \psi_o^*(x) \cdot \psi_o(y) = \lim_{t \to \infty} h(t,x,y)$ is the projector onto the kernel of Δ.

And the related quantity

$$\zeta(s) = \sum_{\lambda \neq 0} \lambda^{-s} = tr \int_M dx\, \zeta(s,x,x) \tag{2.7}$$

$\zeta(s)$ is analytic for s large enough, and for s = 0 one has

$$\zeta(0) = a_o + \int_M dx\, P_o(x,x) \quad . \tag{2.8}$$

Now take an elliptic operator D: $\Gamma(E) \longrightarrow \Gamma(F)$ between two bundles E and F. With the adjoint D^* of D one observes that the operators

$$\begin{aligned} \Delta_E &= D^*D \;:\; \Gamma(E) \longrightarrow \Gamma(E) \\ \Delta_F &= DD^* \;:\; \Gamma(F) \longrightarrow \Gamma(F) \end{aligned} \tag{2.9}$$

are non-negative. Furthermore, they have the same eigenvalues λ, which, for $\lambda \neq 0$ even are of the same multiplicity. To see this, one only has to realize that for $\lambda \neq 0$ and

$$\begin{aligned} \Delta_E \varphi_\lambda &= D^*D \varphi_\lambda = \lambda \cdot \varphi_\lambda \\ \Delta_F (D\varphi_\lambda) &= (DD^*) D\varphi_\lambda = D(D^*D)\varphi_\lambda = \lambda D\varphi_\lambda \end{aligned} \tag{2.10}$$

and that $D\varphi_\lambda \neq 0$ because Δ_E and D have the same kernel. Constructing the corresponding heat kernels $h_E(t,x,y)$ and $h_F(t,x,y)$ one notices that the quantity

$$h_E(t) - h_F(t) = \sum_{\lambda_E} e^{-\lambda_E t} - \sum_{\lambda_F} e^{-\lambda_F t} \tag{2.11}$$

actually does not depend on t, because the contributions from the non-vanishing eigenvalues cancel out.
Hence, because $\ker D = \ker \Delta_E$ and $\ker D^* = \ker \Delta_F$

$$h_E(t) - h_F(T) = \dim \ker D - \dim \ker D^* = \operatorname{ind} D \quad . \tag{2.12}$$

Inserting the asymptotic expansion (2.4) and noticing that only μ_0^E and μ_0^F can contribute , one finally obtains

$$\operatorname{ind} D = \operatorname{tr} \int_M (\mu_0^E(x) - \mu_0^F(x))\, dx = a_0^E - a_0^F \tag{2.13}$$

a formula, which expresses the index of D as an integral over curvature quantities. The integrand is identical with the right-hand side of the Atiyah-Singer index theorem (1.7), if the characteristic classes are expressed in terms of curvature quantities. Of course, it is always advantageous to use (1.7) directly rather than going through all the recursion formulae which lead to (2.13), but the insight coming from the relationship of the index to the heat kernel will be useful for the application of the index theorem to current anomalies, which will be described in section 4.
We conclude with two remarks:
a) The coefficients μ_r of eq. (2.4) normally only have a differential geometric meaning. The Atiyah Singer theorem states that the special combination $\mu_0^E - \mu_0^F$ has topological meaning.
b) The relationship of the index problem to the heat transport equation is used for the heat transport proof of the index theorem. Here formula (2.13) is evaluated for a <u>certain set</u> of elliptic operators, and the resulting integrand identified as a characteristic class in terms of curvature quantities. Then K-theory is applied to show that this is sufficient to show that the index of <u>every</u> elliptic operator is the value of a characteristic class and to derive a formula of type eq. (1.7).

3. Number of Parameters of Instanton Solutions

We consider some Euclidean space time manifold M. We require M to be a four dimensional Riemannian manifold, compact and oriented. The common case in the literature [14-17] is $M = S^4$, the four dimensional sphere, but other base manifolds may be of similar interest, especially in quantum gravity.

On M we have a Yang -Mills field, which belongs to a compact

Lie group G and is given by a (locally defined) gauge potential A*), a one-form on M with values in the Lie algebra $\mathfrak{g}$ of G. In condensed notation the Yang-Mills field theory strength is

$$F = dA + \frac{1}{2} A \wedge A \tag{3.1}$$

a $\mathfrak{g}$-valued two-form on M. The wedge product $A \wedge A$ implies a Lie multiplication.

*) In mathematical terms we have a principal G-bundle P over M and a connection on P.

In our notation the Bianchi identity looks like

$$D_2 F \overset{\text{Def}}{=} dF + A \wedge F = 0 \tag{3.2}$$

and the effect of an infinitesimal gauge transformation on A is

$$\delta A = d\chi + A \wedge \chi \overset{\text{Def}}{=} D_0 \chi \tag{3.3}$$

where χ is some $\mathfrak{g}$-valued function on M (more precisely a section of the vector bundle $V_{\mathfrak{g}}$, the vector bundle with fibre $\mathfrak{g}$, associated to the principal bundle P via the adjoint representation of G.

An <u>instanton</u> over M is defined to be a connection A such that the corresponding Yang-Mills field strength is <u>self-dual</u>:

$$*F = F \quad . \tag{3.4}$$

This implies that the Yang-Mills equation

$$D_2 * F = (d + A \wedge) * F = 0 \tag{3.5}$$

is fulfilled.

The question is now what the dimensionality of a local family of instanton solutions is, if one identifies gauge equivalent instanton solutions.

The variation of F due to an infinitesimal variation δA is because of (3.1) given by

$$\delta F = d\delta A + A \wedge \delta A \overset{\text{Def}}{=} D_1 \delta A \, . \tag{3.6}$$

So, an equivalent formulation of our question is: What is the dimension h_1 of the space of variations δA, which have F self-dual and are not of the form (3.3), i.e. not due to a mere gauge transformation?

Denoting the projector onto antiselfdual two-forms by P_- we can say, that we want to compute the quantity

$$h_1 = \ker P_- \circ D_1 / \mathrm{im}\, D_o \quad . \tag{3.7}$$

This quantity is well defined because, indeed $\mathrm{im}\, D_o \subset \ker P_- D_1$, because $D_1 D_o \chi = F \chi$ and hence

$$(P_- D_1) D_o \chi = 0 \tag{3.8}$$

because F was self dual. The index theorem now enters via the complex 16)

$$D: \; 0 \longrightarrow \Lambda^0_{\mathfrak{g}}(M) \xrightarrow{D_o} \Lambda^1_{\mathfrak{g}}(M) \xrightarrow{P_- D_1} \Lambda^2_{-\mathfrak{g}}(M) \longrightarrow 0,$$

which can easily shown to be elliptic.

Here
$$\begin{aligned} \Lambda^0_{\mathfrak{g}}(M) &= \Lambda^0(M) \otimes V_{\mathfrak{g}} \\ \Lambda^1_{\mathfrak{g}}(M) &= \Lambda^1(M) \otimes V_{\mathfrak{g}} \\ \Lambda^2_{-\mathfrak{g}}(M) &= \Lambda^2_-(M) \otimes V_{\mathfrak{g}} , \end{aligned} \tag{3.9}$$

where $V_{\mathfrak{g}}$ was defined above and $\Lambda^0(M)$, $\Lambda^1(M)$, $\Lambda^2_-(M)$ are the zeroth, first and antiselfdual second exterior power of the cotangent bundle of M. The sections of these bundles are zero-forms, one-forms and antiselfdual two-forms respectively.

Now, according to (1.11)

$$\mathrm{ind}\, D = h_o - h_1 + h_2 \tag{3.10}$$

with h_1 given by (3.7) and

$$\begin{aligned} h_o &= \dim \ker D_o \\ h_2 &= \dim \ker (P_- D_1)^* = \dim (\Lambda^2_{-\mathfrak{g}}(M) / \mathrm{im}\, P_- D_1) \end{aligned} \tag{3.11}$$

ind D can be evaluated by the Atiyah-Singer index theorem, and one obtains

$$h_1 = 4kC(G) - 1/2 \dim G\, (\chi - \tau) + h_o + h_2 \quad , \tag{3.12}$$

Here k is the winding number of the Instanton

χ is the Euler characteristic of M

τ is the signature of M

C(G) and dim G depend on the group and are given by the following table [17]:

G	SU(n)	SO(n) n 7	Sp(2n)	G_2	F_4	E_6	E_7	E_8
C(G)	n	n-2	n+1	4	9	12	18	30

The quantities h_o and h_2 vanish in many interesting cases. h_o is non-vanishing only if the connection is actually reducible to a subgroup $H \subset G$.

A sufficient condition for $h_2 = 0$ is that the scalar curvature of M is non-negative and that the Weyl-tensor of M is selfdual. This also guarantees [10] that the local variations are integrable to global instanton solutions.

As an example one may take the well known case

$$M = S^4, \text{ hence } \chi = 2, \quad \tau = 0,$$
$$G = SU(2) \text{ with } C(G) = 2, \dim G = 3 \tag{3.13}$$
$$\text{and } h_o = h_2 = 0 .$$

Then the famous result $\quad h_1 = 8k - 3 \quad$ (3.14)

is reproduced.

4. Index theorem and Current Anomalies

Current anomalies [26] are a characteristic phenomenon, which may arise if a classical field theory is quantized. The invariance of a classical Lagrangian under a continuous group of transformations leads to the conservation of the corresponding Noether current. Quantization affiliates a quantum current to this classical Noether current, and it may happen, that unlike the classical current, the quantum current is not conserved but has an anomalous non vanishing divergence. In other words, it may be impossible to maintain a classical conservation law in the corresponding quantum field theory. Examples of this situation are provided by the anomaly of the axial current in massless fermion theories and by the trace anomaly, the anomaly of the dilatation current. If the invariance of the Lagrangian is explicitly broken already at the classical level, the classical and quantum Ward identities will differ by the anomalous contribution.

In this chapter we shall show [7) 19) 27)] that for the large class of currents, including the axial current the anomalous divergence of the quantum current has a topological meaning reflecting topological properties of the underlying classical theory. The Atiyah Singer theorem will be **crucial** in revealing this topological relevance of the anomaly and will, moreover, provide a powerful tool for calculating the anomalous divergence of a given current.

In the sequel the classical as well as the quantized field theory will always be Euclidean, and the final physical results will be obtained by Wick rotating at the very end.

We consider the following [19)] quite general classical field theory. The Euclidean classical fields are sections of a vector bundle E over a compact Riemannian space time manifold. (In a later section we shall see that the assumed compactness of Euclidean space-time is only a technical assumption.) Denoting the Euclidean field by ψ we furthermore assume that the classical Lagrangean is of the form [19)]

$$L = \langle \psi , D\psi \rangle + \ldots \qquad (4.1)$$

where $\langle .,. \rangle$ is some hermitean scalar product, which we need not specify for the moment and D: $\Gamma(E) \longrightarrow \Gamma(E)$ is an elliptic operator of the first order. The terms omitted in (4.1) may contain additional interactions without derivatives of ψ and contributions from other Euclidean fields. The assumed ellipticity of D just reflects the physical requirement that the kinetic operator is invertible off-shell, D may contain gauge fixing terms to assure this. We could as well have started out with a Lagrangean bilinear in the derivative:

$$L = \langle D\psi , D\psi \rangle \qquad (4.1a)$$

the anomalous divergences which will emerge later on would in fact be identical, but for the definiteness we stay with (4.1).

Next we assume that L be invariant under transformations

$$\psi \longrightarrow e^{\alpha \Gamma_5} \psi \qquad (4.2a)$$

where Γ_5 is a bundle homomorphism (locally a possibly space time dependent matrix) with the additional properties

$$\Gamma_5 D + D\Gamma_5 = 0 \tag{4.2b}$$

$$\Gamma_5^2 = 1 \tag{4.2c}$$

Finally, although this is not necessary, we shall for convenience assume that the elliptic operator D is of the form

$$D = \Gamma_i \nabla_i \tag{4.3}$$

where ∇_i is the covariant derivative in the direction i.

A classical massless spinor field in curved Euclidean space-time possibly coupled to an additional Yang Mills field is an example of the above described situation, and the notations allude to this standard example, but there are other interesting realizations of the general scheme.

The conserved classical Noether current is

$$J_i = \langle \psi, \Gamma_i \Gamma_5 \psi \rangle \quad . \tag{4.4}$$

Γ_5 is involutive, hence, the fields $\psi \in \Gamma(E)$ can be split into even and odd parts under Γ_5:

$$\Gamma_5 \psi_\pm = \pm \psi_\pm \tag{4.5a}$$

$$\Gamma(E) = \Gamma_+(E) \oplus \Gamma_-(E) \tag{4.5b}$$

and (4.2b) implies that D transforms even fields into odd fields and odd fields into even fields. So, by restriction we can define elliptic operators

$$\begin{aligned} D_+ &: \Gamma_+(E) \longrightarrow \Gamma_-(E) \\ D_- &: \Gamma_-(E) \longrightarrow \Gamma_+(E) \end{aligned} \tag{4.6}$$

which are adjoints of one another.

Now we can state the main assertion of this paragraph [19]:
The divergence of quantum current $\tilde{J}_i$, affiliated to J_i has an anomaly A(x), a polynomial in the curvature quantities of the bundles involved, which is given by minus twice the index density of the elliptic operator D_+. (The index density is the curvature polynomial

on the right hand side of the Atiyah-Singer index theorem (1.7), which has to be integrated over M to give the index.)

In the rest of this section we shall have to justify this vital theorem.

In the quantized theory, the classical current (4.4) is replaced by a functional average (vacuum expectation value of the quantum current operator:

$$\begin{aligned}\tilde{J}_i &= \langle\!\langle J_i \rangle\!\rangle = \int D\psi \, J_i \, e^{-S_E} \Big/ \int D\psi \, e^{-S_E} \\ &= \frac{\partial}{\partial A_i} \ln \int D\psi \, e^{-S_E + A_i J_i} \Big|_{A=0} \\ &\stackrel{\text{Def}}{=} \frac{\partial}{\partial A_i} \Gamma(A) \Big|_{A=0}\end{aligned} \tag{4.7}$$

The integration is a functional integration, the derivative with respect to the external source is a functional derivative, $A_i I_i$ is a short hand notation for $\int_M dx I_i(x) A_i(x)$, and the Euclidean action is

$$S_E = \int_M \mathcal{L} \, dx = \int_M \langle \psi, D\psi \rangle \, dx \tag{4.8}$$

The vacuum functional $\Gamma(A)$ is formally given by

$$\begin{aligned}\Gamma(A) &= -\ln \det (D - A_i \Gamma_i \Gamma_5) \quad \text{, (+ln det for fermions)} \\ &= -\ln \det D^{-1} + A_i \operatorname{tr} \Gamma_i \Gamma_5 D^{-1} + \ldots\end{aligned} \tag{4.9}$$

Hence, in one loop order one has formally

$$\tilde{I}_i = \operatorname{tr} \Gamma_i \Gamma_5 D^{-1} \tag{4.10}$$

Here, D^{-1} is the Green's function of the operator D.

As it stands, the expression (4.10) is of course ill-defined, and there are various equivalent procedures to regularize and define it.

On can start out from the expression

$$\tilde{I}_i = \lim_{x' \to x} \operatorname{tr}(\Gamma_i(x) \Gamma_5(x) G(x,x')) \Big|_{\text{regularized}} \tag{4.11}$$

or, graphically

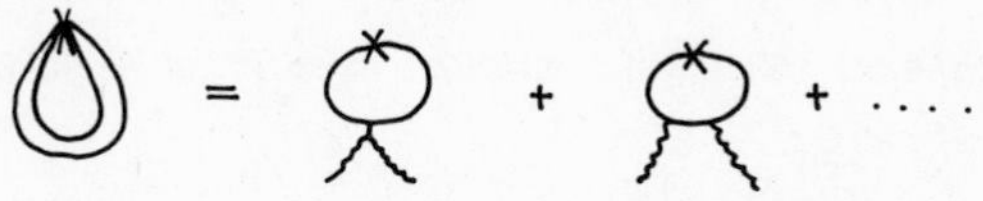

(4.12)

Here the double line represents the propagator of ψ in an external field, the simple line the free ψ propagator, the wavy line the external field and the cross the insertion $\Gamma_i \Gamma_5$.

The task is to calculate $\nabla_i \tilde{I}_i$ in order to find possible anomalies This can be done by [10)]

a) Feynman graph methods [10,28)], where one treats the problem in the momentum space and regularizes the emerging Feynman graphs in a standard way. The anomaly will arise from the lowest graphs depicted in (4.12).

b) Covariant point splitting [19)]. Here one determines the Green's function G for instance, by Schwinger's proper time method and regularizes by dropping terms which are singular in the coincidence limit.

c) ζ - function regularization [27,10,4)].

This method most clearly reveales the relationship between anomalies and the index theorem.

The one-loop vacuum functional is formally given by

$$\Gamma = -\frac{1}{2} \sum_{\lambda} \ln \frac{\lambda^2}{\mu^2} \tag{4.13}$$

where $\{\lambda^2\}$ are the eigenvalues of the non-negative operator $D_A^2 = (D - \Gamma_i \Gamma_5 A_i)^2$, and μ is some regularization mass. Comparing with eq. (2.7) one sees that it is natural to <u>define</u> [4)] Γ by

$$\Gamma = \frac{1}{2} \left[\zeta'(s) + \ln \mu^2 \zeta(s) \right]_{s=0} \quad . \tag{4.14}$$

ζ (s) is the ζ-function of the operator D_A^2. This expression exists for large real part of s, and the value at s = 0 is obtained by analytic combination. For the divergence of $\tilde{I}_i$ one finds (compare 4.11.)

$$\nabla_i \tilde{J}_i(x) = -2 \, \mathrm{tr} \, \zeta(s,x,x) \Gamma_5 \Big|_{A=0} \Big|_{s=0} \qquad \text{(comp. eq. 2.6)}$$

$$= -2 \frac{1}{\Gamma(s)} \int dt \, t^{s-1} \, \mathrm{tr} \, \Gamma_5 \left(h(t,x,x) - P_0(x,x) \right) \Big|_{A=0} \Big|_{s=0} \tag{4.15}$$

$$= -2 \, \mathrm{tr} \, \Gamma_5 \, \mu_0(x) + 2 \, \mathrm{tr} \, \Gamma_5 P_0(x,x) \quad .$$

$\mu_0(x)$ is the coefficient of t^0 in the asymptotic expansion of the heat kernel of the operator D^2.

Now D^2 commutes with Γ_5, and hence, its heat kernel splits into an even and an odd contribution

$$h = h_+ + h_- \ ; \ \Gamma_5 h_{\pm} = \pm h_{\pm} \tag{4.16}$$

$h_{\pm}$ are the heat kernels restricted to $\Gamma_{\pm}(E)$ (compare (4.6)) and belong to the operators

$$\begin{aligned} D_- D_+ &= D_+^* D_+ \quad \text{and} \\ D_+ D_- &= D_+ D_+^* \end{aligned} \tag{4.17}$$

With this splitting we get

$$\nabla_i \tilde{I}_i(x) = -2\mathrm{tr}\{\mu_+(x) - \mu_-(x)\} + 2\mathrm{tr}\{P_{0+}(x,x) - P_{0-}(x,x)\}$$

in evident notation, and looking at (4.17) and (2.13) we see that the anomalous part of the divergence of $\tilde{I}_i$, a curvature polynomial, which is present even if there are no zero modes of D^2, is really given by (-2) times the index density of the operator D_+.

We see, that the anomaly of the current T^i being related to the index of D_+ really has a topological meaning. In contrast to this, for the trace anomaly [4], the anomalous trace of the energy momentum tensor in one loop order the ζ-regularization procedure gives

$$(T_i^{\,i})_{\text{Anomalous}} = -2\,\mathrm{tr}\,\mu_0(x) = -2\,\mathrm{tr}\,(\mu_0^+ + \mu_0^-) \quad . \tag{4.19}$$

This is again a polynomial in the relevant curvature quantities, but this combination does not in general represent a closed differential form and a characteristic class, but has only a metrical meaning. Related to this, the trace anomaly, unlike the anomaly of the axial current is known to be renormalized in higher loop orders.
It is fair to say that we can resume our findings about the topological significance of the anomalies of quantum currents in the following way:

The anomalous part of the divergence of the quantum current $\tilde{I}_i$ is introduced by the regularization procedure, which includes the extraction of the zero modes. The zero modes may show an asymmetry, (different number of even and odd zero modes) which via the Atiyah-Singer index theorem has topological significance.

Apart from this insight into the nature of current anomalies, which turn out to convey a topological message the gain of what we have derived is twofold:

a) We have a powerful tool to compute anomalies quickly by topological methods.
b) We can use our insight to construct new anomalous currents [19], whose anomalies are related to fundamental topological features of the underlying classical theory.
Both points will be illustrated by examples in the next section.

5. Evaluation of Anomalies

We shall now give several realizations [19] of the general framework described by eqs. (4.1) and (4.2). Throughout we shall assume that the space-time manifold M is of even dimension n, compact, oriented and Riemannian.

We denote the Riemannian metric by g. There exist local n-bein tangent vector fields (e_i) $i=1,\dots,n$, such that $(e_1,\dots,e_n)$ defines the orientation of M and

$$g(e_i,e_j) = \delta_{ij}. \tag{5.1}$$

In the standard way we have a unique torsion free Riemannian connection in the tangent bundle, which we write as

$$\begin{aligned} \nabla_x e_i &= \Omega_{ij}(x) e_j \\ \Omega_{ij}(x) + \Omega_{ji}(x) &= 0 \end{aligned} \tag{5.2}$$

with associated Riemannian curvature form R.

From the tangent bundle one can go over to the Clifford bundle C(M) by forming the Clifford algebra of the tangent spaces at all points $m \in M$ with respect to the form $g(m)$.
As local sections of C(M) we shall denote the n-bein fields e_i by γ_i. Because of the orientability of M

$$\gamma_5 = \gamma_1\gamma_2\cdots\gamma_n \tag{5.3}$$

is globally defined.

Our first example is the axial anomaly. Herefore we assume that M has a spin structure (the vanishing of the second Stiefel-Whitney class is a necessary and sufficient condition for this). Then it is possible to construct a spinor bundle $\Delta(M)$ over M, whose sections are $s = 1/2$ spinor fields, and which splits into an even and an odd part under γ_5:

$$\Delta(M) = \Delta^{+}(M) \oplus \Delta^{-}(M) \tag{5.4}$$

On M we have a canonical spinor connection by

$$\hat{\nabla}_X \psi = X\psi + \frac{1}{4}\Omega_{ij}(X)\gamma_i\gamma_j\psi \tag{5.5}$$

and a globally defined Dirac operator D with

$$D\psi = \gamma_i \hat{\nabla}_{e_i}\psi \quad . \tag{5.6}$$

If for the scalar product in the fibres $\langle .,. \rangle$ we take the ordinary scalar product on spinors we have all the data required for the construction (4.1) and (4.2), because really

$$\begin{aligned} D\gamma_5 + \gamma_5 D &= 0 \\ \gamma_5^2 &= 1 \end{aligned} \tag{5.7}$$

So, in this case E is the spinor bundle $\Delta(M)$, the anomalous current is the axial current

$$J_i = \langle \psi , \gamma_i\gamma_5\psi \rangle \tag{5.8}$$

and the anomaly of the axial current is given by the index density of the operator

$$D_{+}: \quad \Gamma\Delta^{+}(M) \longrightarrow \Gamma\Delta^{-}(M) \quad . \tag{5.9}$$

By the methods described in appendix 2, this quantity can be readily determined: it is the $\hat{A}$ genus density [6,25] known from the mathematical literature and given up to terms of degree four by

$$\hat{A} = 1 - \frac{1}{24}p_1 + \dots \quad . \tag{5.10}$$

Here p_1 is the so-called first Pontrjagin class [13], a characteristic class of degree four, which in terms of the Riemannian curvature is given by the closed 4-form [25]

$$p_1 = -\frac{1}{8\pi^2}\mathrm{tr}\, R\wedge R \tag{5.11a}$$

with

$$R_{ij} = \frac{1}{2} R_{ijkl} dx^k \wedge dx^l \quad . \tag{5.11b}$$

So, we can immediately evaluate the gravitational part of the axial anomaly [19], which would, for instance lead to a decay of the π^0-meson into two gravitons. The ordinary Adler-Bell-Jackiw anomaly [26] is also obtained [7], if one adds a gauge degree of freedom of the Fermion fields, which mathematically corresponds to tensorially multiplying with a vector bundle V_G with structural group G:

$$\Delta^{\pm}(M) \longrightarrow \Delta^{\pm}(M) \otimes V_G \tag{5.12}$$

Using the multiplicativity of ch:

$$\mathrm{ch}(E \otimes F) = \mathrm{ch}\, E \cdot \mathrm{ch} F \tag{5.13}$$

and looking at the index theorem (1.7) with
$E = \Delta^+(M) \otimes V$ and $F = \Delta^-(M) \otimes V$ one gets the index density

$$\mathrm{ch} V \cdot \hat{A}(M) = -\dim V \frac{1}{24} p_1 + \frac{\tilde{C}(V)}{4\pi^2} F \wedge F \quad . \tag{5.14}$$

In conventional notation the total axial anomaly on a four dimensional Riemannian manifold is given by

$$\nabla_i J_i = -\frac{\dim V}{384\pi^2} \varepsilon^{abcd} R_{abst} R_{cdst} + \frac{\tilde{C}(V)}{4\pi^2} \varepsilon_{ijkl} \,\mathrm{tr} F_{ij} F_{kl} \tag{5.15}$$

where dim V is the dimension of the representation r(G) under which the spinor field transforms and $\tilde{C}(V)$ is an easily computable number, which only depends on r(G).
For dimension n=4 the two contributions are additive, in higher dimensions also mixed terms will arise.

So we succeed to get the Yang Mills and the gravitational part of the axial anomaly from the index theorem almost without calculation. Especially the evaluation of the gravitational part by conventional methods is quite laborious, and there was a long standing discrepancy about its coefficient in the literature [30].

Next we shall construct two new anomalous currents [19]. For the bundle E over M we take the bundle

$$\Lambda(M) = \bigoplus_{r=0}^{n} \Lambda^r(M) \tag{5.16}$$

The sum of all complexified exterior powers of the cotangent bundle of M. The sections of this bundle $\Lambda(M)$ are just all the complex-valued differential forms on M, not necessarily of homogeneous degree. The dimension of $\Lambda(M)$, and hence the number of complex components of the field ψ is 2^n. For instance, for dimension n=4, ψ has 16 components, namely

1 from 0-forms :	scalar
4 from 1-forms :	vector
6 from 2-forms :	antisymmetric tensor of rank two
4 from 3-forms :	pseudovector
1 from 4-forms :	pseudoscalar

Decomposing into spins we see that for n=4, ψ contains spins not greater than one.

On the cotangent bundle of M we have the Hodge star-involution [11,6], because M is oriented, which in terms of the dual forms e^i of the n-dim fields

$$e^i(e_j) = \delta_{ij} \tag{5.17}$$

is given by

$$* e^{i_1} \wedge \ldots \wedge e^{i_r} = \frac{1}{(n-r)!} \varepsilon_{i_1 \ldots i_r j_1 \ldots j_{n-r}} e^{j_1} \wedge \ldots \wedge e^{j_{n-r}} \quad . \tag{5.18}$$

Hence:

$$* : \Lambda^r(M) \longrightarrow \Lambda^{n-r}(M) \tag{5.18a}$$

The scalar product on the fibres of $\Lambda(M)$ is defined in the following way [6]:

$$\langle \alpha, \beta \rangle = \bar{\alpha} \wedge * \beta \tag{5.19}$$

if α and β are homogeneous of equal degree
$\langle \alpha, \beta \rangle = 0$, if α, β are homogeneous of unequal degree and by linear extension if α or β are nonhomogeneous.
There is also a hermitean form

$$(\alpha, \beta) = \int_M \langle \alpha, \beta \rangle \tag{5.20}$$

which turns out to be positive definite.

On $\Lambda(M)$ we have the ordinary exterior derivative

$$d: \Lambda^r(M) \longrightarrow \Lambda^{r+1}(M) \tag{5.20a}$$

with

$$d^2 = 0 \tag{5.20b}$$

And by

$$(\alpha, d\beta) = (\delta\alpha, \beta) \tag{5.21a}$$

we can define the adjoint δ of d with respect to the hermitean form

$$\delta : \Lambda^r(M) \longrightarrow \Lambda^{r-1}(M) \tag{5.21b}$$

with

$$\delta^2 = 0 \tag{5.21c}$$

which is explicitely given by

$$\delta\alpha = (-1)^{np+n-1} * d * \alpha \quad \text{for } \alpha \in \Gamma\Lambda^p(M) \quad . \tag{5.21d}$$

For the operator D we take the Hodge operator [6) 24)]

$$D = d + \delta$$
$$D: \Gamma\Lambda(M) \longrightarrow \Gamma\Lambda(M) \tag{5.22}$$

which is hermitean and elliptic as can easily be shown.
The kernel of D is spanned by the harmonic differential forms, which are defined by

$$d\omega = \delta\omega = 0 \quad . \tag{5.23}$$

The Hodge decomposition theorem [24)] says that every de Rham cohomology class has precisely one harmonic representative, which implies that the pth Betti number $b_p(M)$ is equal to the number of linearly independent harmonic differential forms of degree p.
Now all other data for the Lagrangean (4.1) are known.
There remains to define the operator Γ_5.
We give two different non-equivalent variants:
a) We define [19)]

$$\Gamma_5^E \omega = (-1)^p \omega \quad , \quad \omega \in \Gamma\Lambda^p(M) \tag{5.24}$$

an operator which is +1 on forms of even degree and -1 on forms of odd degree

b) $\Gamma_5^H \omega = (-1)^{p(p+1)/2 + 1} * \omega$ for $\omega \in \Gamma\Lambda^p(M)$. (5.25)

Here we have assumed that n= dim M = 4l is divisable by four.
One verifies

$$(\Gamma_5^{E,H})^2 = 1$$
$$\Gamma_5 D + D\Gamma_5 = 0 \quad . \tag{5.26}$$

Using the identity

$$d\omega = \sum_i e^i \wedge \nabla_{e_i} \omega \tag{5.27}$$

we can write the Hodge operator D in the form [19]

$$D = (d + \delta) = \Gamma_i \nabla_i \qquad (\nabla_i = \nabla_{e_i}) \tag{5.28a}$$

with

$$\Gamma_i \omega = (e^i \wedge + (-1)^{np+n+1} * e^i \wedge *)\omega \qquad \omega \in \Gamma\Lambda^p(M) \quad . \tag{5.28b}$$

The $2^n \times 2^n$ matrices Γ_i fulfill the same anticommutation rules as the γ_i, and $\{\Gamma_i, \Gamma_5^{E,H}\} = 0$.
The bundle $\Lambda(M)$ splits into even and odd parts under $\Gamma_5^{E,H}$:

$$\Gamma_5^E: \quad \Lambda(M) = \Lambda^e(M) \oplus \Lambda^o(M) \tag{5.29a}$$

$$\Gamma_5^H: \quad \Lambda(M) = \Lambda^+(M) \oplus \Lambda^-(M) \quad . \tag{5.29b}$$

Altogether, the Lagrangean is

$$\mathcal{L} = \langle \omega, \Gamma_i \nabla_i \omega \rangle + \ldots \tag{5.30a}$$

and the conserved classical currents are [19]

$$J_i^{E,H} = \langle \omega, \Gamma_i \Gamma_5^{E,H} \omega \rangle$$

The anomalies of the corresponding quantum currents $\tilde{J}_i^{E,H}$, for which we introduce [19] the names Euler current and Hirzebruch current, are completely determined by the index densities of the operators

$$D_+^E : \Gamma\Lambda^e(M) \longrightarrow \Gamma\Lambda^o(M)$$
$$D_+^H : \Gamma\Lambda^+(M) \longrightarrow \Gamma\Lambda^-(M) \tag{5.31}$$

Let us first discuss D_+^E. The meaning of ind D_+^E is readily identified [6,24]. ind D_+^E = dim ker D_+^E - dim ker D_+^{E*}.

Now ker D_+^E and ker D_+^{E*} are the spaces of harmonic differential forms of even and odd degree respectively. Thus

$$\begin{aligned} \dim \ker D_+^E &= \sum b_{2r}(M) \\ \dim \ker (D_+^E)^* &= \sum b_{2r+1}(M) \end{aligned} \tag{5.31a}$$

and

$$\operatorname{ind} D_+^E = \sum (-1)^p b_p(M) = \chi(M) \quad , \tag{5.31b}$$

the Euler characteristic of M. $\chi(M)$ has another simple meaning [31]: If M is triangulated with m_i i-simplices, then

$\chi(M) = \sum (-1)^i m_i$, independent of the triangulation.

The right hand side of the Atiyah-Singer index theorem for D_+^E gives the so-called Euler class e, a characteristic class of degree n = dim M, which, when integrated over M yields the Euler characteristic $\chi(M)$. Inverting its well known expression [25] in terms of the Riemannian curvature we finally obtain the anomaly [19] for n=4

$$\nabla_i \tilde{J}_i^E = \frac{1}{64\pi^2} \varepsilon^{ikrs} \varepsilon^{uvab} R_{ikuv} R_{rsab} \quad . \tag{5.32}$$

This result has also been checked by comparing it with the result of a covariant point splitting calculation.

We now come to the operator D_+^H:
ker D_+^H and ker D_+^{H*} are the spaces of harmonic forms even and odd under Γ_5^H respectively. Every harmonic form $\alpha \in \Gamma\Lambda^p(M)$ can be written as

$$\alpha = \tfrac{1}{2}(\alpha + \Gamma_5^H \alpha) + \tfrac{1}{2}(\alpha - \Gamma_5^H \alpha) = \alpha_+ + \alpha_- \quad , \tag{5.33}$$

$\alpha_\pm \in \Lambda^\pm(M)$ are harmonic. Hence, for $p \neq 2l$ (remember $n = \dim M = 4l$) the spaces ker $D_+^H \cap \Lambda^p(M)$ and ker $D_+^{H*} \cap \Lambda^p(M)$ are isomorphic, and ind D_+^H gets only contributions from $\Lambda^{2l}(M)$, which is stable under Γ_5^H. On $\Lambda^{2l}(M)$ we have

$$\Gamma_5^H \omega = *\omega \quad . \tag{5.34}$$

In the space of harmonic 2l forms $H^{2l}(M) \subset \Lambda^{2l}(M)$ we introduce a basis $\{\omega_i\}$ such that

$$*\omega_i = \Gamma_5^H \omega_i = \begin{cases} \omega_i & \text{for } 1 \le i \le r \\ -\omega_i & \text{for } r+1 \le i \le r+s = b_{2l}(M) \end{cases} \tag{5.35}$$

Then

$$\mathrm{ind}\, D_+^H = \tau - s \qquad (5.36)$$

This is at the same time the signature of the symmetric quadratic form

$$[\alpha, \beta] = \int_M \alpha \wedge \beta = (\alpha, \Gamma_5 \beta) \qquad (5.37)$$

on $H^{2l}(M)$, which is by definition [6,24,29] the signature of the manifold M.

For the operator D_+^H the index density is given [6,29] by the L genus, a characteristic class whose parts up to degree four are

$$L = 1 + \frac{1}{3} p_1 + \dots \qquad (5.38)$$

It is the content of Hirzebruch's famous signature theorem [29] that the integral over the L class equals the signature of the manifold, which explains the name of Hirzebruch current for J_i^H. Using (5.11) and (5.38) the anomaly of J_i^H is readily obtained [19]. For example for n=4:

$$\nabla_i J_i^H = \frac{1}{48 \pi^2} \varepsilon^{ikrs} R_{ikuv} R_{rsuv} \qquad (5.39)$$

Again, this result has also been verified by a covariant point splitting calculation [19].

It would be straightforward to endow the 2^n component field with an additional gauge degree of freedom, which would just like for the spinor field result in multiplying the index density by ch V_G and yield an easily computable Yang -Mills contribution to the anomalies of J_i^E and J_i^H.

We have succeeded in constructing anomalous currents [19], whose anomalies are related to the most fundamental invariants of the space-time manifold M, the Euler characteristic and the signature. For n=4 these currents incorporate only fields of spin $\leq$ 1. They may play a role in quantum gravity comparable to the importance of the axial current in Yang-Mills theories.

As a last example we shall briefly treat the axial anomaly [28,33,10] of Supergravity [32]. This is a theory, in which a massless spin 3/2 field is coupled to the gravitational field. It is unique in the sense that the usual causality problems, which arise, if a field of higher spin is coupled to the gravitational field are avoided in a consistent way.

The Lagrangean of supergravity is [32)]

$$\mathcal{L} = -\frac{1}{2}\sqrt{g}\,R - \frac{1}{2}\varepsilon^{\lambda\mu\nu\rho}\,\bar{\psi}_\lambda \gamma_\rho \gamma_5 \nabla_\mu \psi_\nu \quad . \tag{5.40}$$

It is invariant under the chiral transformation of the Rarita-Schwinger spinor ψ_μ

$$\psi_\mu \longrightarrow e^{-i\alpha\gamma_5}\psi_\mu \tag{5.41a}$$

which leads to a conserved classical Noether current

$$j^5_\mu = -\frac{i}{2}\varepsilon^{\rho\mu\sigma\tau}\,\bar{\psi}_\rho \gamma_\tau \psi_\sigma \quad . \tag{5.41b}$$

The Rarita-Schwinger field ψ_μ still contains redundant unphysical degrees of freedom, which have to be eliminated by subsidiary conditions like

$$\nabla_\mu \psi_\mu = 0 \quad , \quad \gamma_\mu \psi_\mu = 0 \quad . \tag{5.42}$$

This can be done consistently because the Lagrangean eq. (5.40) has a fermionic gauge invariance under the substitution

$$\psi_\mu \longrightarrow \psi_\mu + \nabla_\mu \chi \tag{5.43}$$

and some of the constraints can be interpreted as gauge fixing conditions.

In a field theoretical treatment a gauge fixing term is added to the supergravity Lagrangean, and the unphysical degrees of freedom are compensated by ghost fields. A careful analysis shows [34)], that the ghost fields are spinor fields [33,34)], two with positive and one with negative chiral charge. This result can be guessed by counting degrees of freedom and by looking at the structure of the subsidiary conditions (5.42). The gauge fixing term helps to render the Euclidean kinetic operator elliptic. Its precise form does not matter for the topological evaluation of the anomaly, because the index theorem (1.7) only depends on the bundles involved.

The contribution of the ghost fields has to be subtracted to get the correct axial anomaly.

In geometrical terms, the Rarita-Schwinger field is a section of the bundle

$$\begin{aligned} R &= R^+ \oplus R^- = \Delta(M) \otimes (TM \otimes \mathbb{C}) \\ &= \Delta^+(M) \otimes (TM \otimes \mathbb{C}) \oplus \Delta^-(M) \otimes (TM \otimes \mathbb{C}) \quad , \end{aligned} \tag{5.44}$$

where $TM \otimes \mathbb{C}$ is the complexified tangent bundle of the compact space-time manifold M, corresponding to the additional four-vector index of the Rarita-Schwinger field ψ_μ.

For the calculation of the axial anomaly one needs the Chern character $ch\, R^+ - ch\, R^-$. Looking at (5.44) and using the multiplicativity of ch one finds

$$ch\, R^+ - ch\, R^- = ch\,(TM \otimes \mathbb{C})\left[ch\, \Delta^+(M) - ch\, \Delta^-(M)\right] \quad , \tag{5.45a}$$

and from an inspection of the index theorem (1.7) one infers that the index density of the elliptic Rarita-Schwinger operator is obtained from the Dirac density by simply multiplying with

$$ch\,(TM \otimes \mathbb{C}) = 4 + p_1 + \cdots \quad . \tag{5.46}$$

(For a derivation of this equality see Appendix II).
The ghost bundle is given by

$$\mathcal{G}^{\pm} = \Delta^{\pm} \oplus \Delta^{\pm} \oplus \Delta^{\mp} \quad . \tag{5.47}$$

In the index theorem, the contributions of two of the ghost spinor bundles will compensate, and the final axial anomaly will be obtained by simply subtracting once the spin 1/2 index density from the Rarita-Schwinger index density. So, finally one has to multiply the spin 1/2 anomaly density by $(3 + p_i + \ldots)$ to arrive at the desired result. Comparing with the spin 1/2 anomaly we find

$$\begin{aligned} \text{ind}\, D_e^+ - \text{ind}\, D_{1/2}^+ &= (3 + p_1 + \ldots)(1 - \tfrac{1}{24}p_1 + \ldots)[M] = \\ (3 + \tfrac{21}{24}p_1)[M] &= \; = -21\ \text{ind}\, D_{1/2}^+ \quad . \end{aligned} \tag{5.48}$$

The axial anomaly of the spin 3/2 field is -21 times [33,10] the corresponding spin 1/2 result. Supposing, one can also add a Yang - Mills degree of freedom in a consistent way, also this contribution to the spin 3/2 axial anomaly can easily be evaluated [10]. The results , normalized to spin 1/2 are given in the following table

Spin	Gravitational part	Yang -Mills part
1/2	1	1
3/2	-21	3

More general spinor fields and additional gauge degrees of freedom have been treated in [43].

This evaluation of the supergravity anomaly especially clearly reveals the power of the topological approach to the anomaly problem. Ref [10] contains a more detailed discussion of the spin 3/2 anomaly including a comparison with other methods for computing anomalies.

We remark here that different values for the spin 3/2 axial anomaly can be obtained by evaluating the relevant Feynman graph [28,10] using noncovariant gauge fixing and assuming that the Adler-Rosenberg method of exploiting "gravitational conservation" is applicable in this case. This, however, turns out not to be true [42].

6. Non Compact Spaces

On noncompact spaces or compact spaces with boundaries the index theorem (1.7) cannot be true without modifications.

To illustrate this we give two examples of noncompact Riemannian manifolds, which can be thought to be obtained by letting a boundary tend to infinity.

a) Hawking's Euclidean Taub-NUT space [35] is a Riemannian space with length element

$$ds^2 = \frac{R+m}{R-m}\,dR^2 + 4(R^2 - m^2)\left\{\sigma_x^2 + \sigma_y^2 + \frac{4m^2}{(R+m)^2}\,\sigma_z^2\right\}, \tag{6.1}$$

$$m \leq R < \infty \quad . \tag{6.1a}$$

Here

$$\begin{aligned} \sigma_x &= \tfrac{1}{2}(-\cos\psi\, d\theta - \sin\theta \sin\psi\, d\varphi) \\ \sigma_y &= \tfrac{1}{2}(\sin\psi\, d\theta - \sin\theta\cos\psi\, d\varphi) \\ \sigma_z &= \tfrac{1}{2}(-d\psi - \cos\theta\, d\varphi) \end{aligned} \tag{6.2a}$$

$$0 \leq \varphi \leq 2\pi \;;\quad 0 \leq \theta \leq \pi \;;\quad 0 \leq \psi \leq 4\pi \tag{6.2b}$$

are the Maurer Cartan forms of the three-sphere S^3 in terms of Euler angles. The singularity of the metric at $R = m$ is only an apparent one and comes only from the polar coordinates. The total topology of the Euclidean Taub-Nut space is simply $\mathbb{R}^4$. Restricting to $m \leq R \leq R_o$ one obtains manifolds with boundary. The boundary surfaces at $R = R_o$ are topological three-spheres with a non-standard metric, which gets more and more asymmetric for $R_o \to \infty$ because the coefficient of σ_z^2 decreases relative to the coefficient $\sigma_x^2 + \sigma_y^2$.

The Riemannian curvature of the metric (6.1) is self dual, which makes the Euclidean Taub-NUT space a good candidate for a gravitational instanton [35]. To show that the single unmodified index theorem cannot hold for this space we compute the Pontrjagin number

$$p_1[M] = -\frac{1}{8\pi^2}\int_M \operatorname{tr} R\wedge R \qquad (6.3)$$

and find

$$p_1[M] = 2 \quad . \qquad (6.4)$$

Eq. (5.10) then tells us that using the index theorem (1.7) this would imply for the index of the Dirac operator

$$\text{"ind } D^+_{1/2}\text{"} = -\frac{1}{12} \qquad (6.5)$$

an absurd fractional result. Hawking presented this situation as a puzzle [35], and we shall soon exhibit the necessary modifications of the index theorem to render this index an integer.

As a second example we take the

b) Eguchi-Hanson space [36], a Riemannian space with metric

$$ds^2 = dR^2(1-(\tfrac{a}{R})^4)^{-1} + R^2\left\{\sigma_x^2 + \sigma_y^2 + (1-\tfrac{a}{R})^4\sigma_z^2\right\} \quad . \qquad (6.6)$$

The singularity at R=a can be removed by identifying antipodal points. The angle ψ then varies only in the interval $0 \leq \psi \leq 2\pi$. The surfaces at $R = R_o$ are therefore not spheres but real projective three-spaces $\mathbb{RP}(3)$, which arise from S^3 by identifying antipodal points. The metric on these projective spaces tends to the standard metric for $R_o \rightarrow \infty$. Again, for (6.6) the Riemannian tensor is self dual, and this time one finds

$$p_1[M] = -3 \qquad (6.7a)$$

and

$$\text{ind } D^+_{1/2} = \frac{1}{8} \quad . \qquad (6.7b)$$

The metric is asymptotically flat, which makes it a still better candidate for a gravitational instanton. Later on we shall deal with more general boundaries than S^3 and $\mathbb{RP}(3)$, namely with so-called Lens spaces [31], which we are now going to describe.

We start out from the three-sphere S^3, which we realize as the following subset of the complex two dimensional space $\mathbb{C}^2$:

$$S^3 = \left\{ (z_1,z_2) \in \mathbb{C}^2 \,\middle|\, |z_1|^2 + |z_2|^2 = 1 \right\} \quad . \tag{6.8}$$

On S^3 we consider the action of a cyclic group G of order m. The action on $\mathbb{C}^2$ and S^3 is determined by specifying the action of a generator $g_o \in G$:

$$g_o(z_1,z_2) = (e^{i\theta_1} z_1, e^{i\theta_2} z_1) \quad . \tag{6.9}$$

Identifying equivalent points under the action of G we arrive at the Lens space

$$L(\theta_1,\theta_2) \overset{\text{Def}}{=} S^3/G \tag{6.10}$$

in particular

$$\mathbb{R}P(3) = L(\pi,\pi) \quad . \tag{6.11}$$

We now state an index theorem of Atiyah, Singer and Patodi [20] for compact Riemannian manifolds with boundary. Our noncompact manifolds are obtained by pushing the boundary to infinity. For definiteness we state it for the Dirac operator. One has [20]

$$\text{ind } D^+_{1/2} = \hat{A}[M] + K - \tfrac{1}{2}(\eta_{1/2}[\partial M] + h_{1/2}) \quad . \tag{6.12}$$

The first term $\hat{A}[M]$ is the same integral over a curvature polynomial which appears on the right hand side of the index theorem for compact manifolds without boundary. This by the way justifies that we restricted ourselves to compact manifolds without boundary in section 4 and 5. The second term K is a local boundary term, an integral over the boundary of a differential form, which depends on the curvature tensor and the second fundamental form. It is the so-called transgression [25] form of the class $\hat{A}$. K vanishes if the manifold is a Riemannian product and in the two examples tends to zero, if R_o of the boundaries $R = R_o$ goes to infinity. We need not further discuss this contribution.

The last contribution

$$\xi_{1/2} = \tfrac{1}{2}(\eta_{1/2}[\partial M] + h_{1/2}) \tag{6.13}$$

is of particular interest. It can be shown to depend only on the boundary ∂M and not on the interior $\overset{\circ}{M}$.

It persists also for $R_o \to \infty$ and will provide the missing fractional part of the index. $\xi_{1/2}$ is a very peculiar spectral invariant of the boundary, defined in the following way:

Introducing normal and tangential coordinates in the neighbourhood of the boundary ∂M on can write the Dirac operator $D^+ : \Delta^+_{1/2}(M) \longrightarrow \Delta^-_{1/2}(M)$ as

$$D^+ = \gamma_n \nabla_n + \gamma_\alpha \nabla_\alpha = \gamma_n (\nabla_n + A) \quad . \tag{6.14}$$

The operator A is hermitean and can be identified with the Dirac operator on the boundary ∂M.

Let

$$h_{1/2} = \dim \ker A, \tag{6.15}$$

be the number of zero modes of the operator A. Denoting the eigenvalues of A by $\{\lambda\}$ one defines

$$\eta_{1/2}(s) = \sum_{\lambda \neq 0} \operatorname{sign} \lambda \, |\lambda|^{-s} \tag{6.16}$$

and

$$\eta_{1/2}[\partial M] = \eta(0) \tag{6.17}$$

in the sense of an analytic continuation.

This strange quantity, obtained from the spectrum of the Dirac operator A on ∂M vanishes if the eigenvalues λ are distributed symmetrically about $\lambda = 0$. It describes an asymmetry of the eigenvalue distribution.

For the Euclidean Taub-NUT space the boundary contribution $\xi_{1/2}$ can be evaluated directly [21,38] diagonalizing the Dirac operator on S^3 with nonstandard left-invariant metric.

The result for the boundary contribution of the boundary at $R_o \to \infty$ is[21,22]

$$\text{Taub-NUT} : \quad \xi_{1/2} = -1/12 \tag{6.18a}$$

and hence

$$\operatorname{ind} D^+_{1/2} = -\frac{1}{12} + \frac{1}{12} = 0 \quad . \tag{6.18b}$$

This vanishing of the index means that no chirality is lost by tunneling of a $s = 1/2$ spinor Taub-NUT background field.

The evaluation of the boundary contribution for the Eguchi-Hanson space is performed [23] with the aid of the G-index theorem. Rather than evaluating $\xi_{1/2}$ on $S^3/G = L(\theta_1, \theta_2)$ with standard metric, the boundary of D^4/G, where $D^4 = \{x \in \mathbb{R}^4 \mid |x| \leq 1\}$, we can undo the identification modulo G and consider the G-index on D^4 with boundary S^3 instead. The origin $O \in \mathbb{R}^4$ is the only fix point of the action of G, and the boundary correction for this case is given by the right hand side of the G index theorem (1.16) and the(vanishing) g-index for the Dirac operator on D^4 with flat metric.

This, in turn, yields the boundary correction of the Lens spaces: 23,39)

$$\xi_{1/2}(\Theta_1,\Theta_2) = \frac{1}{4m} \sum_{k=1}^{m-1} \frac{1}{\sin\frac{k\Theta_1}{2} \cdot \sin\frac{k\Theta_2}{2}} \quad . \tag{6.19}$$

The above sketched procedure is described in detail for the signature operator instead of the Dirac operator in ref. [20]. For more details see ref. [41].

In particular, for the Eguchi-Hanson space we obtain with $\Theta_1 = \Theta_2 = \pi$

Eguchi-Hanson: $\xi_{1/2} = 1/8$ (6.20a)

and

$$\text{ind } D^+_{1/2} = \frac{1}{8} - \frac{1}{8} = 0 \quad , \tag{6.21b}$$

again no chirality loss of a spin 1/2 field in the gravitational instanton field.

This absence of vacuum tunneling does not pertain to the spin 3/2 case [23] as was first conjectured by Hawking and Pope [40], who were able to construct two zero modes of the Rarita-Schwinger operator in a background Eguchi-Hanson metric, both of negative chirality, and none of positive chirality.

The G-index theorem provides a rigorous way to evaluate the index for spin 3/2.

The index of the operator $D^+_R : R^+ \longrightarrow R^-$ (compare section 5) is given [10] by a formula analogous to eq. (6.12):

$$\text{ind } D^+_R = \hat{A}_R[M] - \xi_R[\partial M] \quad , \tag{6.22}$$

where we have already omitted the boundary contribution K, which vanishes for $R_o \longrightarrow \infty$.

In section 5 we had already obtained [10]

$$\hat{A}_R[M] = \{ ch(TM \otimes \mathbb{C})\, \hat{A} \}[M] = \tag{6.23}$$

$$= \frac{5}{6} p_1[M] = -20 \hat{A}[M] \qquad (6.23)$$

$$= 5/2 \qquad \text{for the metric (6.6)} \quad .$$

The boundary correction ξ_R is in an analogous way obtained [23] from the $s = 1/2$ result by multiplying the G-index density with

$$i^* ch_g (TM \otimes \mathbb{C}) = 2 \cos k\Theta_1 + 2 \cos k\Theta_2 \quad . \qquad (6.24)$$

Here i^* means restriction to the fixed point set $M^g = \{0\}$. The result is for Lens spaces [23]

$$\xi_R = \frac{1}{2m} \sum_{k=1}^{m-1} \frac{\cos k\Theta_1 + \cos k\Theta_2}{\sin \frac{k\Theta_1}{2} \sin \frac{k\Theta_2}{2}} \qquad (6.25)$$

$$= -\frac{1}{2} \qquad \text{for the Eguchi-Hanson space} \quad .$$

Hence

$$\text{ind } D_R^+ = -\frac{5}{2} + \frac{1}{2} = -2 \qquad (6.26)$$

and, taking into account the ghost contributions for the Rarita-Schwinger field (compare 5.48)

$$\text{ind } D_R^+ - \text{ind } D_{1/2}^+ = -2 - 0 = -2 \qquad (6.27)$$

in agreement with Hawking and Pope [40].

So, for spin 3/2 chirality does get lost by tunnelling. Boundary contributions for higher spinor fields are evaluated in Ref. [43]. The Atiyah-Singer-Patodi index theorem (6.12) is the appropriate framework [21] to treat broken winding numbers, which appear on non compact manifolds without running into inconsistencies for the chirality bilance. Such fractional configurations may be important in a topological theory of quark confinement.

A more detailed description of the G-index evaluation of boundary terms and a discussion of more general gravitational instanton background fields can be found in Ref. [41].

Appendix I: De Rham Cohomology

In this appendix we shall collect a few fundamental definitions and properties concerning the de Rham cohomology of a manifold, the only kind of cohomology modules which is used in this work.

Let M ba a (smooth) manifold. The linear space of all (smooth) differential p-forms on M is denoted by $\Omega^p(M) = \Gamma \Lambda^p(M)$. Furthermore we write

$$\Omega(M) = \bigoplus_{p=0}^{\dim M} \Omega^p(M) \qquad . \tag{I, 1}$$

On the differential forms we have an exterior derivative

$$d : \Omega^p(M) \longrightarrow \Omega^{p+1}(M) \tag{I, 2}$$

and an associative exterior multiplication

$$\wedge : \Omega^p(M) \times \Omega^q(M) \longrightarrow \Omega^{p+q}(M) \tag{I, 3}$$

with the well-known properties

$$\alpha \wedge \beta = (-1)^{pq} \beta \wedge \alpha \tag{I,4a}$$

$$d(\alpha \wedge \beta) = (d\alpha) \wedge \beta + (-1)^p \alpha \wedge d\beta \tag{I,4b}$$

$$d^2\alpha = 0 \qquad \alpha \in \Omega^p(M) \, , \; \beta \in \Omega^q(M) \, . \tag{I,4c}$$

In addition, d is of course linear and $\wedge$ bilinear.

A differential form ω is called closed if $d\omega = 0$ and exact, if there exists a form φ such that $\omega = d\varphi$. The spaces of closed and exact p-forms on M are denoted by $Z^p(M)$ and $B^p(M)$.
One has, because of (I,4c)

$$B^p(M) \subset Z^p(M) \qquad . \tag{I,5}$$

The pth de Rham cohomology group of M is defined as the quotient vector space

$$H^p(M) = Z^p(M)/B^p(M) \qquad , \tag{I,6}$$

its elements are equivalence classes of closed p-forms, two closed p-forms being equivalent if their difference is an exact p-form. A smooth map f: $M \longrightarrow N$ induces linear maps

$$f^*: \Omega^p(N) \longrightarrow \Omega^p(M) \tag{I,7}$$

such that

$$f^*(\alpha \wedge \beta) = f^*\alpha \wedge f^*\beta \tag{I,8a}$$

$$f^* d\alpha = d f^* \alpha \quad . \tag{I,8b}$$

In terms of coordinates f^* just means composition with the Jacobian matrix of f. Because of (I,8), the map f also induces a linear map between the cohomologies, which we shall also denote by f^*:

$$f^*: \; H^p(N) \longrightarrow H^p(M) \tag{I,9}$$

and one can show that homotopy of two maps f and g implies identity of the associated maps f^* and g^* on the cohomology groups.

We conclude with a few important constructions related to connections on bundles. We shall restrict ourselves to vector bundles.

Let E be a vector bundle over N with standard fibre V and structural group G. Consider a map f: $M \longrightarrow N$. Then one can construct the so-called <u>induced bundle</u> f^*E, a vector bundle over M with the same standard fibre V. We again use the notation f^*, because there is no danger of confusion with the previously introduced mappings f^*. The induced bundle f^*E over M is obtained by taking the fibre of E over f(m) as the fibre over m for f^*E. For details see e.g. ref.[11-13]. Homotopic maps induce equivalent bundles.

Consider now a connection on E with (locally defined) connection form A and curvature $F = dA + 1/2 A \wedge A$. A and F are $\mathfrak{g}$-valued one-and two-forms on N respectively. $\mathfrak{g}$ denotes the Lie algebra of G. Then there is an induced connection on the induced bundle f^*E defined by

$$A^{ind} = f^*A \tag{I,10a}$$

which implies

$$F^{ind} = dA^{ind} + 1/2 A^{ind} \wedge A^{ind} = f^*F \quad . \tag{I,10b}$$

A symmetric r-linear form

$$\begin{aligned} Q: \; & \mathfrak{g} \times \mathfrak{g} \times \ldots \times \mathfrak{g} \longrightarrow \mathbb{C} \\ & (a_1, a_2, \ldots, a_r) \longrightarrow Q(a_1, \ldots, a_r) \end{aligned} \tag{I,11}$$

is called <u>invariant</u>, if

$$Q([b,a],a,\dots,a) = 0 \qquad \text{for all } a,b \in \mathfrak{g} \; . \tag{I,12}$$

Inserting the curvature form F into a symmetric invariant r-linear form one obtains a 2r-form

$$Q_F = Q(F,F,\dots,F) \quad . \tag{I,13}$$

The precise meaning of (I,14) can be described as follows.
Writing F as $F = F_{\mu\nu}(x)\,dx^{\mu}\wedge dx^{\nu}$ with $F_{\mu\nu}(x) \in \mathfrak{g}$
we have

$$Q_F = Q(F_{\mu_1\nu_1},\dots, F_{\mu_r\nu_r})\,dx^{\mu_1}\wedge dx^{\nu_1}\wedge \dots \wedge dx^{\mu_r}\wedge dx^{\nu_r} \quad . \tag{I,13a}$$

We assert that Q_F is closed and, moreover, the cohomology class of Q_F is independent of the specific connection on E and, hence, only depends on E.

For the first assertion we observe (confer def. (3.2), we drop the subscript of D)

$$dQ_F = DQ_F = rQ(DF,F,\dots,F) = 0 \tag{I,14}$$

because of the Bianchi identity $DF = 0$.

To prove the second assertion we have to show that the change δQ_F, due to an infinitesimal change δA of A is an exact form. Indeed

$$\begin{aligned} \delta Q_F &= rQ(\delta F,F,\dots,F) \\ &= rQ(D\delta A,F,\dots,F) \qquad \text{(eq. 3.6)} \\ &= rdQ(\delta A,F,\dots,F) \qquad \text{(invariance of Q)} \; . \end{aligned} \tag{I,15}$$

So, the cohomology class $[Q_F]$ of Q_F is really independent of the connection. This implies that for the induced bundle f^*E we have

$$[Q_{f^*F}] = f^*[Q_F] \tag{I,16}$$

and that $[Q_{f^*F}]$ only depends on f^*E.

Appendix II Characteristic Classes

In this section we confine ourselves to smooth real or complex vector bundles over smooth manifolds. The cohomology will always be de Rham's cohomology which is all we need. For more general cases see e.g. ref. [13].

Consider a vector bundle $E = (\tilde{E}, \pi_E, M_E)$ with total space $\tilde{E}$, base M_E and projection π_E. A characteristic class χ is a function which associates to E a cohomology class of the basis

$$\chi : \quad E \longrightarrow \chi(E) \in H^*(M_E) \tag{II,1a}$$

with

$$H^*(M_E) = \bigoplus_p H^p(M_E) \quad . \tag{II,1b}$$

In addition one requires that χ behaves naturally with respect to induced bundles

$$\chi(f^*E) = f^*\chi(E) \tag{II,1c}$$

Eqs. (II,1) define the notion of a characteristic class.

For instance, the considerations at the end of appendix I and formula (I,16) tell us that for vector bundles with a fixed structure group G every invariant n-linear symmetric form Q defines a characteristic class χ_Q by

$$\chi_Q(E) = Q(F,\ldots,F), \tag{II,2}$$

where F is the curvature form of any connection on the bundle E.

Equation (II,1c) shows that trivial bundles have trivial values for all characteristic classes, because a trivial bundle can always be induced from a bundles whose base is a point.

The most fundamental characteristic classes of <u>complex</u> vector bundles are the Chern classes $c_i(E)$, which are uniquely defined by (II,1) and the following properties:

$$c_i(E) \in H^{2i}(M_E) \tag{II,3a}$$

$$c_o(E) = 1 \quad , \quad c_i(E) = 0 \text{ for } i > \dim E$$

(Here dim E is the dimension of the fibres of E. We define

$$c(E) = \sum_{i \geq 0} c_i(E))$$

$$c(E \oplus F) = c(E)c(F) \tag{II,3b}$$

The multiplication on the right hand side is an exterior multiplication, commutative, because the $c_i(E)$ are forms of even degree. (II,3b) is equivalent with

$$c_i(E \oplus F) = \sum_{k \geq 0} c_k(E)c_{i-k}(F) \quad . \tag{II,3c}$$

A normalization is fixed by prescribing the Chern classes of the canonical one dimensional vector bundle E(n) over the n-dimensional complex projective spaces $\mathbb{C}P(n)$. The fibre over a point of $\mathbb{C}P(n)$ is just the one dimensional complex vector space which represents this point. One defines

$$c(E^{(n)}) = 1 + x \tag{II,3d}$$

where $x \in H^2(\mathbb{C}P(n))$ is the positive integer generator of the de Rham cohomology of $\mathbb{C}P(n)$. Axioms (II,3) in particular guarantee that the integrals of the Chern classes over closed compact surfaces are always integer valued.

Assume now that the complex bundle E is the sum of complex line bundles (one dimensional bundles):

$$E = \bigoplus_{j=1}^{\dim E} E_j \tag{II,4}$$

and call $c_1(E_j) = y_j$. Then (II,3b) implies

$$c(E) = \prod (1 + y_j) \tag{II,5}$$

and $c_k(E)$ is just the kth elementary symmetric polynomial of the quantities y_j. Not every bundle E is splittable into line bundles, but one can prove the splitting principle [6,29], which states that from every complex vector bundle E one can induce a bundle f^*E such that f^*E is splittable and the map $f^*: H^*(M_E) \longrightarrow H^*(M_{f^*E})$ is injective. Hence it suffices to consider splittable bundles. Formally every complex bundle can be treated as a splittable bundle. In terms of such formal splittings one can conveniently define the other characteristic classes which are used in this work. For instance

Todd classes

$$\text{td } E = \sum \text{td}_i E = \prod' \frac{y_i}{1-e^{-y_j}} = 1 + 1/2C_1(E) + 1/12(C_2(E) + C_1^2(E)) + \ldots \tag{II,6}$$

where $\text{td}_i E$ is the homogeneous part of degree i on the right hand side.

Chern character

$$\text{ch } E = \sum \text{ch}_i E = \prod' e^{y_j} \tag{II,7}$$

one verifies at once

$$\text{td } (E \oplus F) = \text{td}E \ \text{td}F \tag{II,8a}$$

$$\text{ch}(E \oplus F) = \text{ch}E + \text{ch } F \tag{II,8b}$$

$$\text{ch}(E \otimes F) = \text{ch}E \cdot \text{ch } F \quad . \tag{II,8c}$$

The last **equation** (II,8c) holds because for line bundles L_k one has

$$c_1(L_i \otimes L_k) = c_1(L_i) + c_1(L_k) \quad . \tag{II,8d}$$

Of particular importance is the case that the complex bundle E arises from complexification of a real bundle:

$$E = E' \otimes \mathbb{C} \tag{II,9}$$

where E' is a real bundle. If E is of even dimension we can always assume [6,29] that the splitting $E = \oplus E_i$ can be chosen such that the classes y_j are pairwise equal up to a sign:

$$y_{2j-1} = - y_{2j} = x_j \qquad (j = 1 \ldots 1/2\dim E) \quad . \tag{II,10}$$

On can furthermore achieve that in terms of the k_j classes x_j the Euler class, which we define for real orientable even dimensional bundles is given by

$$e(E') = \prod' x_j \quad . \tag{II,11}$$

The Pontrjagin classes are defined for real vector bundles E' by

$$p_i(E') = (-1)^i c_{2i}(E' \otimes \mathbb{C}), \text{ hence} \tag{II,12b}$$

$$\sum p_i(E') = \prod (1 + x_i^2) \quad . \tag{II,12b}$$

For convenience we list some of the characteristic classes, which appear in this work in terms of the formal splitting of the complexified tangent bundle $TM \otimes \mathbb{C}$ of an 2r dimensional manifold:

$$\mathrm{ch}\,(TM \otimes \mathbb{C}) = \sum_{i=1}^{r} (e^{x}\ re^{-x}) = 2r + p_1(TM) + \ldots \tag{II,13a}$$

$$e\,(TM) = \prod_{i=1}^{r} x_i \tag{II,13b}$$

$$\mathrm{td}\,(TM \otimes \mathbb{C}) = \prod_{i=1}^{r} \frac{(-x_i^2)}{(1-e^{-x_i})(1-e^{x_i})} \tag{II,13c}$$

$$\mathrm{ch}\,\Lambda^{+}M - \mathrm{ch}\,\Lambda^{-}M = \prod_{i=1}^{r} (e^{-x_i} - e^{+x_i}) \tag{II,13d}$$

$$\mathrm{ch}\,\Delta^{+}M - \mathrm{ch}\,\Delta^{-}M = \prod_{i=1}^{r} (e^{x_i/2} - e^{-x_i/2}) \tag{II,13e}$$

$$L(M) = \prod_{i=1}^{r} \frac{x_i}{\tanh x_i} = 1 + \frac{1}{3} p_1(TM) + \ldots \tag{II,13f}$$

$$\hat{A}(M) = \prod_{i=1}^{r} \frac{x_i/2}{\sinh x_i/2} = 1 - \frac{1}{24} p_1(TM) + \ldots \tag{II,13g}$$

We now define the equivariant Chern character ch_g for the complex vector bundles. We only need it for bundles over fixed point sets of the group element $g \in G$, i.e. for bundles on the base of which g acts trivially and only permutes inside the fibres. Again we can apply a formal splitting [6] $E = \oplus E_j$ and assume furthermore that g acts on E_i simply by multiplication with $e^{i\theta_j}$. Then one has

$$\mathrm{ch}_g\, E = \sum e^{i\theta_j + y_j} \tag{II,14}$$

with $y_j = c_1(E_j)$.

In particular, in terms of the formal splitting of the complexified (co) tangent bundle and for zero dimensional fixed point sets we have

$$\begin{aligned} i^{*}\mathrm{ch}\ g\,(\Lambda_{-1} N^{g} \otimes \mathbb{C}) &= i^{*} \prod_{j=1}^{r} (1 - e^{x_j + i\theta_j})(1 - e^{-x_j - i\theta_j}) \\ &= \prod_{j=1}^{r} (1 - e^{i\theta_j})(1 - e^{-i\theta_j}) \end{aligned} \tag{II,15a}$$

$$i^* \operatorname{ch} g\,(TM\otimes\mathbb{C}) = i^* \sum_{j=1}^{r} (e^{x_j + i\theta_j} + e^{-x_j - i\theta_j}) = \sum_{j=1}^{r} 2\cos\theta_j \qquad \text{(II,15b)}$$

$$i^* \operatorname{ch} g(\Delta^+(M)-\Delta^-(M)) = i^* \prod_{j=1}^{r} (e^{x_j/2+i\theta_j/2} - e^{-x_j/2-i\theta_j/2}) = \prod_{j=1}^{r} (e^{i\theta_j/2} - e^{-i\theta_j/2}) \qquad \text{(II,15c)}$$

i^* here means restriction to the fixed point set M^g.

Finally we note that the Chern classes and, hence all the characteristic classes (II,13), which are polynomials in the Chern classes can be represented by curvature quantities in the form χ_Q of eq. (II,2).

For an s x s matrix X we define the invariant polynomials $Q_j(x)$ by

$$\det\left(1\!\!1 + \frac{i}{2\pi} X\right) = \sum_j Q_j(X)\, t^{s-j} \qquad \text{(II,16)}$$

For a complex vector bundle with fibre dimension s one introduces a connection, whose curvature form F is an s x s matrix of two-forms (on every vector bundle there are infinitely many connections). Then one can prove [25)]

$$c_j(E) = Q_j(F) \qquad \text{(II,17)}$$

independent of F. This formula allows the expression of characteristic classes by curvature quantities, which was used so frequently in our work. The only case not yet covered by eq. (II,17) is the Euler class. For the Euler class of the tangent bundle of an even dimensional orientable Riemannian manifold we have [25)]

$$e(TM) = \frac{(-1)^{n/2}}{2^n \pi^{n/2} (\frac{n}{2})!} \varepsilon_{i_1 \dots i_n} R_{i_1 i_2} \wedge \dots \wedge R_{i_{n-1} i_n} \qquad \text{(II,18)}$$

where R_{ik} is the Riemannian curvature tensor of M.

References

1) M.F. Atiyah, N.J. Hitchin, V.G. Drinfeld, Y.I. Manin, Phys. Lett. 65A , 185 (1978)

2) I.M. Singer, Proceedings of the International Congress of

Mathematicians, Vancouver 1974.
M. Kac, Amer. Math. Monthly 73, 1-23 (1966)

3) H.B. G. Casimir, Proc. Kon. Ned. Akad. Wetenschap 51, 793 (1948), see also T.H. Boyer, Ann. Phys. (N.Y.) 56, 474 (1979) and ref. 5)

4) S.W. Hawking, Comm. Math. Phys. 55, 133 (1977)

5) B.S. De Witt, Physics Reports 19C, 295 (1975)
S.M. Christensen, Phys. Rev. D14, 2490 (1976); D17, 976 (1978)

6) M.F. Atiyah, I.M. Singer, Bull. Amer. Math. Soc. 69, 422 (1969) Amer. Math. 87, 484 and 546 (1968)
M.F. Atiyah, G.B. Segal, Ann. Math. 87, 530 (1968)
R. Palais ed. "Seminar on the Atiyah-Singer Index Theorem", Annals of Math. Studies 57, Princeton University Press.

7) S. Coleman, unpublished
J. Kiskis, Phys. Rev. D15, 2329 (1977)
R. Jackiw, C. Rebbi, Phys. Rev. D16, 1052 (1977)
N.K. Nielsen, B. Schroer, Nuclear Phys. B 127, 493 (1977)
N.K. Nielsen, H. Römer, B. Schroer, Phys. Lett. 70B, 445 (1977)

8) M.F. Atiyah, R. Bott, V.K. Patodi, Inventiones math. 19, 279 (1973)

9) N.K. Nielsen, Nordita preprint 78/24

10) M.T. Grisarn, N.K. Nielsen, H. Römer, P. van Niewenhuizen, Nucl. Phys. B140 477 (1978),
N.K. Nielsen and H. Römer, Proc. Informal Meeting on recent developments in field theory Nov 1977 ICTP/77/152, pp. 27-28.

11) e.g. R.L. Bishop, R.J. Crittenden, "Geometry of Manifolds" Academic Press, New York and London 1964

12) N. Steenrod, " The Topology of Fibre Bundles", Princeton University Press.
D. Husemoller, "Fibre Bundles", Springer New York, Heidelberg, Berlin.

13) J.W. Milnor, J.D. Stasheff, "Characteristic Classes", Annals of Math. Studies Nr.76 Princeton University Press

14) A.S. Schwarz, Phys. Lett. 67B, 172 (1977)

15) R. Jackiw, C. Rebbi, Phys. Lett. 67B, 189 (1977)

16) M. Atiyah, N. Hitchin, I.M. Singer, Proc. Nat. Acad. Sci. USA, 74, 2662 (1977); Proc. R. Soc. Lond. A362, 425 (1978)

17) C.W. Bernard, N.H. Christ, A.H. Gerth, E. Wienberg, Phys. Rev. D16, 2967 (1977)

18) H. Römer, Nuevo Cim. Lett 21. 381 (1978)

19) N.K. Nielsen, H. Römer, B. Schroer, Nucl. Phys. B136, 475 (1978)

20) M.F. Atiyah, V.K. Patodi, I.M. Singer, Bull. Londons Math. Soc 5, 229 (1973), Math. Proc. Cambridge Philos. Soc. 77, 43 (1975);

78, 405 (1975); 79, 71 (1976)

21) H. Römer, B. Schroer, Phys. Lett. 71B, 182 (1977)

22) T. Eguchi, P.B. Gilkey, A.J. Hanson, Phys. Rev. D17, 423 (1978)
C.N. Pope, Nucl. Phys. B141, 432 (1978)

23) A.J. Hanson, H. Römer, Phys. Lett. 80B, 58 (1978)
See e.g.

24) P.B. Gilkey, "Elliptic Operators on Compact Manifolds", Publish or Perish Inc. Boston 1974 and references herin.

25) See e.g. S. Chern, J. Simons, Proc. Nat. Acad. Sci. (USA) 68, 791 (1971); Ann. of Math. (2) 99, 48 (1974) and Milnor, Stasheff, ref. 13)

26) S.L. Adler, Phys. Rev. 177, 2426 (1969)
J.S. Bell, R. Jackiw, Nuovo Cim. 60A, 47 (1969)

27) J.S. Dowker, J. of Phys. A11, 34 (1978)

28) M.T. Grisarn, P. van Nieuwenhuizen, D.Z. Friedemann, Phys. Lett. 71B, 377 (1978)

29) F. Hirzebruch, "Topological Methods in Algebraic Geometry", Grundlagen der Math. Wissenschaften in Einzeldarstellungen Nr. 131, Springer N.Y. 1966.

30) R. Delbowgo, A. Salam, Phys. Lett. 30B, 381 (1972)
T. Eguchi, P.G.O. Freund, Phys. Rev. Lett. 37, 1251 (1976)

31) See e.g. M. Greenberg, "Lectures on Algebraic Topology", W.A. Benjamin, Inc., Amsterdam, New York 1967

32) S. Deser, B. Zumino, Phys. Lett. 62B, 335 (1976)
For a review see e.g. S. Ferrara, Erice Lectures Aug. 1978) CERN TH 2514, an almost complete list of references is contained in "Proceedings of the Supergravity Workshop at Stony Brook, 27.- 29. Sept. 1979, P. van Nieuwenhuizen, D.Z. Friedman eds.

33) S.M. Christensen, M.J. Duff, Phys. Lett 76B, 571 (1978)

34) N.K. Nielsen, Nucl. Phys. B140, 499 (1978); B142, 306 (1978)

35) S.W. Hawking, Phys. Lett. 60A, 81 (1977)

36) T. Eguchi, A.J. Hanson, Phys. Lett. 74B, 249 (1978)

37) V.A. Belinskii, G. W. Gibbons, D.N. Page, C.N. Pope, Phys. Lett. 76B, 433 (1978)

38) N. Hitchin, Advances in Math. 14, 1 (1974)

39) M.F. Atiyah, unpublished

40) S.W. Hawking, C. N. Pope, Nucl. Phys. B146, 381 (1979)

41) G.W. Gibbons, C. N. Pope, H. Römer, Nucl. Phys. B157, 377 (1979)

42) N.K. Nielsen, Nucl. Phys. B151, 536 (1979)

43) H. Römer, Phys. Lett. 83B, 172 (1979)

Topological Concepts in Phase Transition Theory

Mario Rasetti

Istituto di Fisica del Politecnico, Torino, Italy

and

I.N.F.N. e G.N.S.M. Sezioni di Torino, Italy

I. Geometry of Stochastic Processes

The first statements concerning the connection between probability methods - in particular the theory of continuous Markov processes - and the spectral theory of elliptic operators appear in a celebrated paper by Mark Kac [1].

In physics such an idea was especially fertile when it was translated by Feynman [2] into the concept of "integrating along trajectories" as a method for constructing propagators and studying spectra of quantum mechanical observables.

The key point of the method is that a set of suitable measures in the space of trajectories can be constructed from a fundamental solution of a parabolic differential equation (in quantum mechanics the Schrödinger wave equation).

Also implicit in Kac's article is the observation - subsequently more fully appreciated by Singer and McKean [3] and in related further work - that the properties both of the stochastic process and of the solutions of the connected partial differential equation are indeed dependent quite transparently on the global geometrical characterization of the manifold $\mathcal{M}$ on which the process itself takes place.

The link between the usual analytic apparatus of Markov processes [4],[5] and the geometrical and topological properties of $\mathcal{M}$ lies in the fact that the former - at least in the case of "general position" in the sense of Arnol'd [6] - are stably described in terms of an operator, say A, which is a smooth function of the Laplace-Beltrami operator of $\mathcal{M}$[7].

Under suitable, indeed weak, restrictions, the connection remains true when A is a generic elliptic differential operator. Before trying to takle the discussion in fully general terms, let me review briefly the well known example in which $\mathcal{M}$ is a compact d-dimensional Riemannian manifold, and the operator A is simply the Laplacian itself (diffusion process).

$\mathcal{M}$ is characterized by the positive definite metric

$$ds^2 = g_{\alpha\beta}\, du^\alpha du^\beta \tag{1}$$

where u^α ($\alpha = 1,2,\dots,d$) denote the coordinates of the point $x \in \mathcal{M}$. Using the standard notations

$$g = \det g_{\alpha\beta} \tag{2}$$

$$g^{\alpha\gamma} g_{\gamma\beta} = \delta^\alpha_\beta \tag{3}$$

(summation over repeated indices is implied) the volume element at x reads

$$d\sigma_x = \sqrt{g} \prod_{\alpha=1}^{d} du^\alpha \tag{4}$$

and the Laplace-Beltrami operator

$$\Delta = \frac{1}{\sqrt{g}} \frac{\partial}{\partial u^\alpha} \sqrt{g}\, g^{\alpha\beta} \frac{\partial}{\partial u^\beta} \tag{5}$$

It is natural and convenient in the actual calculations, whenever things are studied in proximity of some fixed point $\tilde{x}$, to consider a Euclidean metric $\tilde{g}_{\alpha\beta}$ osculatory to the metric $g_{\alpha\beta}$ at the point $\tilde{x}$ itself. This allows specializing the choice of the coordinates to the so called Riemann's normal coordinates [8].

Such coordinates, say v^{α}, are defined as follows: let v be the shortest distance from $\tilde{x}$ to x, measured naturally along the geodesic line joining them, and let t^{α} be the components (with respect to some fixed Cartesian frame attached to $\tilde{x}$) of a unit vector tangent to this same line at $\tilde{x}$: then by definition simply $v^{\alpha} = v\, t^{\alpha}$.

In terms of the normal coordinates, the metric tensor $g_{\alpha\beta}$ reduces, at the origin $\tilde{x}$, to the Euclidean metric tensor

$$g_{\alpha\beta}(\tilde{x}) = \tilde{g}_{\alpha\beta} = \delta_{\alpha\beta} \quad . \tag{6}$$

Moreover its first-order derivatives $\frac{\partial g_{\alpha\beta}}{\partial v^{\gamma}}$ vanish at $x = \tilde{x}$ which guarantees that indeed the Euclidean space of coordinates v^{α} and metric $\tilde{g}_{\alpha\beta}$ is osculatory to the Riemannian manifold $\mathcal{M}$ at $\tilde{x}$.

At x, the metric tensor $g_{\alpha\beta}$ differs from $\tilde{g}_{\alpha\beta}$ by second and higher order terms in v^{α},

$$g_{\alpha\beta} = \delta_{\alpha\beta} + \frac{1}{3} R_{\alpha\gamma\,\beta\delta}\, v^{\gamma} v^{\delta} + \ldots \quad . \tag{7}$$

Here $R_{\alpha\gamma,\beta\delta}$ is the Riemann-Christoffel curvature tensor at $\tilde{x}$ (naturally expressed in terms of the coordinates v^{α}),

$$R_{\alpha\gamma,\beta\delta} = \frac{\partial}{\partial v^{\delta}} \Gamma_{\alpha,\beta\gamma} + g^{\varepsilon\eta} \Gamma_{\varepsilon,\alpha\beta} \Gamma_{\eta,\gamma\delta} - \{\beta \Leftrightarrow \delta\} \tag{8}$$

$$\Gamma_{\alpha,\beta\gamma} = \frac{1}{2} \left(\frac{\partial g_{\alpha\beta}}{\partial v^{\gamma}} + \frac{\partial g_{\alpha\gamma}}{\partial v^{\beta}} - \frac{\partial g_{\beta\gamma}}{\partial v^{\alpha}} \right) \quad . \tag{9}$$

At the origin $\tilde{x}$ of course these expressions simplify since there $g_{\alpha\beta} = \delta_{\alpha\beta}$ and hence $\Gamma_{\alpha\beta\gamma} = 0$.

In particular contravariant tensors such as

$$R^{\alpha}{}_{\gamma,\beta\delta} = g^{\alpha\varepsilon} R_{\varepsilon\gamma,\beta\delta} \tag{10}$$

or

$$R^{\alpha}{}_{\gamma,}{}^{\beta}{}_{\delta} = g^{\alpha\varepsilon} g^{\beta\eta} R_{\varepsilon\gamma,\eta\delta} \tag{11}$$

have the same components as $R_{\alpha\gamma,\beta\delta}$ at $\tilde{x}$.

$g_{\alpha\beta}$ can then be written as

$$g^{\alpha\beta} = \delta^{\alpha\beta} - \frac{1}{3} \mathcal{R}^{\alpha}{}_{\gamma,}{}^{\beta}{}_{\delta} \, v^{\gamma} v^{\delta} + \dots \tag{12}$$

and

$$g(x) = 1 + \frac{1}{3} \mathcal{R}_{\gamma\delta} v^{\gamma} v^{\delta} + \dots \tag{13}$$

where $\mathcal{R}_{\gamma\delta}$ is the Ricci tensor

$$\mathcal{R}_{\gamma\delta} = \mathcal{R}^{\alpha}{}_{\gamma,\alpha\delta} \quad . \tag{14}$$

In the following we shall also need the Riemannian scalar curvature

$$\mathcal{R} = \mathcal{R}^{\alpha}{}_{\alpha} = g^{\alpha\beta} \mathcal{R}_{\alpha\beta} \quad . \tag{15}$$

In the coordinate system v^{α}, the Laplacian reduces to

$$\Delta = \sum_{\alpha=1}^{d} \left(\frac{\partial}{\partial v^{\alpha}} \right)^2 + \tag{16}$$

$$+ \frac{1}{3} \mathcal{R}^{\alpha}{}_{\gamma,}{}^{\beta}{}_{\delta} v^{\gamma} v^{\delta} \frac{\partial^2}{\partial v^{\alpha} \partial v^{\beta}} + \frac{2}{3} \mathcal{R}_{\alpha}{}^{\beta} v^{\alpha} \frac{\partial}{\partial v^{\beta}} + \dots \quad .$$

If the manifold $\mathcal{M}$ has a boundary $\partial\mathcal{M}$ (which we assume to be a smooth (d-1)-dimensional surface), it is convenient to locate a point x close to $\partial\mathcal{M}$ by a new coordinate set $\{ w, x^{\alpha} \; (\alpha = 2,3,\dots,d) \}$. These are defined as follows. We first associate to x its "projection" μ on $\partial\mathcal{M}$, which is the point $\mu \in \partial\mathcal{M}$ minimizing the geodesic distance from x to μ.

The geodesic line joining x to μ is normal to $\partial\mathcal{M}$ at μ. The first coordinate w of x is but the length of the geodesic segment from x to μ . The remaining (d-1) coordinates x^{α} characterize the position of μ on $\partial\mathcal{M}$.

In this coordinate system (which we denote by (w,)) the element of length[1] takes the form

$$ds^2 = dw^2 + f_{\alpha\beta}(w,\mu) \, dx^{\alpha} dx^{\beta} \tag{17}$$

where $f_{\alpha\beta}(w,\mu)$ are regular functions of all coordinates w and x^{α}, $\alpha = 2,\dots,d$. (Notice that no cross-terms of the form $dw dx^{\alpha}$ enter (17) because the surfaces w = const are perpendicular to the geodesic lines x^{α} = const, $\alpha = 2,3,\dots,d$).

The above choice of coordinates can be further specialized by taking the geometry of $\partial\mathcal{M}$ into account.

One has to recall that a (d-1)-dimensional surface embedded in a d-dimensional space can be characterized by two quadratic forms.

The first of these gives the distance $ds_{\parallel}^2$ between the two points $(\mu, \mu + d\mu)$ on $\partial\mathcal{M}$,

$$ds_{\parallel}^2 = d\mu \cdot d\mu = f_{\alpha\beta}(0,\mu)\, dx^{\alpha} dx^{\beta} \quad . \tag{18}$$

The second form determines the absolute variation Dn_{μ} of the unit normal vector n_{μ} to $\partial\mathcal{M}$ at μ, when μ varies by $d\mu$. The components of Dn_{μ} in our coordinate system are

$$[Dn_{\mu}]_w = \Gamma_{w,w\beta}\, dx^{\beta} = 0 \tag{19}$$

$$[Dn_{\mu}]_{\alpha} = \Gamma_{\alpha,w\beta}\, dx^{\beta} = \frac{1}{2}\frac{\partial f_{\alpha\beta}(0,\mu)}{\partial w}\, dx^{\beta} \tag{20}$$

The second fundamental quadratic form is then

$$-d\mu \cdot Dn_{\mu} = -\frac{1}{2}\frac{\partial f_{\alpha\beta}(0,\mu)}{\partial w}\, dx^{\alpha} dx^{\beta} \quad . \tag{21}$$

Now we can repeat the procedure developed before for a point $\tilde{x}$ interior to $\mathcal{M}$; by considering a fixed point $\tilde{\mu}$ on $\partial\mathcal{M}$ and introducing instead of the (d-1) arbitrary coordinates x^{α} a set of Riemann's normal coordinates y^{α} on $\partial\mathcal{M}$ itself.
The latter can be chosen so as to diagonalize the second fundamental form as well,

$$f_{\alpha\beta}(0,\tilde{\mu}) = \delta_{\alpha\beta} \tag{22}$$

$$\frac{\partial f_{\alpha\beta}}{\partial y^{\gamma}}(0,\tilde{\mu}) = 0 \tag{23}$$

$$-\frac{1}{2}\frac{\partial f_{\alpha\beta}}{\partial w}(0,\tilde{\mu}) = \frac{\delta_{\alpha\beta}}{R_{\alpha}} \tag{24}$$

where R_{α} are the main curvature radii of $\partial\mathcal{M}$ at $\tilde{\mu}$, in terms of which the mean curvature

$$\frac{1}{R} = \frac{1}{d-1} \sum_{\alpha=2}^{d} \frac{1}{R_\alpha} = -\frac{1}{d-1} \frac{\partial}{\partial w} \ln \sqrt{g} \tag{25}$$

can be defined.

In proximity of $\tilde{\mu}$,

$$g_{ww} = g^{ww} = 1 \tag{26}$$

$$g_{w\alpha} = g^{w\alpha} = 0 \tag{27}$$

$$g_{\alpha\beta} = \delta_{\alpha\beta} - \frac{2w}{R_\alpha} \delta_{\alpha\beta} + \dots \tag{28}$$

whence

$$g^{\alpha\beta} = \delta^{\alpha\beta} + \frac{2w}{R_\alpha} \delta^{\alpha\beta} + \dots \tag{29}$$

where $\alpha, \beta = 2,3,\dots,d$ and for the sake of simplicity only terms up to the first order in the coordinates w and y^α have been explicitly written.

(Notice that to this order the eqs. for the metric tensor are exactly the same whether $\mathcal{M}$ is Euclidean or not and at $\tilde{\mu}$

$$ds^2 = dw^2 + \sum_{\alpha=2}^{d} (dy^\alpha)^2 \quad . \tag{30}$$)

To the same order

$$\sqrt{g} = 1 - \frac{d-1}{R} w + \dots \tag{31}$$

and finally

$$\Delta = \frac{\partial^2}{\partial w^2} + \sum_{\alpha=2}^{d} \left(\frac{\partial}{\partial y^\alpha}\right)^2 - \frac{d-1}{R} \frac{\partial}{\partial w} + 2w \sum_{\alpha=2}^{d} \frac{1}{R_\alpha} \left(\frac{\partial}{\partial y^\alpha}\right)^2 + \dots \tag{32}$$

Next order terms would include factors proportional to the second order derivatives and to squares of first order derivatives of the metric tensor taken at $\tilde{\mu}$, and so on for higher order terms.

Now the relevant object to study is the Θ-function of $\mathcal{M}$, namely the Laplace transform of the spectral function of A. In the case under consideration $A = -\frac{1}{2}\Delta$ has a non-negative discrete spectrum of eigenvalues $0 = \lambda_0 < \lambda_1 \le \lambda_2 \le \dots$ and a complete system of corresponding eigenfunctions $\{e_n(x)\}$, $x \in \mathcal{M}$ which are orthonormal with respect to the measure defined by the Riemannian volume element $d\sigma_x$.

The spectral function of A is

$$\mathcal{N}(\lambda) = \sum_{\lambda_n < \lambda} 1 \qquad (33)$$

and thus

$$\Theta(\beta) = \int_0^\infty e^{-\lambda\beta} d\mathcal{N}(\lambda) = \sum_{n=0}^\infty e^{-\beta\lambda_n} \quad . \qquad (34)$$

The reason for considering $\Theta(\beta)$ instead of $\mathcal{N}(\lambda)$ as a primitive object of the theory is that $\mathcal{N}(\lambda)$- which is a sum of step functions - obviously increases by steps of at least one unit with λ (for particular shapes of $\mathcal{M}$ the modes can indeed be many times degenerate, and $\mathcal{N}(\lambda)$ may then increase by much larger steps).

This produces fluctuations in $\mathcal{N}(\lambda)$ of amplitude proportional to $\sqrt{\lambda}$. Such fluctuations give rise to very important physical effects, but mathematically their existence amplifies the difficulty of problems as long as one is interested just in the function $\mathcal{N}(\lambda)$.

The natural way out of this difficulty consists in considering only averaged, smoothed functions, which are sufficient for physical applications. All one has to do is to take proper care that the width of the smoothing function is large enough - both with respect to the distance between eigenvalues and with respect to the period of the fluctuations - that the function $\mathcal{N}(\lambda)$ and its derivative $\rho(\lambda) = \frac{d\mathcal{N}}{d\lambda}$ admit an asymptotic expansion.

A Brownian motion (diffusion process) on $\mathcal{M}$ on the other hand is characterized by a transition function $P(\beta, x, \Gamma)$ - expressing the probability of being at "time" β in a set $\Gamma \subset \mathcal{M}$ starting from a point $x \in \mathcal{M}$ - which is the fundamental solution of the "heat" equation

$$\frac{\partial P}{\partial \beta} + AP = 0 \quad . \qquad (35)$$

By Mercer's theorem, if "initial" conditions $P(0,x,y) = \delta(x-y)$ are assumed P is representable as a uniformly convergent biorthogonal series

$$P(\beta, x, y) = \sum_{n=0}^\infty e^{-\beta\lambda_n} e_n(x) e_n(y) \quad . \qquad (36)$$

It follows that

$$\Theta(\beta) = \int_{\mathcal{M}} P(\beta, x, x)\, d\sigma_x \quad . \qquad (37)$$

The study of the asymptotic behaviour of the transition density P at the "pole" x = y is then equivalent to constructing the decomposition of the Θ function of $\mathcal{M}$[9] (usually referred to as the Minakshisundaram decomposition).

The latter problem has been tackled by several authors (see e.g. Refs. [3] and [9])and the following asymptotic results are known. For large β

$$P(\beta, x, \Gamma) \underset{\beta \to \infty}{\sim} \frac{1}{\sigma(\mathcal{M})} + \mathcal{O}(e^{-\beta\lambda_1}) \tag{38}$$

where

$$\sigma(\mathcal{M}) = \int_{\mathcal{M}} d\sigma_x \tag{39}$$

is the volume of $\mathcal{M}$: i.e. the smallest non-vanishing eigenvalue λ_1 characterizes the speed of convergence of the distribution $P(\beta, x, \Gamma)$ to its final state, characterized by a uniform distribution.

For small β

$$\Theta(\beta) \underset{\beta \to 0}{\sim} \frac{1}{(2\pi\beta)^{d/2}} \left\{ C_0 + \beta C_1 + \beta^2 C_2 + \dots \right\} \quad . \tag{40}$$

The coefficients C_n, due to (16) turn out to be just a global geometrical (and in some instances topological) characterization of $\mathcal{M}$:

$$C_0 = \sigma(\mathcal{M}) \tag{41}$$

equals the volume of $\mathcal{M}$

$$C_1 = -\frac{1}{12} \int_{\mathcal{M}} \mathcal{R}(x)\, d\sigma_x \tag{42}$$

depends on the local Gaussian curvature $\mathcal{R}(x)$ of $\mathcal{M}$ at x.

(Notice that for d = 2, $\frac{1}{2\pi} \int_{\mathcal{M}} \mathcal{R}(x)\, d\sigma_x = E$, the Euler characteristic of $\mathcal{M}$)

$$C_2 = \alpha_d^{(2)} \int_{\mathcal{M}} \mathcal{R}^2(x)\, d\sigma_x \tag{43}$$

$$C_3 = \alpha_d^{(3)} \int_{\mathcal{M}} \mathcal{R}^3(x)\, d\sigma_x + \beta_d^{(3)} \int_{\mathcal{M}} \mathcal{R}\, \Delta \mathcal{R}\, d\sigma_x \quad . \tag{44}$$

The higher C_i's, namely for i=2,3,..., involve integrals of powers of the curvature as well as of its Laplacian, products of the form $\mathcal{R}^{n}(x)[\Delta \mathcal{R}]^{m}$ and $\mathcal{R}^{p}(x)[\Delta \dots \Delta \mathcal{R}]^{q}$, and so on.

The coefficients $\alpha_d^{(i)}, \beta_d^{(i)}, \dots$ $i \geq 2$ on the other hand are universal constants depending only on the dimension of $\mathcal{M}$.

Analogous results hold as well when $\mathcal{M}$ is a manifold-with-boundary.

The connected stochastic process has given boundary conditions, typically Dirichlet boundary conditions $P(\partial\mathcal{M}) = 0$ (in other words it "perishes" on $\partial\mathcal{M}$), or Neumann b.c. $\left.\frac{\partial P}{\partial n}\right|_{\partial\mathcal{M}} = 0$ or mixed b.c. $\left.\left(\frac{\partial P}{\partial n} - \varkappa P\right)\right|_{\partial\mathcal{M}} = 0$, for some given logarithmic derivative $\varkappa$, possibly varying along $\partial\mathcal{M}$.

The main difference with the previous case is that now the local (and again, at least partially, the topological) structure of $\partial\mathcal{M}$ controls the asymptotic expansion of the Θ-function, as could be guessed from eq. (32). Moreover, because of the singularity of $P(\beta,x,x)$ in proximity of $\partial\mathcal{M}$, such an expansion turns out to be, for small β, in steps of $\beta^{1/2}$, as opposed to eq. (40), and it reads

$$\Theta(\beta) \underset{\beta \to 0}{\sim} \frac{1}{(2\pi\beta)^{d/2}} \left\{ \bar{c}_0 + \beta^{1/2} \bar{c}_1 + \beta \bar{c}_2 + \dots \right\} . \tag{45}$$

The coefficients $\bar{c}_i$ have the representation

$$\bar{c}_0 = \sigma(\mathcal{M}) = c_0 \tag{46}$$

$$\bar{c}_1 = \pm \frac{1}{2}\sqrt{\frac{\pi}{2}} \int_{\partial\mathcal{M}} d\sigma_\mu \tag{47}$$

is a measure of the boundary (the $\pm$ signs correspond to Neumann and Dirichlet b.c. respectively; the mixed b.c. would produce additional terms)

$$\bar{c}_2 = c_2 + \frac{d-1}{6} \int_{\partial\mathcal{M}} \frac{d\sigma_\mu}{R_\mu} \tag{48}$$

depends on the local Eulerian curvature of $\partial\mathcal{M}$

$$\bar{c}_3 = \bar{\alpha}_d^{(3)} \int_{\partial\mathcal{M}} \frac{\partial\sigma_\mu}{[R_\mu]^2} + \bar{\beta}_d^{(3)} \int_{\partial\mathcal{M}} \mathcal{R}(\mu)\, d\sigma_\mu \tag{49}$$

involves the Euler characteristic of the boundary. Again the coefficients $\bar{\alpha}_d^{(i)}, \bar{\beta}_d^{(i)}$,..., i $\geq$ 3 are universal non zero algebraic numbers of combinatorial character.

II. Unfolding of Singularities of Stochastic Integrals; ζ Function

In the Feynman sum over histories approach to quantum mechanics on the other hand one considers expressions of the form

$$Z = \int d\{g\} d\{\phi\} \exp[iA(g,\phi)] \tag{1}$$

where $d\{g\}$ is a measure on the space of metrics g of $\mathcal{M}$, $d\{\phi\}$ is a measure normalized on the space of fields ϕ and $A(g,\phi)$ is the action integral. In our case the metric, say g_o, is fixed once for all with a Dirac measure, and we will drop it whenever not strictly necessary.

The functional integration is taken over all the fields ϕ defined over a space-time Σ described in terms of g_o, satisfying a given set of boundary conditions (typically for instance periodic boundary conditions).

Two ways of approaching (1) when compared give an interesting insight into the structure of the problem.

A first approach consists in thinking of the fields as parametrized at each point of space time by a set of parameters $\alpha = \{\alpha_1,\dots,\alpha_k\} \in \mathcal{V} \subset \mathbb{R}^k$.

For purely imaginary time and periodic boundary conditions on ϕ, with period β, eq. (1) can then be written as a locally finite sum over $\mathcal{M}$ of integrals of the form

$$I_{[a,\psi]}(x,\beta) = \left(\frac{\beta}{2\pi}\right)^{k/2} \int_{\mathcal{V}} e^{i\beta\psi(x,\alpha)}\, a(x,\alpha,\beta)\, d\alpha \tag{2}$$

where the phase function ψ is a smooth real-valued function on $\mathcal{M} \times \mathcal{V}$, the manifold $\mathcal{V}$ is open in $\mathbb{R}^k$ for some k and the amplitude $a \in C^\infty(\mathcal{M} \times \mathcal{V} \times \mathbb{R})$.

Also $a(x,\alpha,\beta)$ can be assumed to have an asymptotic expansion in β of the form

$$a(x,\alpha,\beta) \underset{\beta\to\infty}{\sim} \sum_{r=0}^{\infty} a_r(x,\alpha)\, \beta^{\mu-r} \tag{3}$$

with $a_r(x,\alpha) \in C^\infty(\mathcal{M} \times \mathcal{V})$, for some μ.

Indeed it was shown in refs. [10], [11], that the only contributions to $I(x,\beta)$ which are not rapidly decreasing come from the set

$$S_{\psi} = \left\{ (x,\alpha) \in \mathcal{M} \times \vartheta \; ; \; d_{\alpha} \psi(x,\alpha) = 0 \right\} \tag{4}$$

of points where ψ is stationary as a function of α , and an asymptotic expansion for $Z = Z(\beta)$ can then be found, of order β^{ω} for the contribution near $(x,\alpha) \in S_{\psi}$ if $d^2_{\alpha} \psi(x,\alpha)$ is a non-degenerate bilinear form. (Notice that this is where the factor $\left(\frac{\beta}{2\pi}\right)^{K/2}$ in front of the integral is coming from).
At points $(x,\alpha) \in S_{\psi}$ where $d^2_{\alpha} \psi(x,\alpha)$ is degenerate, the procedure may break down, but yet the following analysis is still possible. Suppose a change of integration variables α to $\tilde{\alpha}$ (allowed to depend on x) and a change of x-variables to $\tilde{x}$ (not allowed to depend on α) can be found such that

$$\psi(x(\tilde{x}), \alpha(\tilde{x},\tilde{\alpha})) = \tilde{\psi}(\tilde{x},\tilde{\alpha}) + \vartheta(\tilde{x}) \quad . \tag{5}$$

Let besides

$$b(\tilde{x},\tilde{\alpha},\beta) = a(x(\tilde{x}), \alpha(\tilde{x},\tilde{\alpha}), \beta) \cdot \mathfrak{J}(\tilde{x},\tilde{\alpha}) \tag{6}$$

where

$$\mathfrak{J}(\tilde{x},\tilde{\alpha}) = |\det d_{\tilde{\alpha}} \alpha(\tilde{x},\tilde{\alpha})| \tag{7}$$

is the Jacobian of the change of α -variables. Obviously

$$\mathcal{I}_{[a,\psi]}(x,\beta) = e^{i\beta\vartheta(\tilde{x})} \mathcal{I}_{[b,\tilde{\psi}]}(\tilde{x},\beta) \tag{8}$$

and if $d^2_{\tilde{\alpha}} \tilde{\psi}(\tilde{x},\tilde{\alpha})$ is not degenerate the discussion may safely proceed along the same lines as before.

Eq. (5) is what in the language of Thom defines the equivalence of two functions, ψ and $\tilde{\psi}$, as unfoldings of some function of α .

More precisely two real-valued smooth functions $\psi, \tilde{\psi}$ on $\mathcal{M} \times \vartheta$ are said to be equivalent as unfoldings of a function on ϑ if there exists a diffeomorphism

$$\Pi : (x,\alpha) \longrightarrow (\tilde{x}(x), \tilde{\alpha}(x,\alpha)) : \mathcal{M} \times \vartheta \longrightarrow \mathcal{M} \times \vartheta \tag{9}$$

and a function $\vartheta \in C^{\infty}(\mathcal{M})$ such that

$$\Pi \circ \psi = \tilde{\psi} + \vartheta \quad . \tag{10}$$

In particular a function $\psi \in C^{\infty}(\mathcal{M} \times \mathcal{V})$ is said to be stable as an unfolding if there exists a neighborhood U of ψ in $C^{\infty}(\mathcal{M} \times \mathcal{V})$ for Whitney C^{∞} topology such that ψ and $\tilde{\psi}$ are equivalent as unfoldings of a function on $\mathcal{V}$ for every $\tilde{\psi} \in U$.

Now, if ψ is stable as an unfolding of a function on $\mathcal{V}$, one can prove, using Hironaka's theorem on resolution of singularities and theorems of Duistermaat and Bernshtein [11] [12], the existence of an asymptotic expansion of the form

$$J_{[a,\psi]}(x,\beta) \sim \left(\frac{\beta}{2\pi}\right)^{K/2} \sum_{r=0}^{\infty} \beta^{\mu-r} \sum_{j} \gamma_j(x)\, \beta^{\varepsilon_j} [\ln \beta]^{\eta_j} \tag{11}$$

where $\gamma_j(x)$ are smooth functions of x and the numbers ε_j and η_j are respectively rationals and integers related to monodromy of the singularity.

ψ is stable as an unfolding at (x_o, α_o) if and only if it is locally equivalent to a polynomial of the form

$$\tilde{\psi}(x,\alpha) = f(\alpha) + \sum_{l=1}^{d} x_l f_l(\alpha) \tag{12}$$

where f is a polynomial and the f_l form an arbitrary spanning system of polynomials of degree less than or at most equal to the codimension of f.

The class of equivalence of the germs f of the unfolding ψ and the corresponding f_l has been thoroughly studied by V.I. Arnol'd and his school, who produced a classification of the possible situations [6,10].

An alternative approach to (1) leads, through a similar stationarity technique, to a different asymptotic expansion.

If, as we assumed, time is purely imaginary, a Wick rotation of the time axis will render the argument of the exponential in (1) real, and the integration can be then approximately performed with a saddle point method.

Moreover, if the boundary conditions are once more assumed such that ϕ is periodic in imaginary time over some fixed large enough space boundary with period β, then Z is just the partition function for a canonical ensemble at the temperature $T = \beta^{-1}$.

In order to use a steepest-descend method of integration, let ϕ_o denote the field extremizing the action, and expand the latter in a Taylor series in proximity of such a field:

$$A(\phi) = A(\phi_o) + A_2(\tilde{\phi}) + \dots \tag{13}$$

where

$$\tilde{\phi} = \phi - \phi_0 \tag{14}$$

is the fluctuation of the field, and $A_2(\tilde{\phi})$ is the correction to the extremal action $A(\phi_0)$, quadratic in $\tilde{\phi}$ (higher order terms being at present neglected).
One has

$$\ln Z = i A(\phi_0) + \ln \int d\{\tilde{\phi}\} \exp\left[i A_2(\tilde{\phi})\right] + \ldots . \tag{15}$$

The quadratic term $A_2(\tilde{\phi})$ has in general the form

$$i A_2(\tilde{\phi}) = -\frac{1}{2} i \int_{\Sigma} \tilde{\phi} A \tilde{\phi} \sqrt{-\det g_0}\, dx = -\frac{1}{2}\beta \int_{\mathcal{M}} \tilde{\phi} A \tilde{\phi}\, d\sigma_x \tag{16}$$

where A is a second order differential operator, constructed out of the background field ϕ_0.

Moreover the operator A, if the direction of the wick rotation is suitably chosen, is real, elliptic and self-adjoint.
This implies that it will have a complete spectrum of orthonormal eigenvectors ϕ_n with real eigenvalues λ_n

$$A \phi_n = \lambda_n \phi_n \quad . \tag{17}$$

One can thus express the fluctuation $\tilde{\phi}$ in terms of the set $\{\phi_n\}$

$$\tilde{\phi} = \sum_n (\phi_n, \tilde{\phi}) \phi_n = \sum_n c_n \phi_n \tag{18}$$

whereby the measure $d\{\tilde{\phi}\}$ on the space of all fields $\tilde{\phi}$ can be expressed as

$$d\{\tilde{\phi}\} = \mathcal{N} \prod_n dc_n \tag{19}$$

where $\mathcal{N}$ is some normalization constant.
It follows that

$$\begin{aligned} Z' = Z(\beta) - Z_0 &= \int d\{\tilde{\phi}\} \exp(i A_2(\tilde{\phi})) + \ldots = \\ &\cong \mathcal{N} \prod_n \int dc_n \exp\left(-\tfrac{1}{2}\beta \lambda_n c_n^2\right) = \\ &= \mathcal{N} \prod_n \sqrt{\frac{2\pi}{\beta\lambda_n}} = \frac{\mathcal{N}}{\sqrt{\det\left(\frac{\beta}{2\pi} \cdot A\right)}} \end{aligned} \tag{20}$$

where

$$Z_0 = \exp\left(i A(\phi_0)\right) = \exp\left(-\beta F_0\right) \tag{21}$$

F_o being the unperturbed ground state free energy.
The determinant of the operator A in general diverges because the eigenvalues of A increase without bound, except in the case when one of the eigenvalues (possibly degenerate) is zero. Let's now consider again the heat equation, associated this time with the operator A

$$\frac{\partial}{\partial \beta} P(\beta, x, y) + A P(\beta, x, y) = 0 \quad ; \quad x, y \in \mathcal{M} \tag{22}$$

(Notice that the operator A in (22) is taken to act only on the argument x).
For this equation, according to what was done before, we can construct the Θ-function on $\mathcal{M}$

$$\Theta(\beta) = \mathrm{Tr}\left\{ e^{-\beta A} \right\} \quad . \tag{23}$$

It was noticed by R.T. Seeley [13] first, and successively discussed by M.F. Atiyah, V.K. Patodi and I.M. Singer [14] that in the case we are discussing, the Mellin transform of $\Theta(\beta)$ is rich of information. Indeed

$$\begin{aligned} \frac{1}{\Gamma(s)} \int_0^\infty \beta^{s-1}\, \Theta(\beta)\, d\beta &= \sum_n \lambda_n^{-s} = \\ &= \mathrm{Tr}\left\{ A^{-s} \right\} = \zeta(s) \end{aligned} \tag{24}$$

is but the generalized zeta function of the operator A. The latter has an analytic continuation to the whole s-plane as a meromorphic function of s with isolated simple poles where $s = \frac{1}{2} \dim \mathcal{M} - j$ with $j = 0, 1, \ldots$ is a non-negative integer.
At $s = 0$, $\zeta(s)$ is finite and $\zeta(0)$ can be computed as an explicit integral over the manifold $\mathcal{M}$. The residues of $\zeta(s)$ at the poles are invariants of A and again can be computed as integrals over $\mathcal{M}$ of local quantities depending only on the spectrum of A.

The interesting piece of information however is in the gradient of $\zeta(s)$ at $s = 0$, which reads

$$-\left.\frac{d\zeta}{ds}\right|_{s=0} = \sum_n \ln \lambda_n = \mathrm{Tr}\left\{ \ln A \right\} = \ln \det A \tag{25}$$

whereby one can write

$$\det A = \exp\left(-\frac{d\zeta}{ds}\Big|_{s=0}\right) . \tag{26}$$

Thus in the approximation adopted so far

$$\ln(Z-Z_0) = \frac{1}{2}\left\{\frac{d\zeta}{ds}\Big|_{s=0} - \zeta(0)\ln\left(\frac{\beta}{2\pi}\right)\right\} + \ln\mathcal{N} \tag{27}$$

or equivalently

$$Z(\beta) = Z_0 + \mathcal{N}\exp\left\{\frac{1}{2}\left(\left[\frac{d}{ds} - \ln\left(\frac{\beta}{2\pi}\right)\right]\mathrm{Tr}\left(A^{-s}\right)\right)\Big|_{s=0}\right\}. \tag{28}$$

III. Bifurcation Equation

The discussion of previous sections leads us to a picture in which the partition function on one hand is dominated by the structural stability of some differential operator naturally entering into the theory, through the action integral, and the eigenvalues of such operator on the other hand are strictly related to the global geometric properties of the manifold over which it is defined.

Before we proceed further, it is worth discussing with some more detail a few issues connected with that picture. In a more precise statement the problem we are interested in solving is not a linear problem: indeed the higher order terms so far neglected in the expansion (II.13) play a relevant role, and eq. (II.17) should indeed be written as a nonlinear boundary-value problem

$$A\tilde{\phi} - \lambda\tilde{\phi} + \mathcal{D}\tilde{\phi} = 0 \tag{1}$$

where $\mathcal{D}$ is a nonlinear operator, defined on a subset of the real Hilbert space on which the positive self-adjoint differential operator A is transitive.

The typical nonlinear boundary value problem of the form

$$A\tilde{\phi} + \mathcal{D}\tilde{\phi} = 0 \tag{2}$$

is usually replaced - if A is invertible - by the equivalent equation

$$\tilde{\phi} + A^{-1} \mathcal{D} \tilde{\phi} = 0 \tag{3}$$

which is of the known Hammerstein type.
When the null space of A is non trivial one has however to resort to other methods.

A technique which was introduced with success in connection with several problems in functional analysis by von Neumann [15], Murray [16] and Kato [17] consists in taking the "square root" of the operator A, i.e. in decomposing A as a product $\mathcal{T}\mathcal{T}^*$. This is of course possible since A is positive and self-adjoint, and turns out to be particularly convenient whenever - as in the case we are dealing with - A is an even order differential operator.

In eq. (1) $A - \lambda \mathbb{1}$ of course need not to be a positive operator and the decomposition is not possible in general. Hence we will incorporate the terms $-\lambda \tilde{\phi}$ in the non linear term and consider the resulting equations over a finite-dimensional subspace spanned by the eigenfunctions of A with eigenvalues up to λ.

Let $\mathcal{N}(A)$ and $\mathcal{N}(\mathcal{T}^*)$ denote the null space of A and $\mathcal{T}^*$ respectively, and

$$p = \dim \mathcal{N}(A) = \dim \mathcal{N}(\mathcal{T}^*) \ . \tag{4}$$

If we choose an orthonormal basis of eigenfunctions of A $\{\phi_{01}, \dots, \phi_{0p}, \phi_1, \dots, \}$ with corresponding eigenvalues $\lambda_{01} = \dots = \lambda_{0p} = 0$; $\lambda_i > 0$, $i \geq 1$, with $\lambda_i \leq \lambda_{i+1}$ for $i \geq 1$ and $\lambda_i \to \infty$ as $i \to \infty$, $\mathcal{N}(A)$ is spanned by the set of functions $\{\phi_{0j}, 1 \leq j \leq p\}$.
We now consider, together with A, the differential operator

$$\hat{A} = \mathcal{T}^* \mathcal{T} \tag{5}$$

Clearly $\hat{A}$ is as well positive and self-adjoint, and

$$q = \dim \mathcal{N}(\hat{A}) = \dim \mathcal{N}(\mathcal{T}) \ . \tag{4'}$$

We choose an orthonormal basis $\{\psi_{0j}, 1 \leq j \leq q\}$ for $\mathcal{N}(\hat{A})$.
It is simple matter [18] to show that indeed $\mathcal{N}(A) = \mathcal{N}(\mathcal{T}^*)$ and $\mathcal{N}(\hat{A}) = \mathcal{N}(\mathcal{T})$; and moreover if A is of 2n-th order so is $\hat{A}$.

Also if a set of k $(0 \leq k \leq 2n)$ linearly independent boundary conditions are assigned for ϕ

$$n - k = q - p \quad . \tag{6}$$

Now, for each real number $\lambda > 0$, let

$$\mathcal{E}(\lambda) = \left\{ \tilde{\phi} \mid A\tilde{\phi} = \lambda \tilde{\phi} \right\} \tag{7}$$

and

$$\hat{\mathcal{E}}(\lambda) = \left\{ \tilde{\phi} \mid \hat{A}\tilde{\phi} = \lambda \tilde{\phi} \right\} \quad . \tag{8}$$

Let's assume that $\mathcal{E}(\lambda)$ is not empty, i.e. λ is an eigenvalue of A. For a given arbitrary element $\tilde{\phi}$ of $\mathcal{E}(\lambda)$ let then $\tilde{\psi} = \frac{1}{\sqrt{\lambda}} \tau^* \tilde{\phi}$.

We have

$$\begin{aligned} \hat{A}\tilde{\psi} &= \tau^* \tau \left(\frac{1}{\sqrt{\lambda}} \tau^* \tilde{\phi} \right) = \\ &= \frac{1}{\sqrt{\lambda}} \tau^* (\tau \tau^* \tilde{\phi}) = \frac{1}{\sqrt{\lambda}} \tau^* (\lambda \tilde{\phi}) = \lambda \tilde{\psi} \end{aligned} \tag{9}$$

i.e. $\tilde{\psi} \in \hat{\mathcal{E}}(\lambda)$. In other words the operator $\frac{1}{\sqrt{\lambda}} \tau^*$ maps the eigenspace $\mathcal{E}(\lambda)$ in a one-to-one manner into the eigenspace $\hat{\mathcal{E}}(\lambda)$, whence we may conclude that dim $\mathcal{E}(\lambda) \leq$ dim $\hat{\mathcal{E}}(\lambda)$.

By simple interchange of the roles of A and $\hat{A}$, we can show in the same way that if $\hat{\mathcal{E}}(\lambda)$ is not empty then the operator $\frac{1}{\sqrt{\lambda}} \tau$ maps $\hat{\mathcal{E}}(\lambda)$ in a one-to-one way into $\mathcal{E}(\lambda)$, so that dim $\hat{\mathcal{E}}(\lambda) \leq$ dim $\mathcal{E}(\lambda)$.

It follows that the positive eigenvalues of $\hat{A}$ and A are the same, and they occur with the same multiplicities.
Therefore the functions

$$\psi_i = \frac{1}{\sqrt{\lambda_i}} \tau^* \phi_i \tag{10}$$

are orthonormal eigenfunctions of $\hat{A}$, with eigenvalues λ_i

$$\hat{A}\psi_i = \lambda \psi_i \qquad i \geq 1 \tag{11}$$

$$(\psi_i, \psi_j) = \frac{1}{\sqrt{\lambda_i \lambda_j}} (\tau^* \phi_i, \tau^* \phi_j) = $$
$$= \frac{1}{\sqrt{\lambda_i \lambda_j}} (\tau \tau^* \phi_i, \phi_j) = \sqrt{\frac{\lambda_i}{\lambda_j}} (\phi_i, \phi_j) = \delta_{ij} \tag{12}$$

and the orthonormal basis of eigenfunctions of $\hat{A}$ is given by $\{\psi_{01}, \dots, \psi_{0q}, \psi_1 \dots \psi_k \dots\}$ with corresponding eigenvalues $\hat{\lambda}_{01} = \hat{\lambda}_{02} = \dots = \hat{\lambda}_{0q} = 0$, $\lambda_i > 0$, $i \geq 1$.

Let us denote now by $\mathcal{P}$ the projection operator over the subspace spanned by the first m eigenfunctions of A:

$$\mathcal{P}\tilde{\phi} = \sum_{j=1}^{p} (\tilde{\phi}, \phi_{0j}) \phi_{0j} + \sum_{i=1}^{m} (\tilde{\phi}, \phi_i) \phi_i \tag{13}$$

and by $\mathcal{H}$ the diagonal operator corresponding to A^{-1} defined over the portion of the Hilbert space of A complementing $\mathcal{N}(A)$

$$\mathcal{H} \phi_i = \frac{1}{\lambda_i} \phi_i \qquad i \geq 1 \tag{14}$$

$$\mathcal{H} \tilde{\phi} = \sum_{i=1}^{\infty} \frac{1}{\lambda_i} (\tilde{\phi}, \phi_i) \phi_i \quad . \tag{15}$$

The following identities are straight forward

$$(\mathbb{1} - \mathcal{P}) \mathcal{H} (\mathbb{1} - \mathcal{P}) A \tilde{\phi} = (\mathbb{1} - \mathcal{P}) \tilde{\phi} \tag{16}$$

$$A \mathcal{P} \tilde{\phi} = \mathcal{P} A \tilde{\phi} \quad . \tag{17}$$

Suppose now $\tilde{\phi}$ is a solution of the nonlinear problem (2). Applying $\mathcal{K} = (\mathbb{1} - \mathcal{P}) \mathcal{H} (\mathbb{1} - \mathcal{P})$ to both sides of (2) and using (16) we get

$$(\mathbb{1} - \mathcal{P}) \tilde{\phi} + \mathcal{K} \mathcal{O} \tilde{\phi} = 0 \tag{18}$$

or equivalently

$$\tilde{\phi} + \mathcal{K} \mathcal{O} \tilde{\phi} = \mathcal{P} \tilde{\phi} \quad . \tag{19}$$

Thus any solution of (2) is also a solution of (19). On the other hand, suppose $\tilde{\phi}$ is a solution of (19). Applying A to both sides of (19) [and noticing that $\mathcal{P}(\mathbb{1}-\mathcal{P}) = \mathbb{O}$, whence (16) implies that $A\mathcal{K}\tilde{\phi} = (\mathbb{1}-\mathcal{P})\tilde{\phi} = \mathcal{K}A\tilde{\phi}$] we get

$$A\tilde{\phi} + (\mathbb{1}-\mathcal{P})\,\mathcal{Q}\tilde{\phi} = A\mathcal{P}\tilde{\phi} \tag{20}$$

or, equivalently, by (17)

$$A\tilde{\phi} + \mathcal{Q}\tilde{\phi} = \mathcal{P}(A\tilde{\phi} + \mathcal{Q}\tilde{\phi}) \quad . \tag{21}$$

Hence any solution of (19) is a solution of (2) if and only if

$$\mathcal{P}(A\tilde{\phi} + \mathcal{Q}\tilde{\phi}) = 0 \tag{22}$$

Thus the nonlinear problem (2) is reduced to the equivalent system of eqs. (19) and (22).

Let now Σ_m be the finite dimensional subspace given by $\mathcal{N}(A) \cup \{\phi_j \mid 1 \leq j \leq m\}$, and let $\phi_* \in \Sigma_m$ be an arbitrary element. By definition (13) of course $\mathcal{P}\phi_* = \phi_*$.
If the equation

$$\tilde{\phi} + \mathcal{K}\mathcal{Q}\tilde{\phi} = \phi_* \tag{23}$$

has at least a solution, then for such a solution $\mathcal{P}\tilde{\phi} = \phi_*$.
Hence any solution of (23) is a solution of (19).
In this case eq. (22) reads

$$\mathcal{P}A\tilde{\phi} + \mathcal{P}\mathcal{Q}(\mathbb{1} + \mathcal{K}\mathcal{Q})^{-1}\phi_* = 0 \tag{24}$$

which becomes, using (17)

$$A\phi_* + \mathcal{P}\mathcal{Q}(\mathbb{1} + \mathcal{K}\mathcal{Q})^{-1}\phi_* = 0 \quad . \tag{25}$$

If (25) has a solution $\phi_* \in \Sigma_m$, then any corresponding solution of (23) is a solution to (2).
Eqs. (23) and (25) are the two equations equivalent to eq. (2): they are referred to as the **auxiliary** and the bifurcation equations of the nonlinear boundary value problem respectively, and are of special interest when m = 0, i. e. $\Sigma_m = \mathcal{N}(A)$ (even though m can be any integer ≥ 0).

The auxiliary equation (23) is of the Hammerstein type, and it has been shown under fairly general assumptions [19] - essentially amounting to the requirement that the operator $\mathcal{D}$ be continuous for some topology, and bounded (i.e. mapping bounded sets into bounded sets) - that it has at least one solution. The solution is unique if moreover $\mathcal{D}$ is monotone. Thus everything seems to be indeed controlled by the bifurcation equation (25).
The latter is characterized by the nonlinear operator

$$\mathcal{B} = \mathcal{P}\mathcal{D}(\mathbb{1} + \mathcal{K}\mathcal{D})^{-1} \qquad (26)$$

If m = 0 then the bifurcation equation reduces simply to

$$\mathcal{B}\phi_* = 0 \qquad (27)$$

and it is sufficient that $\mathcal{B}$ be monotone to guarantee the existence of at least one solution. In fact, if $\mathcal{B}$ is strictly monotone (i.e. $(\mathcal{B}\tilde{\phi} - \mathcal{B}\tilde{\psi}, \tilde{\phi} - \tilde{\psi}) = 0$ implies $\tilde{\phi} = \tilde{\psi}$) such a solution is unique.

We have here a hint that, at least for the case m = 0, a pathological situation may occur in the global solution of the eq. (2) and therefore in the analytic properties of $Z(\beta)$, when β is such that the global conditions characterizing $\mathcal{B}$ (notice that these include topological properties, mainly the invariance under homotopy [20]) change.

There is also an evidence that the indicator of such a pathology must have a global geometrical significance, and give a measure of the asymmetry of the spectrum of A, namely emphasize the possibility that A - still remaining self-adjoint and elliptic - be no longer positive.

Following Atiyah this again points out to the ζ function (or its generalizations) as the quantity both indicating and classifying what is going on [21] [22].

IV. Thermodynamic States and their Evolution

We try now to construct a physical setting in which we can embed these concepts.
The physical objects we are dealing with are many component thermodynamical systems, at equilibrium.
The latter are characterized on one hand by a many-body hamiltonian

H, describing the microscopic interactions, an by a Hilbert space $\mathcal{H}$ - on which H is transitive - spanning the manifold variety of finely structured microscopic states; and on the other hand by a set of macroscopic parameters usually a pair of strictly thermodynamical variables, such as temperature, pressure, volume, entropy (indeed the choice of the pair is simply dictated by whose thermodynamical potential - free energy, internal energy, enthalpy or Gibb's potential - stationarity implies equilibrium) and a finite set of external control parameters, such as applied fields of force, chemical potentials and so on defining a much coarser space of thermodynamic states.

The laws of thermodynamics provide just the relationship between such two representations, and our first problem is then stating as rigorously as possible the mathematical framework for this connection which allows the application of the concepts discussed in previous sections.

A hint for doing this in a convenient form comes from Wightman's theorem[23].

This asserts that a unique axiomatic quantum field theory can be constructed when the (equal time) vacuum expectation values of all the n-field operator products are assigned.

Such a theorem provides the ground for the observation, originally formulated by Matsubara[24] years ago that the statistical average of any dynamical observable, represented by an operator $\mathcal{O}$ acting in $\mathcal{H}$ over some ensemble, namely

$$\langle \mathcal{O} \rangle = \frac{\mathrm{Tr}\{\mathcal{O}\exp[-\beta H]\}}{\mathrm{Tr}\{\exp[-\beta H]\}} = \frac{1}{Z(\beta)}\,\mathrm{Tr}\{\mathcal{O}\exp[-\beta H]\} \qquad (1)$$

(notice that if one adopts the grand cancnical ensemble, H in eq. (1) must indeed be assumed to be the total Hamiltonian of the system minus μN , where μ is the chemical potential and N the number operator), has properties strictly similar to the vacuum expectation value of $\mathcal{O}$ in a quantum field theory.

Matsubara's observation was recently revived by Takahashi and Umezawa[25], who proposed a constructive technique whereby a finite temperature quantum field theory can be obtained in which temperature dependent vacuum states - which we denote by $|\beta,\omega\rangle\rangle$, where ω is the set of parameters which, together with the inverse temperature β , characterize the thermodynamic state - realize just the requirement that the expectation value

$$\frac{\langle\langle \beta,\omega | \mathcal{O} | \beta,\omega \rangle\rangle}{\langle\langle \beta,\omega | \beta,\omega \rangle\rangle} = \langle \mathcal{O} \rangle \quad . \tag{2}$$

The construction of $|\beta,\omega\rangle\rangle$ as proposed in ref. [25] has a few ambiguities in it, which are rigorously taken care of in the following approach.

Let's first introduce the differentiable manifold Ω, such that $\omega \in \Omega$.

To each point ω of Ω, we attach a linear frame

$$L(a;\omega) = \Big\{ \langle n,\omega | a_n \;\Big|\; \langle n,\omega| \in \mathcal{H} \, , \, n \in \mathbb{Z} \, ; \; a_n \in \mathbb{R} \, , \sum_{n=0}^{\infty} a_n^2 < \infty \Big\} \tag{3}$$

where the curly brackets denote an ordered set and the vectors in $\mathcal{H}$, $\langle n,\omega|$ are eigenbras of H, with eigenvalues $\{E_n\}$

$$\langle n,\omega | H = E_n \langle n,\omega | \; ; \quad \langle n,\omega | n',\omega \rangle = \delta_{nn'} \quad . \tag{4}$$

One can think of L(a;ω) as a "hyper-vector" whose entries are themselves vectors in $\mathcal{H}$. Therefore it spans a Hilbert space at ω, isomorphic to $\mathcal{H}$, which we denote by $\mathcal{E}_\omega$. Out of all the linear frames L(a;ω) at all points $\omega \in \Omega$ one can construct a product space which is simply the set of pairs

$$L(\Omega) = \{ (\omega, L(a;\omega)) \mid \omega \in \Omega \} \quad . \tag{5}$$

L(Ω) is transitive with respect to a general linear group of transformations in the following sense.

Let

$$\pi : L(\Omega) \longrightarrow \Omega \; : \; L(a;\omega) \longrightarrow \omega \tag{6}$$

be the mapping which maps the linear frame at ω into ω, and let GL($\mathcal{E}_\omega$) denote the linear group of automorphisms of $\mathcal{E}_\omega$.

It is quite evident that GL($\mathcal{E}_\omega$) acts on L(Ω). Indeed, if A $\in$ GL($\mathcal{E}_\omega$) and L(a,ω) is any linear frame, L(aA,ω) is a linear frame as well,

$$L(aA;\omega) = \Big\{ \sum_{m=0}^{\infty} \langle m,\omega | a_m A_{mn} \;\Big|\; n \geq 0 \Big\} = \tag{7}$$

$$= L(a';\omega) = \{ <\widetilde{u,\omega}|a'_u \quad | u \geq 0 \} \tag{7}$$

since of course

$$<\widetilde{u,\omega}\,|\,a'_u = \sum_{m=0}^{\infty} <m,\omega|a_m A_{mu} \in \mathcal{H} \quad . \tag{8}$$

Notice that only the identity element of GL($\mathcal{E}_\omega$) leaves any point of L(Ω) fixed, and hence the action of GL($\mathcal{E}_\omega$) over L(Ω) is effective.

We have then a bundle of frames over Ω, i.e. a principal GL($\mathcal{E}_\omega$)-bundle, identified by the triplet

$$\mathcal{B} = (L(\Omega), \pi, \Omega) \quad . \tag{9}$$

Out of this bundle we may construct another principal bundle by considering any subgroup of GL($\mathcal{E}_\omega$) and defining the corresponding reduced bundle.

The procedure is particularly interesting if one chooses as a subgroup K of GL($\mathcal{E}_\omega$) the subgroup which, in the representation where A $\in$ GL($\mathcal{E}_\omega$) corresponds to a matrix with elements $\{A_{mn}\}$, consists of real diagonal matrices

$$K \ni \varkappa = \{ \varkappa_{m,u} = \varkappa_u \delta_{mu} | \varkappa_u \in \mathbb{R}; u \geq 0; \sum_{u=0}^{\infty} \varkappa_u^2 < \infty \} \tag{10}$$

K of course acts effectively on L(Ω) and the reduced sub-bundle we are looking for is but the principal K-bundle

$$\tilde{\mathcal{B}} = \{ L(\Omega), \tilde{\pi}, L(\Omega)/_K \} \quad . \tag{11}$$

Notice that the quotient space of L(Ω) by the equivalence relation induced by K is actually isomorphic to Ω (K is not a closed subgroup of GL($\mathcal{E}_\omega$)).
Now the reduced map

$$\tilde{\pi} : L(\Omega) \longrightarrow L(\Omega)/_K \tag{12}$$

associates to each K-equivalence class in L(Ω) a point ω in Ω, and can be identified as the canonical map of the bundle.
Eqs. (4) now suggest that

$$H = \sum_{n=0}^{\infty} a_n |n,\omega\rangle\langle n,\omega| = \sum_{n=0}^{\infty} a_n \mathcal{P}_n \tag{13}$$

where $\mathcal{P}_n$ is the projection operator onto the subspace of $\mathcal{H}$ corresponding to the eigenvalue E_n of H.
Obviously $\mathcal{P}_n \mathcal{P}_m = \mathcal{P}_n \delta_{nm}$ whence $[\mathcal{P}_n, \mathcal{P}_m] = 0$ for all n,m and finally $[H, \mathcal{P}_n] = 0, \ \forall\, n \geq 0$.
We can then construct

$$g = \exp\{-\tfrac{1}{2}\beta H + \sum_{n=0}^{\infty} b_n \mathcal{P}_n\} \tag{14}$$

where $b_n \in \mathbb{R}$; $\sum_{n=0}^{\infty} b_n^2 < \infty$ and identify it as an element of the infinite dimensional Lie group G, whose generators are $\{H, \mathcal{P}_n \mid n \geq 0\}$.
K as well on the other hand can be thought of as a Lie group of transformations of $L(\Omega)$ at ω, and in the same representation we can identify

$$\varkappa = \exp\{\sum_{n=0}^{\infty} \varkappa_n \mathcal{P}_n\} \in K \quad . \tag{15}$$

It follows in a straightforward manner that K acts on G

$$\varkappa g = g' = \exp\{-\tfrac{1}{2}\beta H + \sum_{n=0}^{\infty} (\varkappa_n + b_n)\mathcal{P}_n\} \in \mathcal{G} \tag{16}$$

and we may associate to the bundle $\tilde{\mathcal{B}}$ a further fibre-bundle $\Theta(\mathcal{G})$.
The action of K on $L(\Omega) \times G$

$$\varkappa : (x, g) \longrightarrow (x\varkappa, \varkappa^{-1} g)\ ; \ \varkappa \in K,\ g \in \mathcal{G},\ x \in L(\Omega) \tag{17}$$

defines a K-structure on $L(\Omega) \times G$.
Moreover the map by which the above action is restricted to the quotient

$$\hat{\pi} : L(\Omega) \times \mathcal{G} \longrightarrow L(\Omega)\mathcal{G} \tag{18}$$

defines a quotient space

$$\Theta = \hat{\Pi}((L(\Omega) \times \mathcal{G})\,K) \tag{19}$$

which is isomorphic to $(L(\Omega) \times G) \bmod K$.
The commutativity of the diagram

$$
\begin{array}{ccc}
L(\Omega) & \xleftarrow{\hat{P}} & L(\Omega) \times G \\
\pi \big\downarrow & & \big\downarrow \hat{T} \\
\Omega & \xleftarrow[p]{} & \textcircled{H}
\end{array}
\qquad (20)
$$

whereby the map $p: \textcircled{H} \longrightarrow \Omega$ is factorized as the composition $L(\Omega) \times G \xrightarrow{\hat{p}} L(\Omega) \xrightarrow{\pi} \Omega$ by the projection $L(\Omega) \times G \xrightarrow{\hat{\pi}} \textcircled{H}$ shows that $\textcircled{H} = \textcircled{H}(G)$ is indeed a fibre-bundle over Ω (as base space) with standard fibre G (to be considered as a K-module), the manifold of microscopic states $\mathcal{H}$ as a total space and associated principal K-bundle $\tilde{\mathcal{B}}$ (so that K is indeed the structure group of the fibre bundle $\textcircled{H}$).

Finally for every $\omega \in \Omega$, the fibre $p^{-1}(\omega)$ is a space homeomorphic to the group G.

"The points" of such space, i.e. the cross sections of the fibre $p^{-1}(\omega)$, are the objects (hypervectors in our nomenclature)

$$[\![\beta, \omega, \varkappa]\!] = n \, L(\mathbb{1}, \omega) \exp\left(-\tfrac{1}{2} \beta H + \sum_{n=0}^{\infty} \varkappa \, \Theta_n \right) \qquad (21)$$

where $n = n(\beta, \omega)$ is a normalization factor, and $L(\mathbb{1}, \omega) = \{ \langle n, \omega | \mid n \geq 0 \}$ [notice that $L(\mathbb{1}, \omega) = \lim_{a_n \to 1,\, n \geq 0} L(a, \omega) \notin L(\Omega)$ whereas $[\![\beta, \omega, \varkappa]\!] \in L(\Omega)$.
Since the entire previous discussion is not changed, we can include this particular frame in the definition of $L(\Omega)$ replacing the last condition on the r.h.s. of (3) with $\sum_{n=0}^{\infty} a_n^2 \leq \sum_{n=0}^{\infty} 1$]

We will refer to $L(\mathbb{1}, \omega)$ as the infinite temperature frame, since it corresponds to the choice $\beta = 0$ (i.e. $T \longrightarrow \infty$), $\varkappa_n = 0 \;\; \forall n \geq 0$.

The family of sections corresponding to the identity element e_k of K, namely $\varkappa_n = 0$, $\forall n \geq 0$ gives the temperature dependent vacuum states seeked for

$$| \beta, \omega \rangle\!\rangle = [\![\beta, \omega, e_k]\!] \quad . \qquad (22)$$

It is now straightforward checking that these indeed realize the requirement (2). Also, the infinite-temperature state coincides, up to a normalization constant, with the infinite-temperature frame

$$|0\rangle\!\rangle = |0,\omega\rangle\!\rangle = n(0,\omega)\, L(\mathbb{1},\omega) \quad , \quad \omega \in \Omega \quad . \tag{23}$$

The interesting feature of the above constructions which lead us to the identification of the thermodynamic states as a suitable cross section of a fibre-bundle Ⓗ(G) stays in the possibility it fosters of realizing the relationship between any (β , ω) state and another in terms of the usual morphisms of the fibre-bundle.

In our case the topology of K - the group acting as automorphism of the fibre G - is totally disconnected (indeed K, as defined in (15) is immediately seen to be abelian) and the fibre itself is an abelian group.

Thus Ⓗ is indeed a bundle of coefficients [26], and the fibre bundle morphism is derivable in a relatively simple way from the cohomology structure (indeed the cohomology group homeomorphisms) of the base space.

In physical terms the bundle morphism can be easily seen to be a generalized Bogolubov **transformation**.[22]

The latter can be explicitly written (at least as long as "evolution" in inverse temperature is concerned) in the representation adopted so far.

To check this, let's first observe that the thermo**dynamical** state varies indeed with β according to an equation strictly reminiscent of Bloch's equation

$$\frac{\partial}{\partial\beta} |\beta,\omega\rangle\!\rangle = D\, |\beta,\omega\rangle\!\rangle \tag{24}$$

where

$$D = D(\beta,\omega) = -\frac{1}{2}\left(H - \langle H\rangle\right) \tag{25}$$

with

$$\langle H\rangle = -\frac{\partial}{\partial\beta} \ln Z(\beta) \tag{26}$$

representing the internal energy ($Z(\beta)$ is of course the partition function for the given ensemble).

In eqs. (24) to (26), $|\beta,\omega\rangle\!\rangle$ has been assumed normalized to unit norm.

The solution of (24) however is cast into an especially simple form if one gives up normalization requirement (recall that $|\beta,\omega\rangle\rangle$ is - for every β - to be the vacuum state of a suitable quantum field theory, and it is perfectly correct normalizing it in different ways for different β's). It emerges from our previous discussion that, if one fixes the scale of energy in such a way that the physical vacuum has zero energy, $E_o = 0$, then

$$|0,\omega\rangle\rangle = \sqrt{Z(\infty)}\,|\infty,\omega\rangle\rangle \tag{27}$$

(notice that $\beta = \infty$ implies zero temperature).

Thus an interesting representation for the thermodynamic state seems to be given by the vector $\sqrt{Z(\beta)}\,|\beta,\omega\rangle\rangle = |\mathcal{I}\rangle\rangle_\beta$. In such a representation the infinite temperature state is given by $|\mathcal{I}\rangle\rangle_o = \sqrt{Z(0)}\,|0,\omega\rangle\rangle$. By selecting

$$n(\beta,\omega) = [Z(\beta)]^{-1/2} \tag{28}$$

the infinite temperature state turns out to coincide with the unnormalized infinite temperature frame.
The latter has several advantages, the most relevant of which is that it is invariant under a class of non-unitary transformations conserving the number of particles in the system, and is independent of the representation.
Moreover every thermodynamic state can be obtained from it by a transformation of the form

$$|\mathcal{I}\rangle\rangle_\beta = \exp\left\{-\tfrac{1}{2}\beta H\right\}|\mathcal{I}\rangle\rangle_o \tag{29}$$

for which the analogous of eq. (25) is simply

$$\frac{\partial}{\partial\beta}|\mathcal{I}\rangle\rangle_\beta = -\tfrac{1}{2}H|\mathcal{I}\rangle\rangle_\beta \quad . \tag{30}$$

Notice that in terms of $|\mathcal{I}\rangle\rangle_o$, the normalized state $|\beta,\omega\rangle\rangle$ reads

$$|\beta,\omega\rangle\rangle = \exp\left\{-\tfrac{1}{2}S_\beta\right\}|\mathcal{I}\rangle\rangle_o \tag{31}$$

where

$$S_\beta = \beta H + \ln Z(\beta) \tag{32}$$

is an operator whose expectation value $\langle S_\beta \rangle$ is (in units such that $K_B = 1$) the entropy of the physical system.
Also

$$|\mathfrak{I}\gg_0 = \exp\{iB\}\,|0,\omega\gg \tag{33}$$

where B is the Bogolubov transformation connecting the infinite temperature state to the zero temperature one. $\sqrt{Z(\infty)/Z(0)}\,e^{iB}$ is indeed an involutory duality transformation, which is explicitly known for a wide class of statistical systems.
(31) reads then

$$|\beta,\omega\gg = \exp\{-\tfrac{1}{2}S_\beta\}\exp\{iB\}\,|0,\omega> \tag{34}$$

expressing what we referred to as generalized Bogolubov transformation. In general $[H,B] \neq 0$ and (34) is a rather complicated operatorial equation to deal with; however all the features of thermodynamic equilibrium are contained in eqs. (29) and (30) and we may restrict our attention to them assuming as a reference equilibrium state, whereby all other states can be constructed through (29), just $|\mathfrak{I}\gg_0$ as given by (33).
Finally notice that

$$\begin{aligned} Z(\beta) &= {}_\beta\!\ll\mathfrak{I}\,|\,\mathfrak{I}\gg_\beta = {}_0\!\ll\mathfrak{I}\,|\,e^{-\beta H}\,|\,\mathfrak{I}\gg_0 \\ &= Z(\beta)\;{}_0\!\ll\mathfrak{I}\,|\,e^{-S_\beta}\,|\,\mathfrak{I}\gg_0 = <0,\omega|\,\vartheta\,|0,\omega> \end{aligned} \tag{35}$$

where

$$\vartheta = e^{-iB}\,e^{\beta H}\,e^{iB} \tag{36}$$

V. The Functor "index" as Indicator of a Phase Transition

Let's focus now our attention on the operator τ such that

$$e^{-\beta\tau^*\tau} = \frac{1}{\sqrt{Z(\beta)}}\,e^{-\frac{1}{2}\beta H} = e^{-\frac{1}{2}S_\beta} \tag{1}$$

τ, or more precisely its closure $\mathcal{T}$ over the Hilbert space $\mathcal{H}$ when some boundary conditions are assigned, is the object characteriz-

ing in global terms the entire structure.
In general $\mathcal{T}$ is defined over some field $\mathcal{M}$, endowed with a topology induced both by the physical boundary conditions and by the structure of the Lie algebra of the invariance group $\mathcal{G}$ (if any) of the Hamiltonian.
$\mathcal{M}$ is then usually a compact real manifold, transitive under

$$\mathcal{G}(\sigma) = \bigotimes_{x \in \mathcal{M}} \mathcal{G}_x \tag{2}$$

where $\mathcal{G}_x$ is a copy of $\mathcal{G}$ associated to each site $x \in \mathcal{M}$, and σ = volume ($\mathcal{M}$) is the volume occupied by the system. In typical applications $\mathcal{G}$ is either an abelian group (or more precisely a free product of abelian groups) or the semi-direct product of a rotation group $SO(n)$ by an abelian group ("spin" system).
The measure induced as Gibbs measure in such a case is the Haar measure of the homogeneous space $\mathcal{G}/K$, where K is the isotropy subgroup of a point in the unit n-sphere. $\mathcal{T}$ is then given in terms of elements of the group ring.
Moreover we can safely assume $\mathcal{M}$ has always the homotopy type of a finite CW-complex.
Eq. (IV. 30) shows that the problem we are facing is a global one, namely that of giving conditions ensuring the global existence of an algorythm whereby the generalized Bogolubov transformation be reduced to the direct sum of transformations of lower order in $\mathcal{E}_\omega$, $\omega \in \Omega$.
The latter problem in the context described above is a K-theoretical problem[22].
Without entering in the details of the discussion here, let us only recall that its solution points out to be the topological invariants over the reduced ring of bundles in $\mathcal{M}$ as the objects signalling the failure of the required conditions (i.e. the loss of stability of the bundle themselves).
$\mathcal{M}$ is compact and for the sake of simplicity we assume it C^∞ [we pointed out in previous section that the cohomology properties of the manifold control our entire analysis: now the relation between cohomology properties of a field configuration on a continuous ambient space and its analog over a lattice is quite obvious, and it amounts essentially to translating cohomology into homotopy. There is therefore no loss of generality in our assumption].
In such a case $\mathcal{T}$ has the transmission property, namely it maps $C^\infty(E')$ onto $C^\infty(E'')$, where E' and E'' denote the smooth (vector) bundles over $\mathcal{M}$.

In particular $\mathcal{T}$ maps smooth sections of the bundle E' into smooth sections of E''.
Thus $\mathcal{T}$ is in general a pseudo-differential operator. Moreover dim E' = dim E'', and for any <u>local</u> representation the symbol S of $\mathcal{T}$ is invertible (i.e., it is a non-singular matrix).
Hence $\mathcal{T}$ is elliptic.
The theory of elliptic operators[27] guarantees that both the kernel and the cokernel of $\mathcal{T}$ are finite dimensional, so that its Atiyah-Singer index[28]

$$\begin{aligned}\text{index}\,\mathcal{T} &= \dim\ker\mathcal{T} - \dim\operatorname{coker}\mathcal{T} = \\ &= \dim\ker\mathcal{T} - \dim\ker\mathcal{T}^*\end{aligned} \tag{3}$$

- where $\mathcal{T}^*$ denotes the formal adjoint of $\mathcal{T}$ whith respect to some <u>global</u> hermitian inner product in E, say $(\cdot , \cdot)_E$ - is well defined.
When $\partial\mathcal{M}$ is empty ellipticity implies the Fredholm property, hence index $\mathcal{T}$ - which is one (and the most relevant) of the topological invariants we are looking for - is essentially the Euler characteristics.
When $\partial\mathcal{M} \neq 0$, ellipticity does not imply the Fredholm property, which can be ensured only by a suitable choice of the boundary conditions.
Let's then consider the operator

$$T_E = \mathcal{T}^* \mathcal{T} \tag{4}$$

(essentially the operator describing up to an additive factor proportional to the identity, $\frac{1}{2}$ of the Hamiltonian equipped though with the proper boundary conditions) and denote by $\Gamma_\lambda(E)$ the eigenspace of T_E on E associated with the real eigenvalue λ

$$\Gamma_\lambda(E) = \{ \phi \in E \mid T_E \phi = \lambda \phi \} \quad . \tag{5}$$

The countable sequence of such subspaces [for $\mathcal{M}$ compact $\Gamma_\lambda(E) = 0$ except for a discrete set of non negative λ's], gives an orthogonal direct sum decomposition of the Hilbert space $\mathcal{H}(E)$ obtained from E by completion relative to $(\cdot , \cdot)_E$

$$\mathcal{H}(E) = \oplus \Gamma_\lambda(E) \tag{6}$$

Now for $\mathcal{T} : E' \longrightarrow E''$ the Hodge theorem states the two following important facts:

i) for $\lambda > 0$, $\mathcal{T} : \Gamma_\lambda(E') \longrightarrow \Gamma_\lambda(E'')$ is an isomorphism

ii) for $\lambda = 0$, $\Gamma_0(E') \approx \operatorname{Ker} \mathcal{T}$ and
$\Gamma_0(E'') \approx \operatorname{Ker} \mathcal{T}^* = \operatorname{coker} \mathcal{T}$.

It follows that eq. (3) can be written

$$\operatorname{index} \mathcal{T} = \dim \Gamma_0(E') - \dim \Gamma_0(E'') \quad . \tag{7}$$

Let introduce now together with T_E, the operator obtained from it by self-adjoint continuation over the double of $\mathcal{M}$, namely

$$T_{E'}^* = \mathcal{T}\mathcal{T}^* = T_{E''} \quad . \tag{8}$$

In a neighborhood of $\partial\mathcal{M}$, T_E and T_E^* are isomorphic via the symbol s (and such an isomorphism can actually be lifted along the conormal to $\partial\mathcal{M}$ to the entire manifold). Moreover, as we showed in sect. III, having a discrete spectrum with finite multiplicities their nonzero eigenvalues coincide (and $\phi \to \mathcal{T}\phi$, $\phi \in \Gamma_\mu(E)$ defines an isomorphism of the μ eigenspace $\Gamma_\mu(E)$ of T_E into that of T_E^*, with inverse $\phi \to \frac{1}{\mu}\mathcal{T}^*\phi$) and the null space of T_E coincides with that of $\mathcal{T}$, whereas the null space of T_E^* coincides with that of $\mathcal{T}^*$.
The series

$$Z_\beta(T_E) = \sum_\lambda e^{-\beta\lambda} \dim \Gamma_\lambda(E) = \operatorname{Tr}\left\{ e^{-\beta T_E} \right\} \tag{9}$$

and the corresponding for T_E^* converge for almost every $\beta > 0$.
Notice that for any $\beta > 0$, $\exp\{-\beta \mathcal{T}^*\mathcal{T}\}$ is just the fundamental solution of the heat equation (IV.30) giving the "propagator" in (IV.29), so that $\exp\{-\beta\mathcal{T}\mathcal{T}^*\}$ turns out to be the analogous object for the bundle $\mathcal{T}E$.
The Atiyah-Bott-Patodi theorem finally states that (3) or (7) can be rewritten [29]

$$\operatorname{index} \mathcal{T} = \operatorname{Tr}\left\{ e^{-\beta\mathcal{T}^*\mathcal{T}} \right\} - \operatorname{Tr}\left\{ e^{-\beta\mathcal{T}\mathcal{T}^*} \right\} \quad . \tag{10}$$

Let's analyze briefly what kind of information the integer valued functor index $\mathcal{T}$ contributes to our physical picture. First of all index $\mathcal{T}$ is well defined whenever the partition function $Z(\beta')$ (where $\beta' = \frac{1}{2}\beta$) is convergent.

The breaking of such a convergence for a discrete set of "critical" values $\{\beta_c\}$ - which is usually related in statistical mechanics with the appearance of a phase transition - in present picture implies that either index $\mathcal{T}$ is non-uniquely determined or undefined. Also, of course $\beta_c = \beta_c(\omega)$ and if we move ω through Ω , we may expect index $\mathcal{T}$ to jump abruptly (remind that it is integer valued) whenever the point crosses the subset $\Sigma \subset \Omega$ corresponding to the instability of thermodynamical states. Σ is thus identified as the submanifold of structural instability of our system.
This suggests that not only index $\mathcal{T}$ acts as an indicator of the occurence of a phase transition but carries a finer information, through its absolute value in each stability domain, and the amount of the above jump when crossing Σ , concerning both the characterization of each phase and the classification of phase transitions.

This can be reconducted to more familiar physical concepts. Index $\mathcal{T}$ indeed is on one hand a measure of the asymmetry of the spectrum of $\mathcal{T}$, and it accounts for the dimension of ker $\mathcal{T}$ (namely the multiplicity of the zeros eigenvalues of $\mathcal{T}$) - as one can readily check from eq. (7) - and the difference in the asymptotic behaviour of the kernels $e^{-\beta \tau \tau^*}$ and $e^{-\beta \tau^* \tau}$ on the double of $\mathcal{M}$ (the latter can in fact be identified as a suitable Pontrjagin form of the metric on $\mathcal{M}$, hence also the topology of $\mathcal{M}$ is accounted for). Denoting the latter by $\alpha_0(x)$, $x \in \mathcal{M}$ indeed

$$\text{index } \mathcal{T} = \int_{\mathcal{M}} \alpha_0(x)\, d\sigma_x - \frac{1}{2}(p + \zeta(0)) \tag{11}$$

where p = dim ker $\mathcal{T}$ and $\zeta(s)$ is the zeta-function defined in (II.24) (or its generalized form

$$\zeta(s) = \sum_{\lambda \neq 0} \text{sign}\,\lambda \cdot |\lambda|^{-s} \tag{12}$$

to be adopted if $\mathcal{T}$ has negative as well as positive eigenvalues). Thus index $\mathcal{T}$ measures also the stability of the partition function with respect to global changes in the boundary conditions. The latter property is thus a restatement in global terms of Peierl's criterion for the appearance of a phase transiton [30].
On the other hand we saw - up to a trivial factor

$$U_B = \exp\left\{\frac{1}{2} S_\beta\right\} \exp\left\{-\frac{1}{2} S_{\beta'}\right\} \tag{13}$$

is the Bogolubov transormation (indeed an automorphism of the state space) connecting the thermodynamic state at temperature β to the one at temperature β'. The above mentioned analysis in terms of K-theory shows that index $\mathcal{T} > 0$ implies the global existence of a U_B such that $U_B^{-1} \mathcal{T} U_B$ is stably represented as a direct sum of lower order operators.

A jump in index $\mathcal{T}$ has then the following meaning. Moving ω in Ω, the operator $\mathcal{T}$ moves in the bundle of all operators having the same Jordan form - guaranteed by the existence of U_B - realizing an endomorphism of the state space. Corresponding to the decomposition of the space generated by the $\mathcal{T}$'s into bundles, the space Ω decomposes into submanifolds. The exceptional sub-manifold Σ for which the conditions for the existence of U_B fail correspond to a bifurcation of the eigenvalues of $\mathcal{T}$ - in particular the null eigenvalue - as a function of ω, and to nonlinear terms entering into play. Over Σ, which is a finite union of smooth manifolds, each corresponding to a bundle, index $\mathcal{T}$ is undefined.

This picture is itself a characterization in global terms of the mechanism whereby a phase transition is generated, known as Kac's mechanism[1].

Now the final question is: is it worth going through such a complex analysis and try to construct such an object as index $\mathcal{T}$ (and possibly other similar invariants) to study phase transitions?

The answer seems to be positive for two reasons. One is that in principle the method leads to a classification of phase transitions actually exploiting the microscopic properties of the system (as opposed to the Ehrenfest-Tizsa classification, which is phenomenological in its nature and exploits only macroscopic features of the system). The second stays in the so called Atiyah-Singer index theorem[28] which states that

$$\text{index}\ \mathcal{T} = \gamma(\mathcal{T}) \tag{14}$$

where $\gamma(\mathcal{T})$ is the topological index of $\mathcal{T}$, as opposed to index $\mathcal{T}$, often referred to as the analytical index of $\mathcal{T}$. Now eq. (14) should be associated with the fact that the topological index often be computed in ways which do not require the solution of the entire statistical mechanical problem. Indeed it can be written - exploiting the homological features we mentioned before - as an integral over the fundamental cycle of the cotangent bundle of $\mathcal{M}$, and may be expressed in terms of the symbol of $\mathcal{T}$ and the curvature tensor of the manifold $\mathcal{M}$ as

$$\gamma(\tau) = \left\{ ch\, \tau \cdot Td\, \mathcal{M} \right\} \tag{15}$$

where the curly brackets denote the evaluation of the product of cohomologies over the entire manifold, Td is the Todd cohomology class of the tangent bundle of $\mathcal{M}$ and ch the relative Chern class. The problem of classification is then reduced to the determination of the Pontrjagin characteristic classes of the fibre bundle, and the entire structure exhibits thus its intrinsic purely global geometric nature quite explicitly.

VI. An Example: The Ising Model

The two-dimensional Ising model for a system of interacting spins, say on a square lattice, is the ground for a straightforward and natural application of previous discussion. The reason for this is the known connection between the Ising problem and a soluble many-fermion problem proven years ago by Schultz, Mattis and Lieb [31].

The relevant information in their approach, from our point of view, is that, for a lattice, wrapped on a torus i.e. for a system endowed with periodic boundary conditions it is relatively easy to write the partition function $Z(\beta)$ as the expectation value of a suitable operator $\mathcal{V}$ in a"vacuum" state, as in (IV.35).

The latter is properly defined in terms of the operators associated with the spins of a single row, and so $\mathcal{V}$ itself belongs to the ring of the group generated by these (taking advantage as well of the translational degeneracy). The interaction among different rows is automatically included in the definition of $\mathcal{V}$.

This exploits beautifully the advantage of describing the temperature evolution of the system in terms of an entropy operator, which, so to speak, averages over part of the combinatorics. Moreover by a simple canonical transformation $\mathcal{V}$, or more precisely S_β can be written as the exponential of a quadratic form (indeed a direct sum of exact quadratic forms) as required by (V.1).

Let's review briefly some of the relevant steps. To begin with the system is an array of "spin $\frac{1}{2}$" (classical variables σ_{ij} taking on the values $\pm$ 1) on a square lattice Λ of rectangular shape (M columns and N rows so that $1 \leq i \leq N$, $1 \leq j \leq M$; and the lattice has NM sites). M and N are to be considered essentially infinitely large, with their ratio fixed and finite. The interaction is only between nearest neighbours and

$$H_\Lambda = -J_1 \sum_\Lambda \sigma_{nm}\sigma_{n+1,m} - J_2 \sum_\Lambda \sigma_{nm}\sigma_{n,m+1} + \partial H_\lambda \tag{1}$$

where $\sum_\Lambda$ implies that every spin site in the summands belongs to Λ and ∂H_λ is the Hamiltonian interaction between the system and its boundaries. We limit ourselves here to the mentioned periodic b.c., whereby the lattice is framed with an additional virtual row ($i = N+1$, $1 \leq j \leq M$) and column ($j = M+1$, $1 \leq i \leq N$) on which the spins are identical to those in the first row (i=1) and column (m=1) respectively

$$\begin{aligned} \sigma_{n,M+1} &= \sigma_{n,1} \quad , \quad n = 1 \dots N \\ \sigma_{N+1,m} &= \sigma_{1,m} \quad , \quad m = 1 \dots M \end{aligned} \tag{2}$$

so that

$$\partial H_\lambda = -J_1 \sum_{m=1}^{M} \sigma_{Nm}\sigma_{1m} - J_2 \sum_{n=1}^{N} \sigma_{nM}\sigma_{n1} \tag{3}$$

or equivalently Λ is given a toroidal topology, whence

$$H_\Lambda = -J_1 \sum_{n=1}^{N}\sum_{m=1}^{M} \sigma_{nm}\sigma_{n+1,m} - J_2 \sum_{n=1}^{N}\sum_{m=1}^{M} \sigma_{nm}\sigma_{n,m+1} \tag{4}$$

with relations (2) implicit.
J_1 and J_2 are the strengths of interaction.
Each σ_{ij} can be thought of as the z component of a spin operator, and quite naturally one may introduce the corresponding raising and lowering operators

$$\sigma_{nm}^{(\pm)} = \sigma_{nm}^{x} \pm i\sigma_{nm}^{y} \ . \tag{5}$$

The vacuum state, say $|0\rangle$, is defined by

$$\sigma_{Nm}^{(-)} |0\rangle = 0 \qquad , \quad m = 1 \dots M \tag{6}$$

and has the striking property that the partition function

$$Z(\beta) = \langle 0 | \mathcal{V} | 0 \rangle \tag{7}$$

as well as any correlation function, e.g. the two-point correlation function

$$\langle \sigma_{n',m'} \sigma_{n,m} \rangle = \frac{1}{Z(\beta)} \langle 0 | \sigma_{Nm'} \vartheta^{\frac{n'-n}{N}} \sigma_{Nm} \vartheta^{1-\frac{n'-n}{N}} | 0 \rangle \qquad (8)$$

can be expressed as expectation values, with

$$\vartheta = [2 \sinh(2\beta J_1)]^{\frac{NM}{2}} \left\{ \exp\left[\beta J_2 \sum_{m=1}^{M} \sigma_{Nm} \sigma_{N,m+1}\right] \times \right. \qquad (9)$$
$$\left. \times \exp\left[\ln \tanh(\beta J_1) \sum_{m=1}^{M} (\sigma^{(+)}_{Nm} \sigma^{(-)}_{Nm} - \tfrac{1}{2})\right] \right\}^N$$

depending explicitly - as mentioned - only on the spins associated with one row (we selected the N-th row).
By the nonlinear Jordan-Wigner transformation [32] and exploiting the translational degeneracy by the introduction of running wave variables, (9) can be written

$$\vartheta = [2 \sinh(2\beta J_1)]^{\frac{NM}{2}} \exp[-N \sum_{\bar{q}} \varepsilon_{\bar{q}} (a^{\dagger}_{\bar{q}} a_{\bar{q}} - \tfrac{1}{2})] = \qquad (10)$$
$$= Z(\beta) \exp[-N \sum_{\bar{q}} \varepsilon_{\bar{q}} a^{\dagger}_{\bar{q}} a_{\bar{q}}]$$

where $a^{\dagger}_{\bar{q}}$ and $a_{\bar{q}}$ are fermion creation and annihilation operators (whereby $a^{\dagger}_{\bar{q}} a_{\bar{q}}$ has eigenvalues 0 and 1 only) and

$$\cosh \varepsilon_{\bar{q}} = \cosh(2\beta J_2) \cosh[\ln \tanh(\beta J_1)] - \qquad (11)$$
$$- \cos\bar{q} \, \sinh(2\beta J_2) \sinh[\ln \tanh(\beta J_1)] \quad .$$

The reference state we are using is however $|J\rangle\!\rangle_0$, connected to the vacuum $|0\rangle$ through (IV.33) and (IV.35) states that in our representation

$$\langle 0,\omega | \frac{\vartheta}{Z(\beta)} | 0,\omega \rangle = {}_0\langle\!\langle J | e^{-S_\beta} | J \rangle\!\rangle_0 \quad . \qquad (12)$$

A simple inspection of eq. (10) suggests then that for it

$$S_\beta = N \sum_{\bar{q}} \varepsilon_{\bar{q}} a^{\dagger}_{\bar{q}} a_{\bar{q}} \qquad (13)$$

corresponding to a temperature-dependent effective Hamiltonian $H = \ln \vartheta$ (notice that in the representation of the thermodynamical states we adopted, the energy of the asymptotic field in presence of interaction between particles is in general temperature dependent, and so is then the corresponding formal Hamiltonian). Eq. (13) is - as expected - a direct sum of quadratic forms. Such a form makes further calculations much easier. Indeed in this case $\mathcal{T}$ turns out to be a Toepliz operator i.e. a convolution product operator by a smooth function S(z), defined over the unit circle S^1.

It generates therefore a closed path in the complex plane: $\mathcal{T} = \mathcal{T}_S : S^1 \longrightarrow \mathbb{C} - \{0\}$; and the Atiyah-Singer index reduces to the winding number, which is the one and only topological invariant under continuous deformations:

$$W = \text{index}\ \mathcal{T}_S = \frac{1}{2\pi}\left[\arg\{S(e^{i\pi})\} - \arg\{S(e^{-i\pi})\}\right] \tag{14}$$

S(z) is but the phase factor connected with the canonical transformation which diagonalizes ϑ ,

$$S(z) = z \cdot e^{-i\frac{\pi}{4}} \sqrt{\frac{(K_1K_2 z - 1)(K_2 z - K_1)}{(z - K_1K_2)(K_1 z - K_2)}} \tag{15}$$

where

$$K_1 = \exp(-2\beta J_1) \tag{16}$$

$$K_2 = \tanh(\beta J_2) \quad . \tag{17}$$

A straightforward application of (14), which for differentiable S(z) reads

$$W = \frac{1}{2\pi i} \oint \frac{dS}{S} \tag{18}$$

gives

$$W = \left\{ \begin{array}{ccc} 1 & ; & \beta > \beta_c \\ 0 & ; & \beta < \beta_c \end{array} \right\} \tag{19}$$

β_c being just Onsager's [32] critical inverse temperature, solution of the equation

$$k_1 = k_2 \quad . \tag{20}$$

In correspondence of $\beta = \beta_c$, w is undefined.

References

(1) M. Kac, Proc. Second Berkley Symp. Math. Stat. Prob., 189 (1950)

(2) R.P. Feynman, Revx, Mod. Phys. 20, 267 (1948)

(3) H.P. McKean and I.M. Singer, J. Diff. Geom. 1, 43 (1967)
ibid 5, 233 (1971)

(4) A,N. Kolmogorov, Uspekhi Mat. Nauk. 5, 5 (1938)

(5) W. Feller, Uspekhi Mat. Nauk. 8,232 (1940)
H.P. McKean, Stochastic Integrals, Academic Press, New York, 1969

(6) V.I. Arnol'd, Russian Math. Surveys 28 5, 19 (1973)

(7) E.B. Dynkin, Markov Processes, Springer Verlag, Berlin, 1965

(8) E. Cartan, Lecons sur la Geométrie des Espace de Riemann, Gauthier-Villars, Paris, 1928

(9) S. Minakshisundaram and A. Pleijel, Canadian J. Math. 1, 242 1949
M. Berger, P. Gauduchon and E. Mazet, Springer Verlag, Lecture Notes in Math. 194 (1971)

(10) V.I. Arnol'd, Funct. Anal. and its Appl. 6, 61 (1972)
ibid 6. 3 (1972)
ibid 7, 75 (1973)

(11) J. Duistermaat, Comm. Pure Appl. Math. 27, 207 (1974)

(12) I.N. Bernshtein, Funct. Anal. and its Appl. 5, 89 (1971)

(13) R.T. Seeley, Proc. Symp. Pure Math. Amer. Math. Soc. 10, 288 (1967)
Amer. J. Math. 91, 963 (1969)

(14) M.F. Atiyah, V.K. Patodi and I.M. Singer, Bull. London Math. Soc. 5, 229 (1973)

(15) J. von Neumann, Ann. Math. 33, 294 (1932)

(16) F.J. Murray, Trans Amer. Math. Soc. 37, 301 (1935)

(17) T. Kato, Math. Ann. 126, 253 (1953)

(18) J. Locker, Trans. Amer. Math. Soc. 203. 175 (1975)

(19) R. Kannan and J. Locker, J. Diff. Equations 28, 60 (1978)

(20) S. Fucik, J. Necas, J. Soucek and V. Soucek, Springer Verlag Lecture Notes in Math. 346 (1973)

(21) M.F. Atiyah, International Congress of Mathematicians, Moscow, 7 (1966)

(22) V. de Alfaro and M. Rasetti, Fortschritte der Physik, 26, 143 (1978)
M. Rasetti, Global Approach to Phase Transitions, in "Fundamental Problems in Statistical Mechanics", N.N. Bogolyubov jr., and A.M. Kurbatov, eds.; J.I.N.R. Publ., Dubna, U.S.S.R., 1977
M. Rasetti, Structural Stability in Statistical Mechanics, Springer Tracts in Mathematics, W. Güttinger ed., Berlin, 1979 (in press)

(23) A.S. Wightman, Phys. Rev. 101, 860 (1956)

(24) T. Matsubara, Progr. Theor. Phys. 14, 351 (1955)

(25) Y. Takahashi and H. Umezawa, Coll. Phenomena 2, 55 (1975)

(26) N.E. Steenrod, Ann. Math. 44, 610 (1943)

(27) L. Hörmander, Comm. Pure Appl. Math. 18, 501 (1965)
R.T. Seeley, Trans. Amer. Math. Soc. 117, 167 (1965)

(28) M.F. Atiyah and I.M. Singer, Bull. Amer. Math. Soc. 69, 422 (1963)
Ann. Math. 87. 484 (1968), ibid. 87, 546 (1968), ibid. 93, 139 (1971), ibid. 93 199 (1971), Publ. Math. Inst. Hautes Etudes Sci. (Paris) 37 (1969)

(29) M.F. Atiyah, R. Bott and V.K. Patodi, Inventiones Math. 19, 279 (1973)

(30) R. Peierls, Proc. Cambridge Phil. Soc. 32, 477 (1936)
R.B. Griffith, Phys. Rev. 136A, 437 (1964)

(31) T.D. Schultz, D.C. Mattis and E.H. Lieb, Revs. Mod. Phys. 36, 856 (1964)

(32) P. Jordan and E. Wigner, z. Physik 47, 631 (1928)

(33) L. Onsager, Phys. Rev. 65, 117 (1944)

Life without T_2

(Some remarks on certain locally metrizable spaces which need not to be Hausdorff) [1]

by

R.Z. Domiaty, Graz,
Austria

Introductory Remark: This note is a small survey report. In the first part a motivation is given to introduce a special class of locally metrizable spaces, called FM-spaces (definition 2). These spaces seem to be an acceptable generalization of the class of metric spaces, because many of the standard geometrical concepts can be carried over to them - but we will not touch these topics here. In the second part, a result on the existence and structure of certain maximal metrizable subspaces of FM-spaces will be formulated (theorem 1-3). No proofs will be given here, they can be found in forthcoming extensive reports concerning the general theory of locally metrizable spaces [3], [4].

Finally it should be stated, that the motivations for this investigation came from the work of R. Geroch and P. Hajicek on non-T_2 manifolds and from the work of R. Penrose on the structure of space-time.

Notation. Topological spaces are denoted by (X, "X"): X is a set and "X" a topology on X. The sets in "X" are called open subsets. For every $Q \subseteq X$ we denote by $\overline{Q}$ the closure of Q and by ∂Q the boundary of Q with respect to "X"; if Q is considered as a subspace, Q is assumed to carry the usual subspace topology, denoted by "X"$|$Q. For all here not explicitely defined topological notions we refer to [18]. By card Q we denote the cardinality of the set Q; $\aleph_0$ is the cardinality of the set of natural numbers. $\emptyset$ denotes the empty set. The phrase "iff" means "if and only if".

1) This work was supported by the Alexander von Humboldt-Stiftung, Bonn, Germany.

Acknowledgement. I want to thank Prof. Dr. E. Binz (University of Mannheim) and Prof. Dr. K. Bleuler (University of Bonn) for their valuable comments and their patience during the many personal discussions on topics in mathematical physics in the summer-semester 1978.

Introduction. With the appearence of J.L. Kelley's famous book [15] on topology 1955, the interest in non-T_2 spaces was activated. T_2 stands here for the well known Hausdorff-separation axiom. A topological space is said to be T_2, if any two district points can be separated by disjoint neighbourhoods.

At the first moment, life without T_2 seems to be unpleasant: Compact subsets need not to be closed, compact spaces may fail to be normal and paracompactness does no longer imply the existence of a partition of unity subordinate to a locally-finite open covering. Therefore the question arises, why there should be any interest in these rather pathological types of spaces. Before we go into details we want to point to a trivial but important and often overlooked fact: The property of a topological space to be a T_2-space is a global and not a local property. I.e., a space can be locally T_2 (this means, that every point is contained in an open neighbourhood which is T_2 as a subspace) without being a T_2-space. To demonstrate this, let us recall a simple example of a non-T_2 manifold.

The Haefliger-Reeb-Line (multiplication of the origin 0 of the real line $\mathbb{R}$; see [14], [1], 39 , [7], and [3], example 1). Put

$$X := \mathbb{R} \cup S,$$

$S := \{\omega_i\}_{i\in I}$ being a set such that $\mathbb{R}\cap S = \emptyset$, and define an atlas $\mathcal{A} := \{(\varphi_j, \mathbb{R})\}_{j\in J}$, $J := I \cup \{*\}$ (with $* \notin I$), by

$$\varphi_i : \mathbb{R} \longrightarrow X; \quad \varphi_i(x) := \begin{cases} x, \text{ if } x \neq 0 \\ \omega_i, \text{ if } x = 0 \end{cases} \qquad (i \in I)$$

and

$$\varphi_* : \mathbb{R} \longrightarrow X; \quad \varphi_*(x) := x.$$

Then $\mathcal{A}$ defines a C^ω-manifold structure on X. If we denote the underlying manifold-topology on X by "X", then (X, "X") is separable, connected (hence path-connected), locally euclidean (hence locally compact, locally arc-connected and locally metrizable), T_1 and satisfies the first axiom of countability. The following additional properties can be shown:

(1) (X,"X") is T_2 iff card S = 0

(2) (X,"X") is paracompact iff card S $< \aleph_0$

(3) (X,"X") satisfies the second axiom of countability iff card $S \leq \aleph_0$

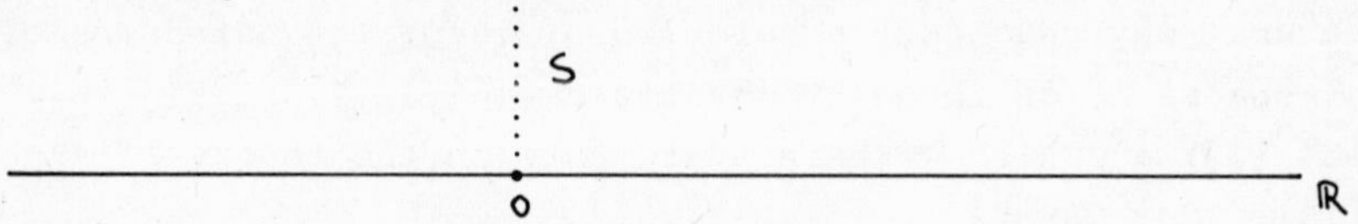

It can easily be seen, that locally X is a T_2-manifold. Hence the violation of the global validity of the T_2-property cannot be discovered by local observations!

At the end of this introduction we want to quote a fundamental remark from a paper of A. Wilanski [21], p. 161. There he writes: "In some situations T_2 may be needed or convenient. In such cases one specializes; but this is no reason to use T_2 as a blanket assumption from the beginning, any more than prospective use of the Čech compactification justifies dealing only with completely regular spaces from the beginning."

Space-Times. The most important models for space-times are Lorentz manifolds (M,g). M is a real n-dimensional ($n \geq 3$), paracompact, connected, C^∞-manifold without boundary. g is a globally defined C^∞-tensor field on M of type (0,2) which is nondegenerate and has signature n-2 everywhere. In general it is assumed, that

(*) M is T_2

([13], 14 , 38 , 57f). But there seems to be no significant physical motivation for the assumption (*) (see [16], 97f, [9], 99f or [13], 57: "M is taken to be T_2 since this seems to accord with normal experience")! Therefore, in what follows, we give up the assumption of (*).

Remark on the paracompactness of M. If (*) is satisfied, by a folklore-theorem in differential topology, the paracompactness of M is a consequence of the other assumptions ([20], 8-53; [8]). Hence the manifold topology is metrizable. In this case, the property "paracompactness" is equivalent to the sharper property "hereditary paracompactness" (i.e. every open, hence any, subspace is paracompact). Because in this paper we give up (*), we demand instead of the paracompactness of M the following sharper assumption to be satisfied:

M is hereditarily paracompact.

A genuine mathematical motivation for this assumption will be given in the next section.

Non-T_2 space-times arise in some places in general relativity in a quite natural way. We mention here two of them: The extension of Taub-NUT-spaces leads to non-T_2 manifolds ([13], 170-178; [16], 97; [11];[12];[2]) and also certain statements in the theory of singularities ([13], 283; 289-292).

FM-Spaces. Now we want to introduce a class of topological spaces, which contains the space-times defined in the preceding section and is at the same time "as close as possible to the class of metrizable spaces" without necessarily being T_2. All our definite geometrical knowledge of space-time is indeed of a local nature. Hence in the first step we have to localize the concept of metrizability. This leads to the class of locally metrizable spaces (see [18], 88;[3]).

Definition 1. (Let (X,"X") be a topological space.

1. (X,"X") is called locally metrizable, abbreviated: (X,"X") $\in$ LM, if every point $p \in X$ has an open neighbourhood U_p such that (U_p, "X"$|U_p$) is metrizable.
2. Let M be a subset of X. M is called a T_2-subset of (X,"X"), if $M \in$ "X" $- \{\emptyset\}$ and (M, "X"$|$M) is T_2; it is called a maximal T_2-subset of (X,"X"), if it is a T_2-subset of (X,"X") which is not a proper subset of any other T_2-subset of (X,"X").

By the lemma of Zorn we get

Lemma 1. Every (connected) T_2-subset of a topological space is contained in at least one (connected) maximal T_2-subset.

An immediate consequence of lemma 1 is the following

Corollary: Every LM-space can be covered by a family of maximal T_2-subsets.

The class of locally metrizable spaces is too general for our purpose. Hence in the second step we have to confine this class in an appropriate way. This motivates the following

Definition 2. A topological space is called a FM-space (from the german word "fast-metrisierbar") if it is locally metrizable and hereditarily paracompact.

Remarks. 1. That FM-spaces are "nearly metrizable" can be seen from the well known metrization-theorem of J. Nagata and Yu. Smirnow([18], 212): A topological space (X, "X") is metrizable iff it is a T_2- and FM-space. In other words: Every metrizable space is a FM-space, and if a FM-space is not metrizable, then the only "missing" property is T_2. This fact motivates the nomenclature in definition 2.

2. Every subspace of a FM-space is again a FM-space.

3. A subspace of a FM-space is metrizable iff it is T_2.

4. As the example of the "long line" (see [19], A 10;[3]) shows, T_2 LM-spaces need not to be metrizable! Hence in general LM-spaces, T_2-subsets are not necessarily metrizable as subspaces. It seems to be an interesting question to characterize those LM-spaces which have the property that all T_2-subsets are metrizable.

5. The HAEFLIGER-REEB line is a FM-space, iff card $S < \aleph_0$.

Maximal metric charts in FM-spaces. We start with

Definition 3. Let (X, "X") be a topological space. A subset $M \subseteq X$ is called a metric chart in (X, "X"), abbreviated $M \in$"K"(X,"X"), if $M \in$ "X" $- \{\phi\}$ and (M,"X"| M) is metrizable; M is called a maximal metric chart in (X,"X"), abbreviated $M \in$ "M"(X,"X"), if $M \in$ "K"(X,"X") and M is not a proper subset of any other metric chart in (X,"X"); M is called a connected (resp. maximal connected) metric chart in (X,"X"), abbreviated $M \in$ "K_c"(X,"X") (resp. $M \in$ "M_c"(X,"X")), if $M \in$ "K"(X,"X") and M is a connected set (resp. $M \in$ "K_c"(X,"X") and M is not a proper subset of any other connected metric chart).

For FM-spaces, the T_2-subsets are exactly the metric charts. Together with lemma 1 we can state a result which is analogous to one which was originally formulated by R. Geroch for a certain class of non-T_2 manifolds (see [11]), section 2) and the following remark 4.)

Theorem 1. In a FM-space every metric chart (resp. connected metric chart) is contained in a maximal metric chart (resp. maximal connected metric chart).

Remarks. 1. Notice, that a maximal metric chart is in general not uniquely determined and that $M \in$ "M_c"(X,"X") does not necessarily mean that $M \in$ "M"(X,"X"), i.e., "M_c"(X,"X") $\nsubseteq$ "M"(X,"X") in general.

2. In LM-spaces which are not paracompact, theorem 1 may be wrong, as the example of the "long line" shows.

3. The HAEFLIGER-REEB line with card S = $\aleph_0$ is an example of a LM-space, which is not a FM-space but has the property that every metric chart is contained in a maximal metric chart.

4. Here we want to give some comments on the above mentioned result of G. Geroch and P. Hajicek. According to a remark in [11], both obtained independently a theorem which is analogous to theorem 1, but they start from a somewhat different topological presupposition: (X,"X") is assumed to be a (not necessarily T_2) n-dimensional differentiable manifold satisfying the second axiom of countability (i.e., (X,"X") has a countable base). The corresponding natural generalization for locally metrizable spaces would be the class of LM-spaces satisfying the second axiom of countability, abbreviated: LM_2-spaces. Again, every subspace of a LM_2-space is a LM_2-space. But FM-spaces and LM_2-spaces are different classes of spaces. On the one hand the HAEFLIGER-REEB line with card S = $\aleph_0$ is a LM_2-space which is not a FM-space, and on the other hand, every non-separable Banachspace is an example of a FM-space which is not a LM_2-space. From topological point of view, this is not surprising, because without additional assumptions, even for LM-spaces, the topological properties "second axiom of countability" and "hereditarily para-compact" are independent. Theorem 1 might fail to be true for LM_2-spaces, because T_2 LM-spaces satisfying the second axiom of countability may not be metrizable! But if (X,"X") is a locally-euclidean LM_2-space, then (X,"X") is locally compact in the following stronger sense: Every point has a neighbourhoodbase consisting of compact sets. Hence every open subset of X is locally compact. Because locally compact T_2-spaces are regular ([18],88), we can apply Urysohn's metrization theorem ([18], 196) to T_2-subsets of (X,"X"), and so obtain theorem 1 for locally-euclidean LM_2-spaces. But this is the version of theorem 1 discovered by R. Geroch and P. Hajicek.

Now arises the problem to characterize the maximal metric charts. To do this, we want to make use of a relation, which was introduced by R. Geroch in an unpublished paper (see [11]).

Definition 4. Let (X,"X") be a topological space. Two points $p,q \in X$ with $p \neq q$ are called in relation $*$, abbreviated $p * q$, if for every neighbourhood U of p and every neighbourhood V of q, we have

$$U \cap V \neq \emptyset$$

(i.e., if p and q cannot be separated by disjoint neighbourhoods). Further, for every $Q \subseteq X$ we denote by Q the set

$$Q^* := \{ x \in X \mid \exists q \in Q : x * q \} .$$

Trivially, $M \subseteq X$ is a T_2-subset of (X,"X") iff $M \cap M^* = \emptyset$. Further we have

Lemma 2. Let (X,"X") be a topological space and $G \in$"X". Then G is a T_2-subset of (X,"X") iff $\overline{G^*} \subseteq \partial G$.
In the case of FM-spaces we can show

Theorem 2. Let (X,"X") be a FM-space.
1. $G \in$ "M"(X,"X") $\Longrightarrow \partial G = \overline{G^*}$
2. $G \in$ "K_c"(X,"X"), $\partial G = \overline{G^*} \Longrightarrow G \in$ "M_c"(X,"X")

We want to note, that there are examples which show, that neither propositon 1. nor 2. can be reversed. But for a special class of FM-spaces both propositons can be put together, to obtain a result analogous to one in [11].

Theorem 3. Let (X,"X") be a locally connected FM-space and $G \in$ "K_c"(X,"X"). Then $G \in$ "M_c"(X,"X") iff $\partial G = \overline{G^*}$.

Concluding Remarks: With the here accomplished considerations one has obtained a rough insight into the basic topological structure of a locally-connected FM-space (X,"X"). Lemma 1. implies that X can be covered by a family $\mathcal{Z} := \{G_\lambda\}$ of connected maximal metric charts, each G_λ having the property $\overline{G_\lambda^*} = \partial G_\lambda$.

An easy calculation shows, that the topology "X" is completely determined by the topology of the subspaces (G_λ,"X"$| G_\lambda$): A subset $Q \subseteq X$ is in "X" iff for every λ we have $G_\lambda \cap Q \in$ "X"$|G_\lambda$.
In other words: "X" can be recovered from the subspaces (G_λ, "X"$| G_\lambda$). In the case that (X,"X") is moreover a manifold, we can take over all locally-defined concepts developed in the theory of T_2-manifolds. For local analysis and local differential geometry the "T_2-property" is irrelevant ([10], 3;[17], 21). To sum up, life without T_2 is not as unpleasant as it seemed to be at the first sight, at least in LM-spaces.

We close with the remark, that a "geometrization" of FM-spaces is possible, because the topology of a FM-space can be defined by special real distance functions (see [4],[5] and [6]for further details).

References.

[1] F. Brickell - R.S. Clark, Differentiable Manifolds, London 1970.

[2] C.J.S. Clarke, Space-time singularities. Commun. Math. Phys., 49 (1976) 17-23.

[3] R.Z. Domiaty, A survey on locally metrizable spaces. Proc. Meeting of General Topology, Trieste, Sept. 4-18th, 1978 (in print).

[4] R.Z. Domiaty, Locally metrizable spaces II. Math. Colloq. Univ. Cape Town, 12 (1979) (in print)

[5] R.Z. Domiaty, The Hausdorff separation property for space-time. Eleftheria (Athens), 2 (1980) (in print).

[6] R.Z. Domiaty, Zur inneren Geometrie quasimetrischer Räume. Colloquium in memory of W. Rinow, Greifswald 1980 (in print).

[7] D.B. Gauld, Topological properties of manifolds. Am. Math. Month., 81 (1974) 633-636.

[8] R. Geroch, Spinor structure of space-time in general relativity, i., J. Math. Phys., 9 (1968) 1739 - 1744

[9] R. Geroch, Space-time structure from a global viewpoint, p. 71 - 1o3 in: General Relativity and Cosmology, Ed.: B.K. Sachs, New York 1971.

[1o] N.J. Hicks, Notes on Differential Geometry. London 1971-

[11] P. Hajicek, Causality in non-Hausdorff space-times. Commun. Math. Phys., 21 (1971) 75-84

[12] P. Hajicek, Bifurcate space-times. J.Math.Phys., 12 (1971) 157-160-

[13] S.W. Hawking - G.F.R. Ellis, The large Scale Structure of Space-Time. Cambridge 1973.

[14] A. Haefliger - G. Reeb, Variétés (non separées) à une dimension et structures feuilletées du plan'. Enseign. Math., 3(1957) 107-125.

[15] J.L. Kelley, General Topology. New York 1955

[16] W. Kundt, Global theory of space-time, p. 93-133 in: Proc. Thirteenth Biennial Seminar, Canadian Math. Congr. Ed.: R. Vanstone. Montreal 1972.

[17] S. Lang, Differential Manifolds. Reading, Mass. 1972.

[18] J. Nagata, Modern General Topology, Amsterdam 1968

[19] M. Spivak, Differential Geometry, Vol. I. Boston 1970

[20] M. Spivak, Differential Geometry, Vol. II. Boston 1970

[21] A. Wilansky, Life without T_2. Am. Math. Month., 77(1970) 157-161; 728

Affine Model of Internal Degrees of Freedom in a Non-Euclidean Space

by

Jan J. Slawianowski

Institute of Fundamental Technological Research
Polish Academy of Sciences
Warsaw

0. Introduction:

In some earlier papers we have presented the basic ideas concerning the concept of affinely-rigid body and its applications in physics and elasticity [12] - [18]. By affinely-rigid body we understand a discrete or continuous system of material points in which all affine relations are frozen (similarly, as in the usual, metrically-rigid body all metrical relations i.e. distances are invariant).

Now, we would like to present a few remarks about the geometric structure of analytical mechanics of material points endowed with affine internal degrees of freedom and moving in an non-Euclidean space. This is simply the mathematical model of test affinely-rigid bodies which are so small that one can treat them as injected into a tangent space rather than into the manifold itself.

The physical space is assumed to be endowed with an affine connection and Riemannian metric. At the beginning there is no relationship between these structures. This is justified by the fact that from the purely logical and also physical, operational point of view the concept of the parallel transport and that of the distance (length of vectors) are independent. Hence, it is reasonable to develop the theory without any relationship between them and to investigate the meaning of any particular linking only on the stage of final results.

For example, we show that the special case of covariantly-constant metric structure (Riemann-Cartan space) is especially suited to the affine model of internal degrees of freedom. This seems to contradict to some current opinions according to which the Riemann-Cartan geometry is distinguished, privileged only from the purely metrical point of view, as the only one which enables us to avoid difficulties when introducing spinors.

Our main purpose is to describe the coupling between spatial geometry (curvature and torsion) and internal degrees of freedom (affine spin i.e. generator of affine rotations of the frame attached to the material point). We remain on the non-relativistic although spatially-non-Euclidean level. This simplifies all calculations. However, this non-relativistic restriction does not matter when the general structure of the coupling between geometry and internal degrees of freedom is concerned.

For what to do things like that? We have in view some possible applications and further developments in relativistic theory of continua, in particular in relativistic theory of elastic continua. Small parts of such a continuum behave as affinely-rigid bodies i.e. as material points with extra attached linear frames.

The affine model of collective degrees of freedom has been used in astrophysics by Bogoyavlenskij, Novikov, Dyson, Chandrasekhar [3][4][5][6][8] (cosmic scale of physical phenomena) and also in nuclear physics (microscale) [19][20] . Some people suspect also that internal degrees of freedom of elementary particles should have something to do with the linear group $GL(3,\mathbb{R})$ (let us mention only that $SU(3)$, $SL(3,\mathbb{R})$ are different, opposite real forms of the same complex Lie group) [7][9][11][21].

Besides, it is obvious that affine model of internal degrees of freedom of heavy matter is strongly suggested by the very model of the physical space as a differential manifold with affine connection (which is simply some geometrical object in the bundle of linear frames i.e. in the manifold of degrees of freedom of test affinely-rigid bodies).

1. Affinely-rigid body in a flat space

Let us summarize the basic concepts concerning affinely-rigid body in a flat space. We do this in a maximally simplified way. For details and rigorous theory cf. [12][14][15][17][18].

The motion of affinely-rigid body splits naturally into the

translational (orbital) motion of the body as a whole and the relative motion of constituents with respect to the centre of mass.

The configuration space of the relative motion in n-dimensional physical space can be identified with $GL(n,\mathbb{R})$ - the real linear group. This identification is achieved by fixing some reference configuration and affine system of coordinates. Obviously, only special cases $n = 1,2,3$ are interesting from the physical point of view. When the body is rigid in the metrical sense, the configuration space becomes the special orthogonal group $SO(n,\mathbb{R}) \subset GL(n,\mathbb{R})$.

The configuration space $Q = GL(n,\mathbb{R})$ carries three natural groups of transformations:

1. Abelian translations:
   $GL(n,\mathbb{R})$ is an open subset of the linear space $L(n,\mathbb{R})$ consisting of all $n \times n$ matrices. Hence, we have the following local group of transformations.

$$\xi \in L(n,\mathbb{R}) : \quad GL(n,\mathbb{R}) \ni \varphi \longrightarrow \varphi + \xi \quad . \tag{1.1}$$

2. Right regular group translations:

$$A \in GL(n,\mathbb{R}) : \varphi \longrightarrow \varphi A \quad . \tag{1.2}$$

When the body is extended, such mappings describe the transformations of the material itself [12] i.e. affine permutations of material points of the body.

3. Left regular group translations:

$$A \in GL(n,\mathbb{R}) : \varphi \longrightarrow A\varphi \quad . \tag{1.3}$$

Such mappings describe spatial, Eulerian rotations and homogeneous deformations of the body in the physical space [12] [17].
Relative i.e. internal motions are described by smooth curves:

$$\mathbb{R} \ni t \longrightarrow \varphi(t) \in GL(n,\mathbb{R}) \quad .$$

Generalized velocity at the time instant t becomes:

$$\xi = \frac{d\varphi}{dt} \in L(n,\mathbb{R}) \quad .$$

There is no need to use here the sophisticated language of tangent bundles because Q is an open submanifold in a linear space $L(n,\mathbb{R})$.

As usually, when dealing with degrees of freedom with a group-theoretical background (Lie group as a configuration space [1]), instead of generalized velocities we employ non-holonomic quasi-velocities which are elements of the corresponding Lie algebra.

Lie algebra of $GL(n,\mathbb{R})$ is canonically $L(n,\mathbb{R})$, hence:

Affine velocity is defined as:

$$\Omega = \frac{d\varphi}{dt}\varphi^{-1} \quad . \tag{1.4}$$

Co-moving affine velocity:

$$\hat{\Omega} = \varphi^{-1}\frac{d\varphi}{dt} = \varphi^{-1}\Omega\varphi \quad . \tag{1.5}$$

When φ remains orthogonal during the motion (gyroscopic constraints imposed) the above objects are skew-symmetric and become the usual angular velocity respectively in spatial and co-moving representation. Hence, affine velocities are nothing alse but the affine generalization of the angular velocity of gyroscopic motion. They obey the following transformation rules:

$$\varphi \longrightarrow \varphi A : \Omega \longrightarrow \Omega \quad , \hat{\Omega} \longrightarrow A^{-1}\hat{\Omega}A \tag{1.6}$$

$$\varphi \longrightarrow A\varphi : \Omega \longrightarrow A\Omega A^{-1}, \quad \hat{\Omega} \longrightarrow \hat{\Omega} \quad . \tag{1.7}$$

This is simply the special case of the adjoint transformation rule for elements of Lie algebra.

When using Hamiltonian formalism we employ the canonical momenta $\pi \in L(n,\mathbb{R})$ which are identified with linear functionals on virtual velocities $\xi \in L(n,\mathbb{R})$ according to the formula:

$$\langle \pi, \xi \rangle = \mathrm{Tr}\,(\pi\xi) \quad . \tag{1.8}$$

Replacing generalized velocities by Lie-algebraic objects Ω, $\hat{\Omega}$ induces the dual replacing of canonical momenta π by Lie-algebraic quasimomenta $\Sigma, \hat{\Sigma} \in L(n,\mathbb{R})$ which are linear functionals on the space of all possible Ω-s, $\hat{\Omega}$-s respectively.

Affine spin is defined as:

$$\Sigma = \varphi\pi \quad . \tag{1.9}$$

The co-moving affine spin takes on the form:

$$\hat{\Sigma} = \pi\varphi = \varphi^{-1}\Sigma\varphi \quad . \tag{1.10}$$

From the very definition, Σ, $\hat{\Sigma}$ satisfy the following contragradient rules:

$$\langle \Sigma, \Omega \rangle = \mathrm{Tr}(\Sigma\Omega) = \mathrm{Tr}(\pi\xi) = \langle \pi, \xi \rangle = \mathrm{Tr}(\hat{\Sigma}\hat{\Omega}) = \langle \hat{\Sigma}, \hat{\Omega} \rangle \tag{1.11}$$

When gyroscopic constraints are imposed ($\varphi \in SO(n,\mathbb{R})$), then $\Sigma, \hat{\Sigma}$ become skew-symmetric and are nothing else but the internal angular momentum in the spatial and co-moving description respectively.

Components of the affine spin Σ are Hamiltonian generators of left affine deformations. Namely, for arbitrary $A \in L(n,\mathbb{R})$ the quantity $F[A] = Tr(A\Sigma)$ generates the group:

$$\varphi \longrightarrow e^{At}\varphi \quad , \pi \longrightarrow \pi e^{-At} \quad . \tag{1.12}$$

Similarly, the co-moving components of affine spin $\hat{\Sigma}$ generate right affine transformations. Namely, $\hat{F}[A] = Tr(A\hat{\Sigma})$ is a Hamiltonian generator of the group:

$$\varphi \longrightarrow \varphi e^{At} \quad , \pi \longrightarrow e^{-At}\pi \quad . \tag{1.13}$$

Equations of motion of affinely-rigid body are equivalent to the balance laws for the affine spin, similarly, as gyroscopic equations are balance laws for the usual spin.

To be able to develop the dynamical theory (equations of motion) we have to postulate some metric tensor i.e. kinetic energy form on the configuration space. It could seem that the only physical possibility is to calculate the kinetic energy of extended affinely-rigid body as a sum of classical kinetic energies of its constituents (material points), taking only into account the affine constraints. However, such a "macroscopic" view, correct in continuous elasticity is no more justified on the microscopic level when trying to describe internal collective degrees of freedom of small objects like nuclei or elementary particles. Besides, there is no reason to proceed in such a way when the affine collective modes of small objects arise according to the adiabatic decoupling scheme rather than according to the constraints mechanism (this is probably the case in the theory of nuclei). Hence, it seems more natural to review some possible kinetic energy forms which are distinguished from the group-theoretical point of view, but without "deriving" them. Of course, in practical applications any particular choice has to be derived or at least justified by experimental data.

The three aforementioned groups of transformations on $GL(n,\mathbb{R})$ (1.1, 1.2, 1.3) give rise to the three natural classes of kinetic energies.

1. Kinetic energies invariant under local abelian translations. When T is invariant under all local mappings: $\varphi \longrightarrow \varphi + \alpha$, then it has to be a quadratic form of $\xi = \frac{d\varphi}{dt}$ with constant coeffi-

cients. This is still a very wide class of kinetic energies. To restrict it in a reasonable way we impose the additional requirement of the invariance under left regular group translations by elements of $SO(n,\mathbb{R})$ i.e. under mappings: $\varphi \longrightarrow U\varphi$, $U \in SO(n,\mathbb{R})$. Such an additional restriction is suggested by the Euclidean geometry of the physical space. The only possibility is the following one:

$$T = \frac{1}{2} \mathrm{Tr} \left(\frac{d\varphi^T}{dt} \frac{d\varphi}{dt} J \right) \tag{1.14}$$

where J is a constant, symmetric and positively definite matrix. Such a formula can also be "derived" in a classical way for the extended, macroscopic affinely-rigid body. J becomes then the co-moving quadrupole momentum of the mass distribution in the reference configuration [12] [17] [18].
The metric tensor corresponding to T does not give rise to the invariant dynamical system on $GL(n,\mathbb{R})$ in the Arnold-Hermann sense [1] because it fails to be invariant either under right or left regular group translations.

2. Kinetic energies invariant under right regular translations.
T is invariant under all mappings (1.2) if and only if it is a quadratic form of $\Omega = \frac{d\varphi}{dt} \varphi^{-1}$ with constant coefficients. Imposing the stronger restriction of the left orthogonal invariance ($\varphi \longrightarrow U\varphi$, $U \in SO(n,\mathbb{R})$, one gets:

$$T = \frac{\mu}{2} \mathrm{Tr} \left(\Omega^T \Omega \right) + \frac{\lambda}{2} \left(\mathrm{Tr}\, \Omega \right)^2 \tag{1.15}$$

where μ, λ are non-negative constants (μ must be strictly positive). For our purposes it will be sufficient to restrict ourselves to the first term only because within the non-Euclidean framework the second one does not influence the structure of the coupling between internal degrees of freedom and geometry. When endowed with the metric tensor corresponding to such a form of T, $GL(n,\mathbb{R})$ becomes a non-flat, curved Riemannian manifold. Such a Riemannian metric, being invariant under right regular translations defines some right-invariant dynamical system on $GL(n,\mathbb{R})$ in the Arnold-Hermann sense.

$\hat{\Sigma}^i{}_j = \frac{\partial T}{\partial \hat{\Omega}^j{}_i}$ become then constants of motion because they generate the symmetry group-right regular translations.

3. Kinetic energies invariant under left group translations.
The invariance under all mappings (1.3) implies T to be a quadratic form of $\hat{\Omega} = \varphi^{-1}\frac{d\varphi}{dt}$ with constant coefficients. There are two natural types of stronger restrictions which lead to two distinguished types of metric:
a) T in the neighborhood of identity (φ = I) asymptotically approaches (1.14). Then:

$$T = \frac{1}{2}\,\mathrm{Tr}\,(\hat{\Omega}^T\hat{\Omega}J) \qquad (1.16)$$

where J is again a constant, symmetric and positive matrix.
b) Assume that T is additionally invariant under right orthogonal group translations: $\varphi \longrightarrow \varphi U$, $U \in SO(n,\mathbb{R})$. Then:

$$T = \frac{\mu}{2}\,\mathrm{Tr}\,(\hat{\Omega}^T\hat{\Omega}) + \frac{\lambda}{2}\,\mathrm{Tr}\,(\hat{\Omega})^2 . \qquad (1.17)$$

Metric tensors corresponding to such a form of T are left-invariant, hence they also give rise to invariant dynamical models on $GL(n,\mathbb{R})$ in the Arnold sense. Curvature tensors are non-trivial. $\Sigma^i{}_j = \frac{\partial T}{\partial \Omega^j{}_i}$ are constants of motion.

When investigating the coupling between internal degrees of freedom and geometry we will imploy rather (1.16). In fact (1.17) arises from (1.16) when putting $J = \mu I$ (and here it is less general) and extra adding the term $\frac{\lambda}{2}(\mathrm{Tr}\,\Omega)^2 = \frac{\lambda}{2}(\mathrm{Tr}\,\hat{\Omega})^2$, but the latter does not influence the structure of the coupling between internal motion and geometry.

Let us notice that the model (1.16) (as any left-invariant) does not presuppose any metric structure in the physical space which is assumed only to carry affine geometry. Hence, it could seem completely non-physical because we know that the space-time is endowed with the Minkowskian metric tensor. Nevertheless, this affine, amorphous model could be a good exercise towards investigating some hypothetic theories where the metric tensor is no more one of fundamental physical fields but rather the secondary, phenomenological manifestation of something more physical (the programme of anti-geometro-dynamics), like some fermion fields, tetread field a.s.f.
Important remark:
There is no kinetic energy form for which simultaneously the invariance under left and right regular translations on $GL(n,\mathbb{R})$

could hold. In this sense the "spherical" rigid body has no affine counterpart. The reason is that in contrary to $SO(n,\mathbb{R})$, $GL(n,\mathbb{R})$ is non-compact and it is no more possible to construct the positive-definite metric tensor from its first and second order Casimir invariants. In fact, the combination :

$$a\,\mathrm{Tr}(\Omega^2) + b(\mathrm{Tr}\Omega)^2 = a\,\mathrm{Tr}(\hat{\Omega}^2) + b(\mathrm{Tr}\hat{\Omega})^2$$

is never positive definite.

2. The test affinely-rigid body in a non-Euclidean space. Degrees of freedom and kinematics.

Let us assume the physical space M to be a smooth n-dimensional manifold. In a manifold there is no more possibility to define globally affine degrees of freedom of extended bodies. Nevertheless, the affine model of internal degrees of freedom is still well-defined. Namely, we consider the motion of material points in M with extra attached linear frames. In principle such a model admits two physical interpretations: a) as something primary, b) as a symbolic way to describe some collective modes of extended but very small composed objects (e.g. molecules moving in a strong gravitational field).

The configuration space Q of the problem is the principal fibre bundle FM of linear frames over M: $Q = FM$. Of course, $\dim Q = n(n+1)$.

Let $\pi : FM \longrightarrow M$ denote the canonical projection of FM onto M. The action of the structural group $GL(n,\mathbb{R})$ is denoted as usually:

$$A \in GL(n,\mathbb{R}) \ : \quad e = (e_1 \ldots e_n) \longrightarrow (e_i A^i{}_1 \ldots e_i A^i{}_n) \quad . \tag{2.1}$$

The mechanics described in the section 1 remains valid in any fibre $\pi^{-1}(m)$ which is a free homogenous space of $GL(n,\mathbb{R})$ (and after fixing some standard configuration can be identified with $GL(n,\mathbb{R})$ itself).

We shall also use the singular configuration space $Q_s = F_sM$ which is simply the completion of FM containing also singular frames i.e. linearly dependent n-tuples of vectors. In terms of Whitney sums:

$$Q_s = F_sM = \bigoplus_n TM = \bigcup_{m \in M} \ \mathop{\times}_n T_mM \quad . \tag{2.2}$$

Q_s is a linear bundle over M and Q is an open submanifold in Q_s.

The bundle of singular co-frames will be denoted as:

$$Q^*_s = F^*_s M = \bigcup_{m\in M} \underset{n}{\times} T^*_m M = \underset{n}{\oplus} T^* M \quad . \tag{2.3}$$

${}^*\pi : F^*M \longrightarrow M$ denotes the natural projection. As usually, TM, T^*M denote respectively the tangent and co-tangent bundles over M. We shall also use projections:

$$\theta_A : F_s M \longrightarrow TM \quad \text{where:} \quad \theta_A (e_1 \ldots e_n) = e_A \in TM \quad . \tag{2.4}$$

Motions are described by smooth curves: $\varsigma : \mathbb{R} \longrightarrow F_s M$.
The orbital i.e. translational part of motion is well-defined, namely it is described by the curve:

$$\pi \circ \varsigma : \mathbb{R} \longrightarrow M \quad . \tag{2.5}$$

Unfortunately, that was all what could be done in a bare manifold M (Excepting trivial statements that the analytical mechanics of the problem should be formulated in TFM and T^*FM).

In fact, there is no well-defined splitting of ς into translational and internal part. How to define the amount of change of the internal configuration along some orbital trajectory in M? The only possibility is to endow M with an affine connection Γ.

Let Γ be an affine connection on M described by some $L(n,\mathbb{R})$-valued connection form ω. The components of Γ with respect to some local coordinates x^i on M will be denoted by $\Gamma^i{}_{jk}$.

Even after introducing the affine connection Γ, configurations do not split into translational and internal parts, but infinitesimal behaviours i.e. velocities do. Namely, the tangent vector $\varsigma'(t) \in T_{\varsigma(t)}M$ can be replaced by the following pair:

$v = (\pi \circ \varsigma)'$ i.e. $v^i = \dfrac{dx^i}{dt}$ - translational velocity

$V_A = \dfrac{D(\theta_A \circ \varsigma)}{Dt} = \dfrac{De_A}{Dt}$ $A = 1 \ldots n$ -velocities of internal motion.

D denotes here the covariant differentiation in the Γ-sense.
At a given time instant t, all vectors v, V_A, e_A are attached at the same point $m = \pi(e) = (\pi \circ \varsigma)(t)$ (the position of the centre of mass i.e. orbital configuration).

Remark: $(v = (\pi \circ \varsigma)', \ V_A = \frac{D}{Dt}\theta_A \circ \varsigma)$, $A = 1 \ldots n$ is a non-holonomic quasivelocity. It becomes holonomic if and only if Γ is flat.

The rich structure of the configuration space Q as the bundle of frames over M, endowed with the connection Γ enables us to modify the formalism of analytical mechanics. Usually one uses two kind of canonical state spaces: TQ as a Newtonian state space and T^*Q as a Hamiltonian state space. The canonical Cartan form on T^*Q will be denoted as ω_Q : $\omega_{Q^p} = p \circ T\tau^*_{Qp}$ where τ^*_Q: $T^*Q \longrightarrow Q$ - the natural projection. The Hamiltonian phase space carries the natural symplectic structure: $(T^*Q, d\omega_Q)$.

Affine connection gives rise to splitting kinematical quantities into translational and internal parts. This enables us to introduce the auxiliary covariant state spaces:
The Newtonian state space (velocity state space):

$$P_N = F_sM \oplus TM \oplus F_sM = \bigcup_{m \in M} \times_{(2n+1)} T_mM = \oplus_{(2n+1)} TM . \tag{2.6}$$

The Hamiltonian state space becomes:

$$P = F_sM \oplus T^*M \oplus F^*_sM = \bigcup_{m \in M} \times_n T_mM \times_{(n+1)} T^*_mM . \tag{2.7}$$

The physical meaning of elements of canonical state spaces is as follows:

$$(\ldots e_A \ldots ; v ; \ldots V_A \ldots) \in P_N \tag{2.8}$$

$e = (\ldots e_A \ldots)$ - configuration; $m = \pi(e)$ - spatial location
v - translational velocity; $(\ldots V_A \ldots)$ - internal velocities.

$$(\ldots e_A \ldots ; p ; \ldots p^A \ldots) \in P \tag{2.9}$$

$e = (\ldots e_A \ldots)$ - configuration
p - covariant translational momentum
$(\ldots p^A \ldots)$ - covariant momenta of internal motion.
The connection Γ gives rise to the natural diffeomorphisms:

$$\begin{aligned} \Gamma &: TF_sM \longrightarrow P_N \\ {}^*\Gamma &: TF_sM \longrightarrow P \end{aligned} \tag{2.10}$$

Hence, our state spaces will be manifolds P_N, P. Canonical projections onto the configuration space Q will be denoted as:

$$pr: P_N \longrightarrow F_sM \qquad {}^*pr: P \longrightarrow F_sM, \text{ where:}$$

$$\mathrm{pr}(\ldots e_A \ldots;\ v\ ;\ldots V_A \ldots) = {}^{*}\mathrm{pr}(\ldots e_A \ldots;\ p\ ;\ldots P^A \ldots) = \tag{2.11}$$

$$= e = (\ldots e_A \ldots)$$

The covariant state spaces P_N, P enable us to define affine velocities and affine spin exactly as introduced in the section 1. This would be impossible without the connection Γ, i.e. when forced to work in TF_sM, $T^{*}F_sM$.
Affine velocity in the state $(\ldots e_A \ldots;\ v\ ;\ldots V_A \ldots)$ is a linear mapping $\Omega(e,V) : T_{\pi(e)}M \longrightarrow T_{\pi(e)}M$ defined as follows:

$$V_A = \Omega(e,V)e_A \quad \text{i.e.} \quad \Omega^i{}_j = V^i{}_A \ e^A{}_j \tag{2.12}$$

In co-moving description:

$$\hat{\Omega}^A{}_B = \langle e^A, V_B \rangle = e^A{}_i \, V^i{}_B \tag{2.12a}$$

We shall also employ shorthands $V^i{}_j$ instead of $\Omega^i{}_j$ and $V^A{}_B$ instead of $\hat{\Omega}^A{}_B$.
Affine spin in a Hamiltonian state $(\ldots e_A \ldots;\ p\ ;\ldots P^A \ldots)$ is a linear mapping $\Sigma(e,p) : T_{\pi(e)}M \longrightarrow T_{\pi(e)}M$

defined as:

$$\Sigma(e,P) = e_A \otimes P^A \qquad \text{i.e.} \quad \Sigma^i{}_j = e^i{}_A P^A{}_j \tag{2.13}$$

(hence: $e^A \circ \Sigma(e,P) = P^A$)
When using the co-moving description:

$$\hat{\Sigma}^A{}_B = \langle P^A, e_B \rangle = P^A{}_i \, e^i{}_B \tag{2.13a}$$

We shall also use the simplified, shorthand notation: $P^i{}_j$, $P^A{}_B$ instead of $\Sigma^i{}_j$, $\hat{\Sigma}^A{}_B$ respectively.

Geometrical meaning of quantities $P^A{}_i$, $P^i{}_j$, $P^A{}_B$ is similar as in Euclidean theory: $P^A{}_i$ are generators of abelian translations in fibres of F_sM, $P^i{}_j$ are generators of left affine rotations in internal degrees of freedom and $P^A{}_B$ generate right affine rotations i.e. the action of the structural group $GL(n,\mathbb{R})$.
The term "Hamiltonian generator" is understood here in the sense of the natural symplectic structure $(P, d\omega_\Gamma)$ on P, where ω_Γ is related to the Cartan form of Q through the Γ-mapping:

$$\omega_Q = {}^{*}\Gamma^{*}\omega_\Gamma \tag{2.14}$$

The simplest explicit and effective description of the symplectic structure is that in terms of Poisson brackets for some special functions. In the sequel we shall use the adapted local coordinates $(Q^i, Q^i{}_A, P_i, P^A{}_i)$ on P induced by arbitrary coordinates x^i on M. ($Q^i = x^i \circ \pi \circ {}^*pr$ and the values of remaining coordinates are simply the components of tensorial objects e_A, p, P^A in coordinates x^i). Such coordinates, although especially convienient, are non-canonical:

$$d\omega_\Gamma \neq dP_i \wedge dQ^i + dP^A{}_i \wedge dQ^i{}_A \quad . \tag{2.15}$$

Namely, we have the following formulas:

$$\begin{aligned}
&\{Q^i, Q^j\} = 0 \quad \{Q^i{}_A, Q^j{}_B\} = 0 \quad \{Q^i, Q^j{}_A\} = 0 \quad \{P^A{}_i, P^B{}_j\} = 0 \\
&\{P^A{}_i, Q^j\} = 0 \quad \{P^A{}_i, Q^j{}_B\} = \delta_i{}^j \delta^A{}_B \quad \{P_i, Q^j\} = \delta_i{}^j \\
&\{P^a{}_b, P^c{}_d\} = P^a{}_d \delta^c{}_b - P^c{}_b \delta^a{}_d \quad .
\end{aligned} \tag{2.16}$$

The last equation describes simply the structural relations of the real linear group GL(n,ℝ). All the above equations are exactly the same as in a flat space. Geometry of M is reflected by the remaining part of the basic Poisson brackets:

$$\{P_i, P_j\} = -P^k{}_l R^l{}_{kij} \quad . \tag{2.17}$$

Hence, in some sense the affine spin $P^k{}_l$ is a "charge" which "feels" the curvature tensor R. The coupling of internal degrees of freedom (through $P^a{}_b$) with the curvature is related to the non-integrability of infinitesimal actions generated by P_i - s .

When (M, Γ) is a Riemann-Cartan space with the metric g, then the g-symmetric part of $P^a{}_b$ does not contribute to (2.17). Hence, in this special case the kinematical coupling of internal degrees of freedom with geometry is completely described by the angular momentum itself. Besides, we have the following Poisson bracket-relations:

$$\begin{aligned}
&\{P_i, P^A{}_j\} = P^A{}_k \Gamma^k{}_{ji} \quad \{P_i, Q^j{}_A\} = -Q^k{}_A \Gamma^j{}_{ki} \\
&\{P_i, P^k{}_j\} = -P^l{}_j \Gamma^k{}_{li} + P^k{}_l \Gamma^l{}_{ji} \quad .
\end{aligned} \tag{2.18}$$

This means that P_i - s generate parallel transports along the x^i-th coordinate lines.

The co-moving objects enable us to describe the above Poisson brackets in more geometric terms:

$$\{ f\circ {}^*pr, g\circ {}^*pr \} = 0 \text{ - "positions do commute"} \tag{2.19}$$

$$\{ P_A, f\circ {}^*pr \} = (Df)_A \circ {}^*pr \tag{2.20}$$

where Df denotes the Γ -covariant differential of f (the horizontal part of df) and $(Df_e)_A$ is a value of Df_e on the horizontal lift of e_A to $e = (e_1 \ldots e_n) \in F_sM$. The equation (2.20) means that $P_A = P_i Q^i{}_A$ generates the parallel transport along the A-th standard horizontal vectorfield on F_sM.

$$\{ P^A{}_B, f\circ {}^*pr \} = (E^A{}_B \cdot f)\circ {}^*pr = \langle df, E^A{}_B \rangle \circ {}^*pr \tag{2.21}$$

where $E^A{}_B$ are Killing vectorfields of the structural group $GL(n,\mathbb{R})$. Locally:

$$E^A{}_B = Q^i{}_B \frac{\partial}{\partial Q^i{}_A} \quad .$$

The system of basic Poisson brackets in a co-moving representation is completed by the following equations:

$$\{ P^A{}_B, P^C{}_D \} = P^C{}_B \delta^A{}_D - P^A{}_D \delta^C{}_B \tag{2.22}$$

$$\{ P_C, P^A{}_B \} = - P_B \delta^A{}_C \tag{2.23}$$

$$\{ P_A, P_B \} = - P^D{}_C R^C{}_{DAB} + 2P_C S^C{}_{AB} \tag{2.24}$$

where $R^C{}_{DAB}$, $S^C{}_{AB}$ are co-moving components of the curvature and torsion tensors in a moving frame $(\ldots e_A \ldots)$. It is remarkable that the spatial counterpart of (2.24) i.e. (2.17) does not involve the torsion. (2.22) are the structural relations of $GL(n,\mathbb{R})$ and (2.23) means that P_A behave like vectors under $GL(n,\mathbb{R})$. However, P_A, $P^A{}_B$ do not integrate to the n-dimensional affine group, because, as shown by (2.24) the P_A - s do not commute according to the coupling of motion with spatial geometry.

Let us finish this section with the following conclusion: On the purely kinematical level, internal degrees of freedom are coupled to the curvature (through the affine spin), and translational ones to the torsion (through the linear momentum). This is just the meaning of (2.24).

3. Equations of motion.

The dynamical coupling between internal degrees of freedom and geometry.

In the aforementioned scheme of the kinematical coupling between degrees of freedom of our object and the spatial geometry there is no physics at all: This is simply the geometry of the bundle of linear frames with connection. The physical coupling is reflected by equations of motion, especially by the deviation from the geodesic motion caused by internal phenomena.

We start with a Hamiltonian system the dynamical model of which is based on the Lagrangian $L = T - V$ where V is some potential energy (depending on configurations only) and T - the kinetic energy based on the appropriate metric tensor G on FM. The simplest way to derive equations of motion is to use the Hamiltonian formalism and Poisson brackets given in the section 2. This is just the method we have used here.

Classification of metric tensors G provides us with a key to classify dynamical models with FM-degrees of freedom. We shall consider three most natural classes of G-s which arise as a globalization of metric tensors on $GL(n,\mathbb{R})$ as described in the section 1 onto the manifold of frames FM. All of them are compatible with Γ in the sense that horizontal and vertical subspaces are G-orthogonal.

1. G induced by the metric tensor g on M, isotropic and invariant under abelian translations (the global counterpart of (1.14)). This means that besides of the connection Γ, M is endowed with a Riemannian metric g. T is assumed to be invariant under abelian translations in fibres of F_sM as vector spaces and under local g-rotations in tangent spaces, generated by the "laboratory" components of spin: $S^a{}_b = P^a{}_b - P_b{}^a$. The only possibility is:

$$T(e,v,V) = \frac{1}{2}\langle G_e, (v,V)\otimes(v,V)\rangle =$$

$$\frac{M}{2}\langle g_{\pi(e)}, v\otimes v\rangle + \frac{1}{2}\langle g_{\pi(e)}, V_A\otimes V_B\rangle J^{AB} = \qquad (3.1)$$

$$\frac{M}{2} g_{ij}v^i v^j + \frac{1}{2} g_{ij} V^i{}_A V^j{}_B J^{AB}$$

where M is a positive constant (mass) and J^{AB} some symmetric and positive matrix (internal inertia). Without any physical restrictions we can put J to be diagonal.

Now, let us notice that after introducing the metric g, the manifold M becomes endowed with two affine connections: Γ and the natural Christoffel connection $\{\ \}$ induced by g. This enables us to introduce the tensor field:

$$\mathcal{K}^{i}{}_{jk} = \left\{ {i \atop j\,k} \right\} - \Gamma^{i}{}_{jk} . \tag{3.2}$$

$\mathcal{K}$ is algebraically built of Γ hence it contains also the information about the torsion tensor S of Γ. In the sequel, all symbols of the covariant differentiation are to be understood in a Γ-sense.

The resulting equations of motion, derived with the help of Poisson brackets given in the section 2 have the following form:

$$M \frac{Dv^{a}}{Dt} = -M\mathcal{K}^{a}{}_{bc} \cdot v^{b}v^{c} + P^{d}{}_{c}R^{c\ a}{}_{d\ b}\, v^{b} + F^{a} + \mathcal{K}_{cb}{}^{a} \frac{De^{b}{}_{A}}{Dt} \frac{De^{c}{}_{B}}{Dt} J^{AB}$$

$$e^{i}{}_{A} \frac{D^{2}e^{j}{}_{B}}{Dt^{2}} J^{AB} = -e^{i}{}_{A}\, g^{jk} \cdot \frac{Dg_{kl}}{Dt} \cdot \frac{De^{l}{}_{B}}{Dt} + N^{ij} \tag{3.3}$$

where:

$$N^{ij} = -e^{i}{}_{A} \frac{\partial V}{\partial e^{k}{}_{A}} g^{kj} \tag{3.4a}$$

is the affine momentum of forces (the first virial, or hyperforce) affecting the internal degrees of freedom, and

$$F^{a} = - g^{ab} \left(\frac{\partial V}{\partial x^{b}} - N^{m}{}_{j} \Gamma^{j}{}_{mb} \right) \tag{3.4b}$$

is the translational force. The covariant vector $\tilde{F}$ with components $F_{a} = g_{ab} F^{b}$ is related to the covariant exterior differential of V:

$$\tilde{F}(e) = - DV_{e} \circ \mathrm{lift}_{e}$$

where lift_{e}: $T_{\pi(e)}M \longrightarrow T_{e}\, F\, M$ is the Γ-horizontal lift.

The translational part of (3.3) proves that the dynamical coupling between degrees of freedom of the body and spatial geometry is in agreement with the kinematical one. Namely, the affine spin $P^{d}{}_{c}$ is a "charge" which "feels" the curvature R. Even when there are no forces F, the coupling between internal phenomena and R prevents the body to move along the geodesics. The translational momentum $p = Mv$ "feels" the torsion because it is coupled to $\mathcal{K}$.

Such a model predicts the exchange of the kinetic energy between orbital and internal degrees of freedom even in an interaction-free

case. The reason is that g is non-invariant under parallel transports. Such exchange of energy between two kinds of degrees of freedom disappears in a Riemann-Cartan space, where: $\nabla g = 0$.
$\mathcal{K}$ becomes then the "contorsion":

$$\mathcal{K}^k{}_{ij} = - S^k{}_{ij} + S_j{}^k{}_i - S_{ij}{}^k \tag{3.5}$$

and:
$$R^i{}_{jkl} = - R_j{}^i{}_{kl} \quad . \tag{3.6}$$

The last equation implies that it is only spin $S^i{}_j$ what interacts with the curvature. The symmetric i.e. deformative part of the affine spin becomes decoupled from the geometry. Namely, we get the following, simplified form of equations of motion:

$$M \frac{Dv^a}{Dt} = 2M\, v^b v^c\, S_{cb}{}^a + \frac{1}{2} S^d{}_c\, R^c{}_d{}^a{}_b\, v^b + F^a \tag{3.7a}$$

$$e^i{}_A \frac{D^2 e^j{}_B}{Dt^2} J^{AB} = N^{ij} \quad . \tag{3.7b}$$

The second equation can be interpreted as a balance law for affine spin:

$$\frac{DP^{ij}}{Dt} = \Omega^i{}_k \cdot P^{kj} + N^{ij} = \frac{De^i_A}{Dt} \cdot \frac{De^j_B}{Dt} \cdot J^{AB} + N^{ij}$$

$$= N^{ij} + 2 \frac{\partial T_{int}}{\partial g_{ij}} \tag{3.8}$$

where the last expression means that the non-conservation of the symmetric part of P^{ij} even in the interaction-free case is due to the explicit dependence of T and G on the spatial metric g. Hence, the kinematical affine symmetry of degrees of freedom is broken by the kinetic term even before introducing interactions.

When imposing gyroscopic constraints:

$$\langle g, e_A \otimes e_B \rangle = g_{ij}\, e^i{}_A\, e^j{}_B = \delta_{AB} \tag{3.9}$$

we get the theory of the test rigid body in a manifold (cf. [10]). Our equations of motion consist then of the balance of $P^a = Mv^a$ and $S^{ij} = P^{ij} - P^{ji}$. The last balance law is simply the skew-symmetric part of (3.8):

$$\frac{DS^{ij}}{Dt} = 2\, N^{[ij]} = N^{ij} - N^{ji} \quad . \tag{3.10}$$

When passing to the co-moving description of (3.7) we get the affine counterpart of the famous Euler equations for the gyroscope:

$$M\frac{dv^A}{dt} = -M\hat{\Omega}^A{}_B\, v^B + 2Mv^B\cdot v^C\cdot S_{CB}{}^A + \frac{1}{2M}S^C{}_D\cdot R^D{}_C{}^A{}_B\cdot p^B + F^A$$

$$\frac{d\hat{\Omega}^B{}_C}{dt}\cdot J^{CA} = -\hat{\Omega}^B{}_D\hat{\Omega}^D{}_C\cdot J^{CA} + N^{AB} \qquad (3.11)$$

or, equivalently:

$$\frac{dp^A}{dt} = -p^B\cdot J_{BC}\cdot P^{CA} + \frac{2}{M}p^B\cdot p^C\cdot S_{CB}{}^A + \frac{1}{M}S^D{}_C\cdot R^C{}_D{}^A{}_B\cdot p^B + F^A$$

$$\frac{dP^{AB}}{dt} = -P^{AC}\cdot J_{CD}\cdot P^{DB} + N^{AB} \quad . \qquad (3.12)$$

2. G induced by the metric tensor g on M, isotropic and invariant under structural group.

This is simply the global counterpart of (1.15). We put $\lambda = 0$ because the trace does not influence the coupling of internal motion with geometry. Then:

$$T(e,v,V) = \frac{M}{2}\langle g_{\pi(e)}, v\otimes v\rangle + \frac{\mu}{2}\langle g_{\pi(e)}, (\Omega\otimes\Omega)\cdot\tilde{g}_{\pi(e)}\rangle \qquad (3.13)$$

where $\tilde{g}$ is reciprocal to g and μ is a positive constant (scalar measure of internal inertia). In coordinates:

$$T = \frac{M}{2}g_{ij}\cdot v^i v^j + \frac{\mu}{2}g_{ij}\,\Omega^i{}_k\,\Omega^j{}_l\cdot g^{kl} \quad . \qquad (3.13a)$$

The equations of motion become:

$$M\frac{Dv^a}{Dt} = -M\,\mathcal{K}^a{}_{bc}\,v^b v^c + P^k{}_l\cdot R^l{}_k{}^a{}_i\,v^i + F^a +$$

$$+\frac{1}{\mu}P^i{}_k\,P^j{}_l\left(g_{ij}\,\mathcal{K}^{kla} - g^{kl}\,\mathcal{K}_{ij}{}^a\right) \qquad (3.14)$$

$$\frac{dP^A{}_B}{dt} = N^A{}_B \quad . \qquad (3.15)$$

The "internal part" (3.15) is a balance law for the co-moving components of affine spin. It becomes conservation law in the interaction-free case because the theory is invariant under the structural group $GL(n,\mathbb{R})$.

In spatial, laboratory terms:

$$\frac{DP^k{}_l}{Dt} = N^k{}_l - P^k{}_j \Omega^j{}_l + \Omega^k{}_i P^i{}_l \tag{3.16}$$

or, equivalently:

$$\frac{D\Omega^{jk}}{Dt} = -\mu\cdot\Omega^{ik}\Omega_i{}^j + \mu\Omega^{ji}\Omega^k{}_i + N^{kj} \ . \tag{3.17}$$

It is important that when (M, Γ, g) is a Riemann-Cartan space, then the translational equation (3.14) becomes identical with (3.7). Hence the coupling of degrees of freedom with geometry and the corresponding deviation of the trajectory from the geodesic form is in a Riemann-Cartan space the same for both models of the kinetic energy (3.1), (3.13).

3. Amorphous connection-space without fixed metric in M.
This will be the global FM-counterpart of (1.16). The kinetic energy of translational-less motion is then invariant under affine mappings acting separately in tangent spaces of M.

Hence, we use the same formula as in (3.1) but instead of the metric tensor g we substitute the tensor γ induced in T_mM by the configuration $e = (e_1 \ldots e_n)$, $m = \pi(e)$.

$$\gamma = \eta_{AB}\, e^A \otimes e^B \quad \text{i.e.:} \quad \gamma_{ij} = \eta_{AB}\, e^A{}_i\, e^B{}_j \tag{3.18}$$

where η is some numerical metric, e.g. $\eta_{AB} = \delta_{AB}$.
Then:

$$\begin{aligned} T(e,v,V) &= \frac{M}{2} \langle \gamma(e),\, v \otimes v \rangle + \frac{1}{2} \langle \gamma(e),\, V_A \otimes V_B \rangle \cdot J^{AB} = \\ &\frac{M}{2}\, \eta_{AB}\, v^A\, v^B + \frac{1}{2}\, \eta_{AB}\, \hat{\Omega}^A{}_C \hat{\Omega}^B{}_D \cdot J^{CD} \ . \end{aligned} \tag{3.19}$$

Equations of motion become:

$$M\frac{Dv^a}{Dt} = M\left(\Omega^a{}_b + \Omega_b{}^a\right)v^b + 2M\, v^b \cdot v^c \cdot S_{cb}{}^a + P^c{}_d \cdot R^d{}_c{}^a{}_b \cdot v^b + F^a \tag{3.20}$$

$$\frac{DP^{ab}}{Dt} = P^{ac}\left(\Omega_c{}^b + \Omega^b{}_c\right) - M\, v^a \cdot v^b + N^{ab}$$

or, lowering indices by means of $\gamma(e)$:

$$M\frac{Dv_a}{Dt} = 2M\, v^b \cdot v^c \cdot S_{cba} + P^c{}_d \cdot R^d{}_{cab} \cdot v^b + F_a \tag{3.21}$$

$$\frac{D}{Dt} P^{a}{}_{b} = N^{a}{}_{b} - M v^{a}{}_{\cdot} v_{b} \quad . \tag{3.21}$$

What is important, in this amorphous, metric-free space the structure of the coupling between degrees of freedom (in particular-internal ones) and geometry, responsible for the geodesic deviation, is exactly the same as in previous models (3.1) and (3.13) when applied to a Riemann-Cartan space.

Within amorphous framework we are still able to define gyroscopic constraints:

$$\frac{D\gamma_{ab}}{Dt} = 0 \quad \text{i.e.} \quad \Omega^{(ab)} = 0 \quad . \tag{3.22}$$

Equations of internal motion become then:

$$\frac{DS^{ab}}{Dt} = 2 \frac{DP^{[ab]}}{Dt} = 2 N^{[ab]} = N^{ab} - N^{ba} \quad . \tag{3.23}$$

However, in contrary to the theory of test gyroscopes in a manifold with a fixed metric g, gyroscopic constraints (3.22) are non-holonomic. Probably, they become holonomic if and only if (M, Γ) has a Riemann-Cartan structure.

Remark: In a model (3.19) the metric concepts (γ) become merely the algebraic manifestation of the"physical" matter represented here by the moving frame i.e. material point with extra attached affine degrees of freedom. Hence, this is the simplest, academic model of the anti-geometrodynamical physics.

Conclusions

Let us summarize briefly the above results:

1. In the theory of test affinely-rigid bodies in a non-Euclidean manifold the affine internal degrees of freedom are coupled to the curvature tensor through the affine spin which becomes some kind of the tensorial "charge" which "feels" the curvature. Translational degrees of freedom are coupled to the torsion (through the linear momentum).
2. The Riemann-Cartan model of spatial geometry is distinguished and geometrically priviliged because:

a) In a Riemann-Cartan space the couplings between internal degrees of freedom and spatial geometry for both g-models (3.1) and (3.13) become the same and independent on the details of the model like

J^{AB}, μ. The same concerns the coupling between translational motion and geometry.

b) The Riemann-Cartan pattern is strongly suggested by the amorphous theory (model (3.19)) where the general structure of the coupling between degrees of freedom and spatial geometry is just as for the Riemann-Cartan space in models (3.1), (3.13).

c) In a Riemann-Cartan space the coupling between internal degrees of freedom and geometry is "theory-invariant" - it does not depend on any particular dynamical model. This provides us with the possibility to "measure" the geometry by means of "geodesic deviation" of test affinely-rigid bodies (like molecules P_4, As_4, CH_4) even when we do not know the details of the dynamical theory.

References

1. V.I. Arnold: Inst. Fourier (Grenoble) 16, 1966, 319
2. V.I. Arnold: "Mathematical Methods of Classical Mechanics" (in Russian), Moscow, 1974, Nauka.
3. O.I. Bogoyavlenskij: Pisma w Astronomiceskij Journal 1 no. 9, 1975, 22 (in Russian).
4. O.I. Bogoyavlenskij: Prikladnajy Matematika i Mekhanika 40 no 2, 1976 270 (in Russian)
5. O.I. Bogoyavlenskij: Doklady Akademii Nauk SSSR 232 nr. 6, 1977m 1289
6. S. Chandrasekhar: "Ellipsoidal Figures of Equilibrium", New Haven and London, Yale University Press, 1969
7. Y. Dothan, M. Gell-Mann, Y. Ne´eman: Phys. Lett. 17, 1965, 148
8. F.J. Dyson: J. Math. and Mech. 18 nr. 1, 1968, 91
9. M. Gell-Mann: Phys. Rev. Lett. 14, 1965, 77
10. H.P. Künzle: Comm. Math. Phys. 27, 1972. 23
11. H. Römer, K. Wespfahl: Annalen der Physik, 22, 1969. 264
12. J.J. Slawianowski: Archives of Mechanics 26, 1974, 569
13. J.J. Slawianowski: Bull. de l'Academie Polonaise des Sciences, serie des sciences Techniques 23, 1975, 17
14. J.J. Slawianowski: Archives of Mechanics, 27, 1975, 93
15. J.J. Slawianowski: "Geometry of Phase Spaces" (in Polish), Polish Scientific Publishers, Warsaw, 1975
16. J.J. Slawianowski: Bull. de l'Academie Polonaise des Sciences. serie des sciences techniques 23, 1975, 43
17. J.J. Slawianowski: Int. J. Theor. Phys. 12, 1975, 271
18. J.J. Slawianowski: Rep. Math. Phys. 10, 1976, 219

19. L. Weaver, L.C. Biedenharn, Nucl. Phys. A185, 1972, 1

20. L. Weaver, L.C. Biedenharn, R.Y. Cusson: Ann. Phys. 77, 1973, 250

21. K. Westpfahl: Annalen der Physik 20, 1967, 113.

Jet Bundles and Weyl Geometry

J.D. Hennig

Institut für Theoretische Physik
Technische Universität Clausthal
D-3392 Clausthal

Abstract

Describing the projective structure $\mathcal{P}$ (given by the set of "freely falling particles") and the conformal (light cone) structure $\mathcal{C}$ of space time via subbundles of second order frame bundles, we investigate the existence and uniqueness of a Weyl geometry compatible with $\mathcal{P}$ and $\mathcal{C}$.

We first review some basic notions concerning the fibre bundle description of geometric structures on differentiable manifolds and then apply this formalism to the central step in the axiomatic approach to space time geometry presented by Ehlers, Pirani and Schild in (1). For a more detailed version of our lecture, cf. (2)

1. Geometric objects and G-structures

The coordinate free generalization of the classical definition of a "geometric object" on $\mathbb{R}^n$ leads in a natural way to the concept of higher order frame bundles and their associated bundles. E.g., a vector field v on an open subset $U \subset \mathbb{R}^n$ usually is given by n functions

$$v^\mu : U \longrightarrow \mathbb{R} \quad , \quad \mu = 1,\dots,n$$

with the transformation rule $\bar{v}^\mu(m) = (\partial \bar{x}^\mu / \partial x^\nu)|_m \, v^\nu(m)$, $m \in U$.

In contrast, the global, coordinate independent description of a vector field w on an n-dimensional manifold M is defined by means of a cross section

$$w : M \longrightarrow TM$$

of the tangent bundle TM of M (similarly for all "first order" geometric objects, like tensor fields, densities etc., i.e. objects, whose transformation properties only depend on first order partial derivatives of coordinate transformations).

TM may be identified with the fibre bundle "associated" with the bundle PM of linear frames over M via the standard representation of $GL(n,\mathbb{R})$ on $\mathbb{R}^n$. This yields the following possibility of introducing "higher order" geometric objects (transforming with respect to higher order partial derivatives of coordinate transformations, as e.g. linear connections).

Obviously, in a first step one has to generalize in an appropriate way the notion of a linear frame ('1-frame') $e(m) = (e_1(m),\ldots,e_n(m))$ at a point $m \in M$. This can be done if one interpretes e(m) as an equivalence class of coordinate systems $\varphi_\alpha : M \supset U_\alpha \longrightarrow \mathbb{R}^n$ around m with $\varphi_\alpha(m) = 0$ and first order contact in m.

Correspondingly, a <u>k-frame</u> $e^{(k)}(m)$ in $m \in M$ is an invertible k-jet $j^k(\varphi)$ of coordinate systems with source m and target $o \in \mathbb{R}^n$ (3). The set P^kM of all k-frames for all $m \in M$ then can be given a natural C^∞ bundle structure, fibred over M by the canonical projection $\pi^k : e^{(k)}(m) \longrightarrow m$. In particular, $P^1M = PM$.

In order to define P^kM as a principal fibre bundle, we have to introduce a structure group $G^k(n)$ for P^kM in analogy with the structure group $G^1(n) := GL(n,\mathbb{R})$ of PM. The jet interpretation of k-frames over M suggests to choose for $G^k(n)$ the set of all k-frames in $0 \in \mathbb{R}^n$ (classes of 'coordinate transformations' $\psi_\beta : \mathbb{R}^n \supset U_\beta \longrightarrow \bar{U}_\beta \subset \mathbb{R}^n$, $\psi_\beta(0) = 0$, with k-order contact in $0 \in \mathbb{R}^n$). The natural composition of jets then defines a (Lie-) group structure on $G^k(n)$ and a canonical C^∞ left action of $G^k(n)$ on P^kM, thus yielding P^kM as a principal fibre bundle over M with structure group $G^k(n)$.

Now, let $\phi : G^k(n) \times F \longrightarrow F$ be a C^∞ left action on a finite dimensional manifold F. Define in the usual manner the bundle $P^kM \times^\phi F$ over M, 'ϕ-associated' with P^kM (4). I.e., take the induced left action on $P^kM \times F$ and consider the factorizing 'principal map' $\nu : P^kM \times F \longrightarrow P^kM \times^\phi F$ onto the set of $G^k(n)$-orbits in $P^kM \times F$. There is a canonical projection $\pi^k_\phi : P^kM \times^\phi F \longrightarrow M$, uniquely determined through the following commutative diagram, and a

C^∞ structure on $P^kM \times^\phi F$, such that the latter becomes a C^∞ fibre bundle over M,

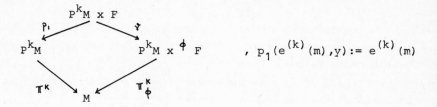

$$\begin{array}{ccccc} & & P^kM \times F & & \\ & \swarrow^{p_1} & & \searrow^{\gamma} & \\ P^kM & & & & P^kM \times^\phi F \\ & \searrow_{\pi^k} & & \swarrow_{\pi^k_\phi} & \\ & & M & & \end{array} \quad , \; p_1(e^{(k)}(m),y) := e^{(k)}(m)$$

We define a <u>geometric object of order k and type (ϕ,F)</u> to be a C^∞ cross section of $P^kM \times^\phi F$

$$\sigma : \; M \longrightarrow P^kM \times^\phi F \; .$$

σ is called irreducible if ϕ is transitive.

For given F and transitive ϕ let us fix now a point $y_o \in F$. Then we obtain a one to one correspondence between the set of all (irreducible) geometric objects of order k and type (ϕ,F) and the set of all <u>G-structures of order k</u>, i.e. the set of all principal subbundles of P^kM with structure group G, where G is the isotropy subgroup of $G^k(n)$ with respect to ϕ and y_o. The correspondence is given by the map $\rho^{y_o} := \gamma \circ i^{y_o}$

$$\begin{array}{ccccc} & & P^kM \times \{y_o\} & & \\ & \nearrow^{i^{y_o}} & & \searrow^{\gamma} & \\ P^kM & & \xrightarrow{\quad \rho^{y_o} \quad} & & P^kM \times^\phi F \\ & \searrow_{\pi^k} & & \nearrow_{\sigma} & \\ & & M & & \end{array} \quad , \; i^{y_o}(e^{(k)}(m)) := (e^{(k)}(m), y_o) ,$$

which defines for each cross section σ of $P^kM \times^\phi F$ the G-structure $(\rho^{y_o})^{-1}(\sigma(M)) \subset P^kM$.

In this way, geometric structure of quite different type can be described in a uniform, global and coordinate free manner via cross sections σ in fibre bundles, associated with the bundles P^kM, where k determines for σ the order of 'infinitesimal' dependence of its geometric transformation rule on changes of coordinate systems. Moreover, if σ is irreducible, it may be represented by a G-subbundle of P^kM, with suitable $G \subset G^k(n)$. Hence, for these objects, P^kM plays the role of a 'universal' bundle, which

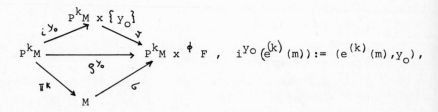

(i) contains the information of all types of 'infinitesimal geometries' of order k on M, given through the closed (isotropy) subgroups G of $G^k(n)$, and

(ii) by its global topological properties determines all possibilities of pointwise fitting together infinitesimal geometries of the same type in order to get global differentiable structures on M.

Examples for irreducible geometric objects and G-structures, respectively, of order $k = 1$ are Riemannian or pseudo-Riemannian structures ($G = O(p,q)$, $p + q = n$), conformal structures ($G = CO(p,q) = O(p,q) \times D$, D being the group of dilatations in $\mathbb{R}^n$), symplectic structures ($G = Sp(n)$), orientations ($G = GL^+(n,\mathbb{R})$) or parallelizations ($G = \{$unit element δ in $GL(n,\mathbb{R})\}$). Examples for non-irreducible objects are non-constant vector fields with zeros.

Since we are interested here in projective structures and Weyl structures (which are defined by means of linear connections), we have to discuss in particular second order objects, due to the fact, that connections contain first and second order partial derivatives in their characteristic transformation rule:

2. Second order structures

The structure group $G^2(n)$ of P^2M may be realized as $G^2(n) = \{ (a,b)/a = (a^\mu_\nu) \in GL(n,\mathbb{R});\ b = (b^\mu_{\nu\rho}),\ b^\mu_{\nu\rho} = b^\mu_{\rho\nu} \in \mathbb{R},\ \mu,\nu,\rho = 1,\dots,n \}$ with the multiplication law

$$(a^\mu_\nu, b^\mu_{\nu\rho})\ (\bar{a}^\mu_\nu, \bar{b}^\mu_{\nu\rho}) = (a^\mu_\alpha \bar{a}^\alpha_\nu,\ a^\mu_\alpha \bar{b}^\alpha_{\nu\rho} + b^\mu_{\alpha\beta}\, \bar{a}^\alpha_\nu\, \bar{a}^\beta_\rho)\ .$$

This reflects nothing else than the chain rule, if one interpretes a^μ_ν, $b^\mu_{\nu\rho}$ as the partial derivatives (in $0 \in \mathbb{R}^n$) $\partial\bar{x}^\mu/\partial x^\nu$, $\partial^2\bar{x}^\mu/\partial x^\nu \partial x^\rho$ of coordinate transformations $\psi : \mathbb{R}^n \ni (x^\mu) \longrightarrow (\bar{x}^\mu) \in \mathbb{R}^n$, $\psi(0) = 0$.
$G^2(n)$ turns out to be a semidirect product $G^1(n) \circledS B$ of $G^1(n) = GL(n,\mathbb{R}) = \{(a^\mu_\nu, 0)\}$ [1) and the invariant abelian subgroup $B = \{ (\delta^\mu_\nu, b^\mu_{\nu\rho}) \}$.

As an example for an irreducible second order geometric object we may take each symmetric linear connection Γ on M. The transformation formula for Γ then implies $G = G^1(n)$ for the corresponding isotropy subgroup of $G^2(n) = G^1(n) \circledS B$ (with suitably chosen y_o in the standard fibre F of $P^2M \times_\Phi F$; cf. section 1). Thus Γ can be

1) We simply do not distinguish between $G^1(n)$ and $G^1(n) \circledS \{0\} \subset G^2(n)$.

represented by a $G^1(n)$-subbundle of P^2M.

A <u>projective structure</u> on M is an equivalence class $[\Gamma]$ of symmetric linear connections, where Γ and Γ' are said to be equivalent, if their autoparallels coincide as unparametrized curves. $[\Gamma]$ is another example of an irreducible second order object, and it may be described by a J_{proj} -subbundle of P^2M with

$$J_{proj} := G^1(n) \otimes H_{proj} \subset G^2(n),$$

$$H_{proj} := \{(\delta^\mu_\nu, b^\mu_{\nu\rho}) / b^\mu_{\nu\rho} = \tau_\nu \delta^\mu_\rho + \delta^\mu_\nu \tau_\rho \ , \ (\tau_\sigma) \in \mathbb{R}^{n*}\}.$$

This J_{proj} -structure can also be obtained as the union of all $G^1(n)$-structures in P^2M equivalent to Γ .

We continue this list of examples with a second order description of conformal structures [1]. Consider a pseudo-Riemannian structure g of signature (p,q) and, for a given coordinate system, its local representation by functions $g_{\mu\nu}$. Then $g_{\mu\nu}$ and the partial derivatives $g_{\mu\nu,\rho}$ together constitute an irreducible second order object and correspond to an O(p,q)-structure in P^2M, $O(p,q) \subset G^1(n) \subset G^2(n)$ (we call a second order O(p,q)-structure 'prolonged', if it is generated by a first order O(p,q)-structure in this way; there are O(p,q)-subbundles in P^2M which are not prolonged). Similarly to the case of projective structures, we take now the union of all O(p,q)-structures in P^2M, conformally equivalent to g. This gives a $J^{p,q}_{conf}$ -structure, where

$$J^{p,q}_{conf} := CO(p,q) \otimes H^{p,q}_{conf} \subset G^2(n),$$

$$H^{p,q}_{conf} := \{(\delta^\mu_\nu, b^\mu_{\nu\rho}) / b^\mu_{\nu\rho} = \tau_\nu \delta^\mu_\rho + \delta^\mu_\nu \tau_\rho - \eta_{\nu\rho}\eta^{\lambda\mu}\tau_\lambda, \ (\tau_\sigma) \in \mathbb{R}^{n*}\}^{2)}$$

Thus we may characterize $[g]$ in a unique manner by a $J^{p,q}_{conf}$ -subbundle of P^2M.

Let us restrict for the following to the physically relevant case of a 4-dimensional, connected, paracompact manifold M. Denote by $S[O(3,1)]$, $S[J_{conf}]$, $S[GL(4,\mathbb{R})]$, $S[J_{proj}] \subset P^2M$ the second order G-structures generated, as indicated above, by g, $[g]$, Γ and $[\Gamma]$, where g is a given Lorentz metric on M and Γ its Levi-Civita connection. The well known expression of $\Gamma^\mu_{\nu\rho}$ in terms of $g_{\mu\nu}$ and $g_{\mu\nu,\rho}$ then implies $S[O(3,1)] \subset S[GL(4,\mathbb{R})]$.

1) Notation: If g stands for a given pseudo-Riemannian structure on M, the class of all g', conformally equivalent to g, i.e. $g' = \Omega g$ with positive function Ω, will be denoted by $[g]$. $[g]$ is called the <u>conformal structure</u> induced by g.

2) $(\eta_{\mu\nu}) = (\eta^{\mu\nu})$ diagonal with $\eta_{\mu\mu} = 1$ for $\mu = 1,\dots,p$ and $\eta_{\mu\mu} = -1$ for $\mu = p+1,\dots, n=p+q$.

Thus we get

$$(1)\qquad \begin{array}{ccc} S[O(3,1)] & \subset & S[J_{conf}] \\ \cap & & \\ S[GL(4,\mathbb{R})] & \subset & S[J_{proj}] \end{array} \subset P^2M$$

as the G-structure version of the statements: $[g]$, Γ are induced by g, and $[\Gamma]$ is induced by Γ .

3. The Weyl structure

Our aim now is to reverse the discussion leading to the inclusions (1) and to start with independently given conformal (signature (3.1)) and projective structures $\mathcal{C}$ and $\mathcal{P}$ on M. This is physically motivated by the possibility, at least in principle, to determine the conformal and the projective structure of space time by measuring the world lines of photons (giving the light cones on M, hence $\mathcal{C}$) and of freely falling point-particles (forming a sufficiently large set $\mathcal{F}$ of unparametrized autoparallels of a linear connection on M in order to fix $\mathcal{P}$); cf. (1).

In their axiomatic formulation of space time geometry Ehlers, Pirani and Schild propose a compatibility condition for $\mathcal{C}$ and $\mathcal{P}$, which expresses the possibility to approximate the space time path of a photon by those of freely falling particles. Using this condition the authors give a coordinate dependent proof for the existence of a unique Weyl structure on M, which induces $\mathcal{C}$ and $\mathcal{P}$.

(2) Compatibility of $\mathcal{C}$ and $\mathcal{P}$: Each $m \in M$ has a neighbourhood U_m in M such that in U_m the timelike cone C_m coincides with the union of all $p \cap U_m$, $p \in \mathcal{F}_m$ [1].

As we shall indicate below (cf. (2), for the proofs), condition (2) may be translated into the coordinate free bundle language. The unique existence of the Weyl structure then turns out to be a simple consequence of algebraic properties of the structure groups J_{proj} and J_{conf}:

1) If p is the unparametrized world line of a freely falling particle passing through $m \in M$, we write $p \in \mathcal{F}_m$.
For $m \in M$ (and in a sufficiently small neighbourhood of m) $C_m \subset M$ denotes the timelike cone in m, i.e. the interior of the double light cone with vertex m.

Let us call an unparametrized null geodesic of $\mathcal{C}$ a 'null geodesic path' (cf. (5)) and an unparametrized autoparallel of $\mathcal{P}$ a '$\mathcal{P}$-path'. By continuity arguments, it is easy to show that each null geodesic path of $\mathcal{C}$ is a $\mathcal{P}$-path, if (2) is fulfilled.
We call $\mathcal{C}$ and $\mathcal{P}$ then <u>null path compatible</u>. This, however, is a condition of second order contact of null geodesic paths and $\mathcal{P}$-paths. Hence we may express it in terms of second order G-structures.
We get

<u>Lemma A</u> Let $S[J_{conf}]$, $S[J_{proj}] \subset P^2M$ be the second order structures corresponding to given conformal and projective structures $\mathcal{C}$, $\mathcal{P}$ on M. Then $\mathcal{C}$ and $\mathcal{P}$ are null path compatible if and only if

(3) $$S[J_{conf}] \cap S[J_{proj}] \neq \emptyset$$ 1)

<u>Definition</u> A symmetric linear connection Γ on M together with a conformal structure $\mathcal{C}$ on M is called a <u>Weyl structure</u>, if for each two points m, m' $\in$ M and each curve c, connecting m, m', the Γ-parallel-transport along c maps light like directions in m onto light like directions in m'.

This condition for Γ and $\mathcal{C}$, too, is a second order compatibility condition, hence a property of second order G-structures:

<u>Lemma B</u> Let $S[GL(4,\mathbb{R})]$, $S[J_{conf}] \subset P^2M$ be the second order structures belonging to a given symmetric linear connection Γ and a conformal structure $\mathcal{C}$ on M. Then $\mathcal{C}$, Γ constitute a Weyl structure if and only if
$$S[GL(4,\mathbb{R})] \cap S[J_{conf}] \neq \emptyset .$$

Having translated all relevant terms into the language of G-structures, it is trivial now to compare the intersection conditions in lemma A and B in order to show the existence and uniqueness of a Weyl structure on M:

Note first, that $S \cap S' \neq \emptyset$ for two principal subbundles S, S' (with structure groups G, G') of a principal fibre bundle P implies

1) We say, two subjects A, B of a fibre bundle E over M have a 'non-trivial intersection', $A \cap B \neq \emptyset$, if for each $m \in M$ $A \cap B \cap E_m \neq \emptyset$, where E_m denotes the fibre in E over m.

$S \cap S'$ to be also a principal subbundle of P with structure group $G \cap G'$ (an analogous statement for arbitrary C^∞ fibre bundles E is not true; the intersection of two subbundles of E need not be a submanifold at all).

Hence, given $\mathcal{C}$, $\mathcal{P}$ on M or, equivalently, $S[J_{conf}]$, $S[J_{proj}] \subset P^2M$, condition (2) implies eq. (3), wich means

$$S[J_{conf}] \cap S[J_{proj}] = S[CO(3,1)]$$

because of $J_{conf} \cap J_{proj} = CO(3,1)$ ($S[CO(3,1)]$ being a $CO(3,1)$-subbundle in P^2M).
The action of $GL(4,\mathbb{R}) = G^1(4) \subset G^2(4)$ on $S[CO(3,1)]$ yields a $GL(4,\mathbb{R})$-structure $S[GL(4,\mathbb{R})] \subset P^2M$, hence determines a symmetric linear connection Γ on M. Obviously

$$(4) \qquad \begin{aligned} &S[GL(4,\mathbb{R})] \subset S[J_{proj}] \qquad \text{and} \\ &S[GL(4,\mathbb{R})] \cap S[J_{conf}] \neq \emptyset \quad , \end{aligned}$$

i.e. Γ belongs to $\mathcal{P}$ and $(\Gamma, \mathcal{C})$ defines a Weyl structure on M.

Moreover, Γ is given uniquely. For, let $\bar{S}[GL(4,\mathbb{R})] \subset P^2M$ be another $GL(4,\mathbb{R})$-structure fulfilling (4). This implies $S[CO(3,1)] = S[J_{conf}] \cap S[J_{proj}] \subset \bar{S}[GL(4,\mathbb{R})]$.
The invariance of $\bar{S}[GL(4,\mathbb{R})]$ under the action of $GL(4,\mathbb{R})$ then yields $\bar{S}[GL(4,\mathbb{R})] = S[GL(4,\mathbb{R})]$.

Proposition

Consider a conformal and a projective structure $\mathcal{C}$ and $\mathcal{P}$ on M. Assume the existence of a distinguished class $\mathcal{F}$ of $\mathcal{P}$-paths (called 'freely falling particles'), such that $\mathcal{C}$, $\mathcal{P}$ are compatible in the sense of condition (2). Then there is exactly one symmetric linear connection Γ belonging to $\mathcal{P}$ (i.e., $\mathcal{P} = [\Gamma]$), which, together with $\mathcal{C}$, constitutes a Weyl structure on M.

4. Concluding remarks

The physical implication of this proposition is straightforward, as pointed out in (1). There is a well-known necessary and sufficient condition for the existence of a Lorentz metric g in $\mathcal{C}$ ($\mathcal{C} = [g]$), which induces Γ as its Levi-Civita connection (6) (or, equivalently, for the existence of a prolonged $O(3,1)$-structures in $S[J_{conf}] \cap S[GL(4,\mathbb{R})] = S[CO(3,1)]$). This condition states, that two identical 'gravitational clocks' in $m \in M$, transported to m' along different curves, should show the same time rate in m'. Or, in mathematical terms, Γ -parallel-transport of a (timelike) vector v from m to m' should change the length of v (measured with respect to an arbitrary metric $\bar{g}$ in $\mathcal{C}$) independently of the curve connecting m and m', thus defining a scaling function Ω all over M. Then $g = \alpha\, \Omega^{-1}\, \bar{g}$, i.e. g is unique up to a constant positive factor α (a prolonged $O(3,1)$-subbundle in $S[CO(3,1)] \subset P^2M$ is unique up to the action of an element $\alpha\delta$ in the dilatation subgroup $D \subset G^1(4)$), and we recover the inclusions (1).

Hence, starting with the structures $\mathcal{C}$ and $\mathcal{P}$ on space time M and assuming the validity of the compatibility (2) and the integrability condition described above (which, in principle, both can be tested by experiment only using the concepts of light rays and freely falling particles), it is possible to derive the 'physical' Lorentz metric on M up to a constant factor.

Of course, our considerations in section 3 can easily be repeated in local coordinates, thus yielding a simple derivation (different to that one proposed in (1.1)) of the Weyl structure. Our main purpose, however, was to indicate by an instructive example, how the formalism of G-structures treats 'geometric objects' different in type and order such as tensor fields, connections, projective structures etc. in a completely uniform, coordinate free manner, and how interrelations between these structures may reduce to simple set- and group-theoretical properties of principal bundles and their structure groups.

References

(1) J. Ehlers, F.A.E. Pirani, A. Schild, in General Relativity (ed. L. O'Raifeartaigh, Oxford (1972)
F.A.E. Pirani, Symposia Mathematica XII, 119, 67 (1973)
J. Ehlers, in Relativity, Astrophysics and Cosmology, (ed. W. Israel), D. Reidel, Dordrecht-Holland (1973)

(2) J.D. Hennig, G-structures and Space Time Geometry I, ICTP, Trieste preprint IC/78/46 , and
G-Structures and Space Time Geometry II, in preparation

(3) S. Kobayashi, Transformation Groups in Differential Geometry, Springer, New York, (1972)
J. Dieudonné, Treatise on Analysis III, Academic Press, New York (1972)

(4) S. Kobayashi, K. Nomizu, Foundations of Differential Geometry I, Interscience, New York (1963)
W. Greub, S. Halperin, R. Vanstone, Connections, Curvature and Cohomology II, Academic Press, New York (1973)

(5) F.A.E. Pirani, A. Schild, in Perspectives in Geometry and General Relativity, (ed. B. Hoffmann), Bloomington (1966)

(6) H. Weyl, Mathematische Analyse des Raumproblems, Berlin (1923)

Line Fields and Lorentz Manifolds

by

W.H. Greub, University of Toronto

Abstract:
The first result of this paper is that the total Gaussian curvature of a compact Lorentz manifold vanishes (Theorem I, sec.8). The argument in the proof is different from those given in [1] and [2]. In fact, we make extensive use of the existence of a global line field on a Lorentz manifold, rather than reducing the problem to the classical case of a Riemannian manifold.

In §4 the index of a line field at an isolated singularity is defined in intrinsic terms. Theorem II in sec. 12 expresses the index sum of a line field on a compact Riemannian 4-manifold in terms of the total Gaussian curvature. A consequence of this result is that every compact 4-manifold which admits a line field, also admits a vector field without zeros. (Theorem III, sec. 13). Finally, in sec. 15 it is shown that the notions of orientability and time-orientability on a Lorentz manifold are independent.

§1. Basic concepts

In this introductory paragraph we summarize a number of basic definitions and theorems about smooth manifolds and vector bundles. For details the reader is referred to [3] and [4].

1. Manifolds.

All manifolds, maps and functions are assumed to be of class C^∞. Let M be a manifold and let $S(M)$ denote the ring of functions on M. A vector field is a derivation in $S(M)$. Thus a vector field is an operator $X\colon S(M) \longrightarrow S(M)$ which is linear over the function ring and satisfies

$$X(f\cdot g) = X(f)\cdot g + f\cdot X(g) \qquad f,g \in S(M) .$$

The vector fields on M form an $S(M)$-module denoted by $X(M)$. A diffeomorphism $\varphi: M \longrightarrow N$ induces an isomorphism $\varphi_*: X(M) \xrightarrow{\cong} X(N)$.

A p-form on M is a skew symmetric p-linear (over $S(M)$) map

$$\Phi : \underbrace{X(M) \times \ldots \times X(M)}_{p \text{ times}} \longrightarrow S(M) .$$

The module of p-forms on M is denoted by $A^p(M)$. A map $\varphi: M \longrightarrow N$ induces a homomorphism $\varphi^*: A^p(M) \longrightarrow A^p(N)$. The exterior derivative of a p-form Φ will be denoted by $\delta\Phi$.

An n-manifold M is called <u>orientable</u>, if there exists an n-form Δ such that $\Delta(x) \neq 0$, $x \in M$. Two such forms are called equivalent, if the function f determined by $\Delta_2 = f\cdot\Delta_1$, is positive. An <u>orientation</u> of M is an equivalence class of n-forms under this relation. Thus a connected orientable manifold has exactly two orientations.

A <u>pseudo-Riemannian metric</u> of signature s on a manifold is a symmetric tensor field g of degree 2 which makes every tangent space into a pseudo-Riemannian vector space with signature s. The inner product of two vector fields X and Y on such a manifold will be denoted by (X, Y).

On an oriented pseudo-Riemannian manifold M there is a unique n-form Δ_M which represents the orientation and satisfies the Lagrange identity

$$\Delta_M(X_1,\ldots,X_n)\cdot\Delta_M(Y_1,\ldots,Y_n) = (-1)^{\frac{1}{2}(n-s)} \det((X_i,Y_j)) \quad , X_i,Y_j \in X(M) .$$

It is called the <u>positive normed n-form</u> on M.

A pseudo-Riemannian manifold is called <u>Riemannian</u>, if the inner product is positive definite ($s = 0$). A <u>Lorentz manifold</u> is a pseudo-Riemannian manifold of dimension 4 and signature 0.

2. Fibre bundles

A <u>fibre bundle</u> is a collection (E, π, M, Y) where E, M and Y are smooth manifolds and $\pi: E \longrightarrow M$ is a smooth map onto M with the following property: There exists an open covering $\{U_\alpha\}$ of M and a family of diffeomorphisms

$$\psi_\alpha : U_\alpha \times Y \xrightarrow{\cong} \pi^{-1}(U_\alpha)$$

such that

$$\pi\,\psi_\alpha(x,y) = x \qquad x \in U_\alpha\,,\ y \in Y\ .$$

Thus, for every $x \in U_\alpha$, ψ_α restricts to a diffeomorphism $\psi_{\alpha,x}: Y \longrightarrow Y_x$ where $Y_x = \pi^{-1}(x)$. Y_x is called the fibre over x.

A cross-section in a fibre bundle is a smooth map: $\sigma: M \longrightarrow E$ such that

$$\pi \circ \sigma = id\ .$$

A vector bundle ξ is a fibre bundle where $Y = F$ and $Y_x = F_x$ are vector spaces of fixed dimension k and all maps $\psi_{\alpha,x}: F \xrightarrow{\ \cong\ } F_x$ are linear (and hence linear isomorphisms). Vector bundles will be denoted by

$$\xi = (E, \pi, M, F)\ .$$

A typical example is the tangent bundle $\tau_M = (T_M, \pi_M, M, \mathbb{R}^n)$. The cross-sections in a vector bundle form a module over $S(M)$ and denoted by $\mathrm{Sec}\,\xi$. Thus, $\mathrm{Sec}\,\tau_M = \mathcal{X}(M)$. Next, let $(\dot{T}_M, \pi_M, M, \dot{\mathbb{R}}^n)$ denote the fibre bundle obtained by removing the zero-cross-section in the tangent bundle. Identifying each vector $h \in \dot{T}_x(M)$ with all vectors λh $(\lambda \neq 0)$ we obtain a new fibre bundle $(P_M, \tilde{\pi}_M, M, \mathbb{R}P^{n-1})$ called the projective bundle over M. The canonical projections $\rho_x : \dot{T}_x(M) \longrightarrow P_x(M)$ determine a smooth map $\rho: \dot{T}_M \longrightarrow P_M$ which makes the diagram

$$\begin{array}{ccc} \dot{T}_M & \xrightarrow{\ \rho\ } & P_M \\ & {\scriptstyle \pi_M}\searrow \quad \swarrow{\scriptstyle \tilde{\pi}_M} & \\ & M & \end{array}$$

commute.

A line field on M is a cross-section in the projective bundle. Thus a line field σ assigns to every point $x \in M$ a 1-dimensional subspace $\sigma(x)$ of $T_x(M)$. If $\varphi: M \xrightarrow{\ \cong\ } N$ is a diffeomorphism, then a line field σ on M determines a line field $\tau = \varphi_* \sigma$ on N. A local lifting of a line field over an open subset U is a vector field X in U such that $\rho_x X(x) = \sigma(x)$, $x \in U$. Every point $a \in M$ has a neighbourhood U such that a line field σ lifts over U. In fact, consider the point $\sigma(a) \in P_M$ and choose a neighbourhood V of $\sigma(a)$ such that the bundle $\dot{T}_M \longrightarrow P_M$ admits a cross-section s over V. Now choose U such that $\sigma(U) \subset V$ and define X by

$$X(x) = s(\sigma(x))\ , \qquad x \in U\ .$$

If X and Y are liftings of σ over U, then we have $Y = f \cdot X$ where $f(x) \neq 0$, $x \in U$.

3. Bundle valued differential forms

A ξ-valued p-form in a vector bundle is a skew symmetric p-linear map

$$\Phi : \underbrace{X(M) \times \dots \times X(M)}_{p \text{ times}} \longrightarrow \operatorname{Sec} \xi \ .$$

The module of ξ-valued p-forms will be denoted by $A^p(M;\xi)$ and we shall write

$$\sum_{p=0}^{n} A^p(M;\xi) = A(M;\xi) \ , \ n = \dim M.$$

Note that $A^0(M;\xi) = \operatorname{Sec}\xi$. Moreover, if ξ is the product bundle $M \times \mathbb{R}$ then a ξ-valued p-form is just an ordinary p-form on M.

Let ξ^* denote the dual vector bundle and denote the scalar product of two cross- sections $\sigma^* \in \operatorname{Sec}\xi^*$ and $\sigma \in \operatorname{Sec}\xi$ by $\langle \sigma^*, \sigma \rangle$. Then every pair of forms $\Phi \in A^p(M;\xi^*)$ and $\Psi \in A^q(M;\xi)$ determines an (ordinary) differential form of degree p+q by

$$\langle \Phi, \Psi \rangle (X_1, \dots, X_{p+q}) = \frac{1}{p!\,q!} \sum_\sigma \varepsilon_\sigma \langle \Phi(X_{\sigma(1)}, \dots, X_{\sigma(p)}), \Psi(X_{\sigma(p+1)}, \dots, X_{\sigma(p+q)}) \rangle .$$

It is called the scalar product of Φ and Ψ.

Now consider for every r the vector bundle $\Lambda^r \xi$ whose fibre at x is the space $\Lambda^r T_x(M)$. Then the exterior product of two forms $\Phi \in A^p(M;\Lambda^r\xi)$ and $\Psi \in A^q(M;\Lambda^s\xi)$ is the (p+q)-form with values in $\Lambda^{r+s}\xi$ given by

$$(\Phi \wedge \Psi)(X_1, \dots, X_{p+q}) = \frac{1}{p!\,q!} \sum_\sigma \varepsilon_\sigma \, \Phi(X_{\sigma(1)}, \dots, X_{\sigma(p)}) \wedge \Psi(X_{\sigma(p+1)}, \dots, X_{\sigma(p+q)})$$

where the $\wedge$ on the right hand side denotes the $\wedge$-product of a cross-section in $\Lambda^r\xi$ with a cross-section in $\Lambda^s\xi$.

The product defined above satisfies

$$(1) \qquad \Phi \wedge \Psi = (-1)^{pq+rs} \Psi \wedge \Phi \qquad \begin{array}{l} \Phi \in A^p(M;\Lambda^r\xi) \\ \Psi \in A^q(M;\Lambda^s\xi) \end{array} .$$

The k-th exterior power of Φ is defined by

$$\Phi^k = \underbrace{\Phi \wedge \dots \wedge \Phi}_{k} .$$

Next, let σ^* be a cross-section in the dual bundle. It determines an operator

$$i(\sigma^*): \mathrm{Sec}\,\Lambda^r\xi \longrightarrow \mathrm{Sec}\,\Lambda^{r-1}\xi \qquad r \geq 1$$

given by

$$i(\sigma^*)(\sigma_1,\dots,\sigma_r) = \sum_{j=1}^{r} (-1)^{j-1} \langle \sigma_j^*, \sigma_j \rangle \sigma_1 \wedge \dots \wedge \hat{\sigma}_j \wedge \dots \wedge \sigma_r$$

$$\sigma_j \in \mathrm{Sec}\,\xi \; .$$

In particular,

$$i(\sigma^*)\sigma = \langle \sigma^*, \sigma \rangle \qquad \sigma \in \mathrm{Sec}\,\xi \; .$$

The operator $i(\sigma^*)$ extends in a natural way to an operator

$$i(\sigma^*): A^p(M;\Lambda^r\xi) \longrightarrow A^p(M;\Lambda^{r-1}\xi) \qquad r \geq 1 \, , \, p \geq 0 \; .$$

It satisfies

(2) $$i(\sigma^*)(\Phi \wedge \Psi) = i(\sigma^*)\Phi \wedge \Psi + (-1)^r \Phi \wedge i(\sigma^*)\Psi$$

$$\Phi \in A(M;\Lambda^r\xi) \; , \; \Psi \in A(M;\Lambda^s\xi) \; .$$

This relation implies that

(3) $$\sigma \wedge i(\sigma^*)\Phi = \langle \sigma^*, \sigma \rangle \Phi$$

$$\sigma \in \mathrm{Sec}\,\xi \; , \; \sigma^* \in \mathrm{Sec}\,\xi^* \; , \; \Phi \in A(M;\Lambda^r\xi) \; ,$$

where r is the rank of ξ.
In fact

$$\langle \sigma^*, \sigma \rangle \Phi - \sigma \wedge i(\sigma^*)\Phi = i(\sigma^*)\sigma \wedge \Phi - \sigma \wedge i(\sigma^*)\Phi$$

$$= i(\sigma^*)(\sigma \wedge \Phi) \; .$$

But $\sigma \wedge \Phi \in A(M;\Lambda^{r+1}\xi) = 0$ (ince r is the rank of ξ) and so it follows that $\sigma \wedge \Phi = 0$ whence (3).

4. Exterior covariant derivative.

Recall that a linear connection in a vector bundle is an operator

$$\nabla: \mathrm{Sec}\,\xi \longrightarrow A^1(M;\xi)$$

which satisfies

$$\nabla(\sigma + \tau) = \nabla\sigma + \nabla\tau \qquad \sigma, \tau \in \mathrm{Sec}\,\xi$$

and

$$\nabla(f\cdot\sigma) = \delta f\cdot\sigma + f\cdot\nabla\sigma \qquad f\in S(M)\ ,\ \sigma\in \mathrm{Sec}\,\xi\ .$$

If Z is a vector field on M, the covariant derivative of σ with respect to Z, denoted $\nabla_Z\sigma$, is the cross-section defined by

$$\nabla_Z\sigma = (\nabla\sigma)(Z)\ .$$

It is again a cross-section in ξ.

A linear connection ∇ in ξ induces linear connections in the exterior bundles $\Lambda^{r}\xi$, also denoted by ∇. It is determined by

$$\nabla(\sigma_1,\dots,\sigma_r) = \sum_{i=1}^{r} \sigma_1\wedge\dots\wedge\nabla\sigma_i\wedge\dots\wedge\sigma_r$$

$$\sigma_i\in \mathrm{Sec}(\xi)\ .$$

Now let $\Phi\in A^p(M;\Lambda^r\xi)$. Then an element $D\Phi\in A^{p+1}(M;\Lambda^r\xi)$ is defined by

$$D\Phi(X_0,\dots,X_p) = \sum_{i=0}^{p}(-1)^i\,\nabla_{X_i}\left(\Phi(X_0,\dots,\hat{X}_i,\dots,X_p)\right) + \sum_{i<j}(-1)^{i+j}\,\Phi([X_i,X_j],X_0,\dots,\hat{X}_i,\dots,\hat{X}_j,\dots,X_p)\ .$$

It is called the exterior covariant derivative of Φ. In particular we have

$$D\sigma = \nabla\sigma \qquad \sigma\in \mathrm{Sec}\,\xi$$

and if $\xi = M\times\mathbb{R}$ is the product bundle and ∇ is the ordinary derivative, then D is the classical exterior derivative of a p-form.

To define the curvature of a linear connection, consider the bundle $L(\xi)$ whose fibre at x is the space of linear transformations of $T_x(M)$. Then the curvature for ∇ is the 2-form R with values in $L(\xi)$ given by

$$R(X,Y) = \nabla_X\nabla_Y - \nabla_Y\nabla_X - \nabla_{[X,Y]} \qquad ,\ X,Y\in \mathfrak{X}(M)\ .$$

If σ is any cross-section in ξ, we have the formula

$$(4) \qquad D^2\sigma = R\sigma\ .$$

5. Pseudo-Riemannian connections.

Let M be a pseudo-Riemannian manifold. Then a linear connection in τ_M is called pseudo-Riemannian, if

$$Z(X,Y) = (\nabla_Z X,Y) + (X,\nabla_Z Y) \qquad ,\ X,Y,Z\in \mathfrak{X}(M)\ .$$

It can be shown that there exists a unique such connection with torsion zero. It is called the Levi-Civita connection.

The tensor $g(x)$ defines an isomorphism

$$\gamma_x : T_x(M) \xrightarrow{\cong} T_x(M)^*$$

($T_x(M)^*$ the dual space) given by

$$\langle \gamma_x h, k \rangle = g(x; h, k) \qquad h, k \in T_x(M) .$$

From γ_x we obtain a natural isomorphism

$$Sk\, T_x(M) \xrightarrow{\cong} \Lambda^2 T_x(M)$$

where $Sk\,T_x(M)$ denotes the space of skew linear transformations of $T_x(M)$. This yields an isomorphism

$$\mathrm{Sec}\,(Sk\,\tau_M) \xrightarrow{\cong} \mathrm{Sec}\,\Lambda^2 \tau_M$$

where $Sk\,\tau_M$ denotes the bundle over M whose fibre at x is the space $Sk\,T_x(M)$.

Now consider the curvature R of a pseudo-Riemannian connection. It is a 2-form on M with values in the bundle $Sk\,\tau_M$. Hence R determines under the isomorphism above a 2-form Ω with values in $\Lambda^2 \tau_M$. It is called the curvature form for ∇. (Observe that in classical notation Ω corresponds to the tensor field $R^{ij}{}_{kl}$.

Now formula (4) reads

$$(5) \qquad D^2 X = i(X^*)\Omega \qquad , \quad X^* = \gamma(X) .$$

The Bianchi identity states that the exterior covariant derivative of Ω (with respect to the induced connection in $\Lambda^2 \tau_M$) vanishes,

$$D\Omega = 0 \qquad .$$

Now suppose that M is orientable and of even dimensions n=2m and let Δ_M be the positive normed n-form. Regard Δ_M as a cross-section in $\Lambda^n \tau_M$. Then $\langle \Delta_M, \Omega^m \rangle$ is an ordinary n-form on M . Thus there is a unique scalar function K such that

$$(6) \qquad \langle \Delta_M, \Omega^m \rangle = K \cdot \Delta_M$$

K is called the Gaussian curvature of the pseudo-Riemannian manifold M .

§2. The differential forms Φ_X and Φ_σ

6. The form Φ_X. Let M be a pseudo-Riemannian 4-manifold with curvature form Ω. In this section we shall express the 4-form $<\Delta_M, \Omega \wedge \Omega>$ in terms of the exterior derivative of a certain 3-form Φ_X which will depend on a vector field X.

Let X be a vector field defined in an open subset $U \subset M$ such that $(X(x), X(x)) \neq 0$, $x \in U$. Consider the $\Lambda^4 \tau_M$-valued 3-forms

$$\Psi^0 = \frac{1}{(X,X)} X \wedge DX \wedge \Omega$$

and

$$\Psi^1 = \frac{1}{(X,X)^2} X \wedge (DX)^3$$

where D denotes the exterior covariant derivative (cf. sec.5).

It is easy to check that these forms do not change if X is replaced by $f \cdot X$, where f is a function without zeros.

Lemma 1: The forms Ψ^0 and Ψ^1 satisfy

(a) $$D\Psi^0 = \varepsilon (DX)^2 \wedge \Omega + \frac{1}{2}\Omega^2$$

and

(b) $$D\Psi^1 = 3\varepsilon (DX)^2 \wedge \Omega$$

where $\varepsilon = +1$ if $(X,X) > 0$ and $\varepsilon = -1$ if $(X,X) < 0$.

Proof of (a): Using formula (5) and the Bianchi identity we obtain

$$D\Psi^0 = \varepsilon\left[(DX)^2 \wedge \Omega \wedge X + X \wedge i(X^*)\Omega \wedge \Omega\right] \quad , \; X^* = \gamma(X).$$

But, by (2)

$$i(X^*)\Omega \wedge \Omega = \frac{1}{2} i(X^*)\Omega^2 .$$

Now relation (3), applied with $\Phi = \Omega^2$ yields

$$X \wedge i(X^*)\Omega \wedge \Omega = \frac{1}{2} < X^*, X > \Omega^2 =$$
$$= \frac{1}{2}(X,X)\Omega^2 = \frac{1}{2}\varepsilon\Omega^2 \qquad ((X,X) = \varepsilon = \pm 1)$$

and so (a) follows.

Proof of (b): Since (X,X) is constant, we have for every vector field Z

(7) $$(\nabla_Z X, X) = 0 .$$

This relation implies that

$$i(X^*)\,\nabla_Z X = \langle X^*, \nabla_Z X\rangle = (X, \nabla_Z X) = 0$$

whence

$$i(X^*)\,DX = i(X^*)\,\nabla X = 0$$

and so,

$$i(X^*)(DX)^2 = 0 \quad .$$

Next, observe that, in view of (7), the vector $(\nabla_Z X)(x)$ is contained in the 3-dimensional complement of $X(x)$. Consequently,

$$(\nabla X)^4 = 0 \qquad \text{and so} \qquad (DX)^4 = (\nabla X)^4 = 0.$$

Thus we have

$$D\Psi' = (DX)^4 + 3X\wedge(DX)^2\wedge i(X^*)\Omega = 3X\wedge i(X^*)\left[(DX)^2\wedge\Omega\right] .$$

Applying formula (3) with $\Phi = (DX)^2\wedge\Omega$ yields

$$X\wedge i(X^*)\left[(DX)^2\wedge\Omega\right] = \varepsilon\left[(DX)^2\wedge\Omega\right]$$

and so (b) follows. This completes the proof of Lemma I.

Now set

$$\Psi_X = 2\left(\Psi_0 - \tfrac{1}{3}\Psi_1\right) \quad .$$

Then Lemma I implies that

$$(8) \qquad D\Psi_X = \Omega^2 \quad .$$

Now assume that M is oriented and let Δ_M denote the positive normed 4-form on M. Regard Δ_M as a cross-section in the bundle $\Lambda^4\tau_M^*$. Then Δ_M is parallel with respect to the induced connection,

$$(9) \qquad D\Delta_M = 0 \quad .$$

Contracting the $\Lambda^4\tau_M$-valued 3-form Ψ_X with the cross-section Δ_M we obtain an (ordinary) 3-form

$$\Phi_X = \langle \Delta_M, \Psi_X\rangle \quad .$$

Explicitly,

$$\Phi_X = \frac{2}{(X,X)^2}\langle \Delta_M, (X,X)X\wedge DX\wedge\Omega - \tfrac{1}{3}X\wedge(DX)^3\rangle \quad .$$

Note that $\Phi_{fX} = \Phi_X$ where f is any function without zeros. Now we obtain from relations (8) and (9)

$$(10) \qquad \delta\Phi_X = \langle \Delta_M, \Omega^2\rangle \quad .$$

Proposition I: Let $\tilde{g}$ be a second pseudo-Riemannian metric on M such that $\tilde{g} = g$ in the neighbourhood of a point a. Then the difference

$$\langle \tilde{\Delta}_M, \tilde{\Omega}^2 \rangle - \langle \Delta_M, \Omega^2 \rangle$$

is exact.

Proof: Choose a vector field X in $M - a$ such that $X(x) \neq 0$, $x \in M - a$ (cf. [3], Chapter VIII, §5). Then formula (10) yields

$$\delta \Phi_X = \langle \Delta_M, \Omega^2 \rangle \qquad \delta \tilde{\Phi}_X = \langle \tilde{\Delta}_M, \tilde{\Omega}^2 \rangle \quad ,$$

whence

$$\langle \tilde{\Delta}_M, \tilde{\Omega}^2 \rangle - \langle \Delta_M, \Omega^2 \rangle = \delta (\tilde{\Phi}_X - \Phi_X) \quad \text{in} \quad M - a \; .$$

On the other hand, since $\tilde{g} = g$ in $\mathcal{U}$,

$$\langle \tilde{\Delta}_M, \Omega^2 \rangle - \langle \Delta_M, \Omega^2 \rangle = 0 \quad \text{in} \quad \mathcal{U} \; .$$

Thus the above difference is exact on M.

Remark: Assume that the metric is flat, $\Omega = 0$. Then Φ_X is given by

$$\Phi_X (X_1, X_2, X_3) = - \frac{2}{3} \frac{1}{(X,X)^2} \langle \Delta_M, X \wedge (DX)^3 \rangle$$

Now

$$(DX)^3 (X_1, X_2, X_3) = \sum_\sigma D_{X_{\sigma(1)}} X \wedge D_{X_{\sigma(2)}} X \wedge D_{X_{\sigma(3)}} X =$$

$$= 3! \, D_{X_1} X \wedge D_{X_2} X \wedge D_{X_3} X$$

and so

$$\Phi_X (X_1, X_2, X_3) = - 4 \frac{1}{(X,X)^2} \Delta_M (X, D_{X_1} X, D_{X_2} X, D_{X_3} X) \; . \tag{11}$$

7. The form Φ_σ. Let σ be a line field in W (W a connected open subset of an oriented pseudo-Riemannian 4-manifold M). Choose an open covering $\{\mathcal{U}_\alpha\}$ of W such that σ lifts to a vector field X_α over $\mathcal{U}_\alpha$. Let Φ_α be the 3-form in $\mathcal{U}_\alpha$ corresponding to X_α (see sec. 6). Since $\Phi_{fX} = \Phi_X$ for every function f without zeros, it follows that $\Phi_\alpha = \Phi_\beta$ in $\mathcal{U}_{\alpha\beta}$. Thus the local 3-forms Φ_α define a global 3-form Φ_σ in W. Formula (10) implies that

$$\delta \Phi_\sigma = \langle \Delta_M, \Omega^2 \rangle \; . \tag{12}$$

In particular, if the metric is flat in W formula (11) yields

$$\Phi_\sigma (X_1, X_2, X_3) = - 4 \frac{1}{(X,X)^2} \Delta_M (X, D_{X_1} X, D_{X_2} X, D_{X_3} X) \quad , \tag{13}$$

where X is a local lift of σ.

§3. Lorentz manifolds

8. Let M be a Lorentz manifold with Lorentz metric g_L. Recall that a nonzero tangent vector h at $x \in M$ is called

space-like, if $g_L(x;h,h) > 0$
time-like, if $g_L(x;h,h) < 0$
a light vector, if $g_L(x;h,h) = 0$

A line field σ on M is called time-like, if the vectors in the subspace $\sigma(x) \subset T_x(M)$ are time-like for all $x \in M$.

Proposition II: Every Lorentz manifold admits a time-like vector-field. Conversely, let M be a 4-manifold which admits a line field σ. Then M can be made into a Lorentz manifold such that σ is time-like.

Proof: Suppose M is a Lorentz manifold. Choose any Riemannian metric g_R on M. Then, for every $x \in M$, a linear transformation $\varphi_x : T_x(M) \longrightarrow T_x(M)$ is defined by the equation

$$g_L(x;h,k) = g_R(x;\varphi_x h,k) \qquad h,k \in T_xM \ .$$

Since g_L is symmetric, φ_x is self-adjoint with respect to g_R. Thus φ_x has four real eigenvalues. Since g_L is a Lorentz metric, exactly one eigenvalue will be negative. Let $\sigma(x)$ denote the 1-dimensional subspace of $T_x(M)$ spanned by the corresponding eigenvector. Then the correspondence $x \longrightarrow \sigma(x)$, $x \in M$, defines a time-like line field on M.

Conversely, let M be a 4-manifold with a line field σ. Again choose a Riemannian metric g_R on M. Then every tangent vector $h \in T_x(M)$ can be uniquely decomposed in the form

$$h = h_1 + h_2 \quad \text{where} \quad h_1 \in \sigma(x) \ , \ h_2 \in \sigma(x)^{\perp} \ .$$

Now set

$$g_L(x;h,k) = -g_R(x;h_1,k_1) + g_R(x;h_2,k_2) \ .$$

Then g_L is a Lorentz metric on M, as is easily checked, and σ is a time-like line field.

Theorem I: Let M be an oriented compact Lorentz manifold. Let Δ_L

and Ω_L denote the normed 4-form and the curvature form corresponding to the Lorentz metric. Let K denote the corresponding Gaussian curvature (cf. sec.5). Then

$$\int_M K \cdot \Delta_L = 0 .$$

Proof: In view of Proposition II, M admits a time-like line field σ. Let Φ_σ be the corresponding 3-form on M defined in sec. 7. Then formula (12) shows that

$$\delta \Phi_\sigma = \langle \Delta_L , \Omega_L^2 \rangle .$$

Now Stokes' theorem yields

$$\int_M \langle \Delta_L , \Omega_L^2 \rangle = \int_M \delta \Phi_\sigma = 0$$

and so

$$\int_M K \cdot \Delta_L = \int_M \langle \Delta_L , \Omega_L^2 \rangle = 0 .$$

§4. Index of a line field

9. Index of a line field in $\mathbb{R}^n$. Let $\mathbb{R}^n$ be an oriented Euclidean space of dimension n and let Δ denote the positive normed determinant function. Consider the (n-1)-form Γ in $\dot{\mathbb{R}}^n$ given by

$$\Gamma(x; h_1, \dots, h_{n-1}) = |x|^{-n} \Delta(x; h_1, \dots, h_{n-1}) \qquad x \in \dot{\mathbb{R}}^n, \ h_i \in \mathbb{R}^n.$$

A simple commutation shows that Γ is closed. Moreover, if S^{n-1} is any (n-1)-sphere centered at x = 0, with the orientation induced from $\mathbb{R}^n$, then

$$(14) \qquad \int_{S^{n-1}} \Gamma = k_{n-1}$$

where k_{n-1} denotes the volume of the (n-1)-dimensional unit sphere.

Suppose now that n is even. Then Γ determines a unique (n-1)-form $\bar{\Gamma}$ in $\mathbb{R}P^{n-1}$ such that

$$\rho^* \bar{\Gamma} = \Gamma$$

where $\rho : \dot{\mathbb{R}}^n \longrightarrow \mathbb{R}P^{n-1}$ is the canonical projection.

Since Γ is closed, so is $\bar{\Gamma}$. Moreover, formula (14) implies that

$$2 \int_{\mathbb{R}P^{n-1}} \bar{\Gamma} = \deg \rho \cdot \int_{\mathbb{R}P^{n-1}} \bar{\Gamma} = \int_{S^{n-1}} \rho^* \bar{\Gamma} = \int_{S^{n-1}} \Gamma = k_{n-1}$$

whence

$$(15)\qquad \int_{\mathbb{R}P^{n-1}} \bar{\Gamma} = \frac{1}{2} k_{n-1} .$$

Observe that the (n-1)-form Γ depends heavily on the Euclidean inner product in $\mathbb{R}^n$. However, the index of a line field is independent of the metric. In fact, let $\tilde{\Gamma}_\sigma$ denote (n-1)-form corresponding to a second inner product in $\mathbb{R}^n$. Then, since any two positive definite bilinear forms in $\mathbb{R}^n$ can be deformed into each other, it follows that $\tilde{\Gamma}_\sigma - \Gamma_\sigma$ is exact, and so

$$\int_{\tilde{\tilde{S}}_a^{n-1}} \tilde{\Gamma}_\sigma = \int_{S_a^{n-1}} \Gamma_\sigma .$$

Now let σ be a line-field defined in $\dot{U} = U - a$ where U is a neighbourhood of a point $a \in \mathbb{R}^n$. Then $\Gamma_\sigma = \sigma^* \bar{\Gamma}$ is a closed (n-1)-form in $\dot{U}$. The <u>index</u> of σ at a is defined by

$$j_a(\sigma) = \frac{1}{k_{n-1}} \int_{S_a^{n-1}} \Gamma_\sigma ,$$

where S_a^{n-1} is a sphere in U around a with the orientation induced from $\mathbb{R}^n$.

Observe that, if the orientation of $\mathbb{R}^n$ is reversed, Γ_σ goes into $-\Gamma_\sigma$ and, at the same time, the induced orientation of S_a^{n-1} is reversed. Thus the index does not depend on the orientation of $\mathbb{R}^n$. The index of a line field is a half integer. In fact, formula (15) implies that

$$\int_{S_a^{n-1}} \Gamma_\sigma = \int_{S_a^{n-1}} \sigma^* \bar{\Gamma} = \deg\sigma \cdot \int_{\mathbb{R}P^{n-1}} \bar{\Gamma} = \frac{1}{2} k_{n-1} \cdot \deg\sigma$$

whence

$$j_a(\sigma) = \frac{1}{2} \deg\sigma .$$

Since the mapping degree is an integer, $j_a(\sigma)$ is a half integer.

The <u>index of a vector field</u> is defined in a similar way. Now we can drop the assumption that n is even. In fact, let X be a vector field in $\dot{U}$ such that $X(x) \neq 0$, $x \in \dot{U}$. Regard X as a map $\dot{U} \longrightarrow \dot{\mathbb{R}}^n$ and let Γ_X be the (n-1)-form in $\dot{U}$ given by $\Gamma_X = X^* \Gamma$. Now define the <u>index of X</u> at a by

$$j_a(X) = \int_{S_a^{n-1}} \Gamma_X .$$

Thus $j_a(X)$ is the degree of the map $S_a^{n-1} \longrightarrow S^{n-1}$ determined by the unit vector field corresponding to X and so it is an integer. Again, the index of X at a is dependent of the orientation of $\mathbb{R}^n$. Also observe that

$$j_a(X) = (-1)^n j_a(-X) .$$

Now suppose that a line field σ lifts to a vector field X in $\dot{U}$. Then we have $\sigma = \rho \circ X$ and so $\Gamma_\sigma = \sigma^*\bar{\Gamma} = X^*\rho^*\bar{\Gamma} = X^*\Gamma = \Gamma_X$.

It follows that $j_a(\sigma) = j_a(X)$ and so the index of σ is an integer in this case.

Example 1: Assume that $n \geqslant 4$. Then every line field in $\mathbb{R}^4$ lifts to a vector field, since $\dot{\mathbb{R}}^4$ is simply connected. Thus, if $n \geqslant 4$, the index of a line field is always an integer.

Example 2: Let $\mathbb{C}$ be the complex plane and let $a = 0$. Consider the line field σ determined by the (double valued) map $z \to z^{1/2}$. Then Γ_σ is the 1-form in $\dot{\mathbb{C}}$ given by

$$\Gamma_\sigma(z; Y) = \frac{1}{2} \operatorname{Im} (\bar{z}^{\frac{1}{2}} . z^{\frac{1}{2}} . Y) = \frac{1}{2} \operatorname{Im} \left(\frac{Y}{z}\right) \qquad z \in \dot{\mathbb{C}} , Y \in \mathbb{C} .$$

Now a simple computation shows that

$$\int_{S^1} \Gamma_\sigma = \pi$$

and so $j_a(\sigma) = 1/2$. It follows that the line field σ does not lift to a vector field in $\dot{\mathbb{C}}$.

Next let $\mathbb{C}^+$ denote the compact complex plane and use $\omega = \frac{1}{z}$ as local parameter in a neighbourhood U_∞ of z_∞. Then the line field above, in terms of ω is given by

$$\sigma(\omega) = - \omega^{3/2} .$$

Thus,

$$j_\infty(\sigma) = \frac{3}{2} .$$

Observe that

$$j_0(\sigma) + j_\infty(\sigma) = 2 .$$

Example 3: Let X be the vector field in $\mathbb{C}$ given by

$$X(z) = \frac{1}{z^2} .$$

Then $j_0(X) = -2$ and $j_\infty(X) = 4$. Hence, again $j_0(X) + j_\infty(X) = 2$.

10. The (n-1)-forms Θ_X and Θ_σ. To define the index of a line field σ on a manifold, we shall first associate with σ a certain (n-1)-form Θ_σ. Let M be an oriented Riemannian manifold of even

dimension and let X be a vector field without zeros defined in an open subset $W \subset M$. Then an (n-1)-form Θ_X is defined in W by

$$\Theta_X(X_1,\dots,X_{n-1}) = |X|^{-n}\,\Delta_M(X,\nabla_{X_1}X,\dots,\nabla_{X_{n-1}}X) \quad .$$

Observe that, since n is even, $\Theta_{fX} = \Theta_X$ for every function f without zeros.

A straightforward calculation shows that the exterior derivative of Θ_X is given by

$$\delta\Theta_X = \langle \Delta_M, (DX)^{n-2} \wedge R(X) \rangle \quad .$$

In particular, if the metric is flat, $R = 0$, then Θ_X is closed.

Now let σ be a line field in W. Choose an open covering $\{U_\alpha\}$ of W such that σ lifts to a vector field X_α over U_α and consider the local (n-1)-forms $\Theta_{X_\alpha} = \Theta_\alpha$. Then we have $\Theta_\alpha = \Theta_\beta$ in $U_{\alpha\beta}$. And so the Θ_α determine a global (n-1)-form Θ_σ in W.
If $\varphi : M \xrightarrow{\cong} N$ is an orientation preserving isometry, we have the relation

$$(16) \qquad \Theta_\sigma = \varphi^*\Theta_\tau \qquad \text{where} \qquad \tau = \varphi_*\sigma \quad .$$

<u>Lemma II</u>: Let σ be a line field in an open subset $W \subset \mathbb{R}^n$ and let Θ_σ be the corresponding (n-1)-form defined via the standard connection in $\mathbb{R}^n$. On the other hand, consider the (n-1)-form Γ_σ (see sec. 10). Then $\Theta_\sigma = \Gamma_\sigma$.

<u>Proof</u>: Fix $a \in W$ and let X be a lifting of σ defined in a neighbourhood U of a. Then we have for $x \in U$

$$\Gamma_\sigma(x; h_1,\dots,h_{n-1}) = \bar\Gamma(\sigma x; (d\sigma)_x h_1,\dots,(d\sigma)_x h_{n-1}) =$$

$$= \bar\Gamma(\rho X(x); (d\rho)_x (d\sigma)_x h_1,\dots,(d\rho)_x (d\sigma)_x h_{n-1}) =$$

$$= \rho^*\bar\Gamma(X(x); (d\sigma)_x h_1,\dots,(d\sigma)_x h_{n-1}) =$$

$$= \Gamma(X(x); (d\sigma)_x h_1,\dots,(d\sigma)_x h_{n-1}) =$$

$$= \Theta_X(x; h_1,\dots,h_{n-1}) = \Theta_\sigma(x; h_1,\dots,h_{n-1}) \quad .$$

Thus, $\Gamma_\sigma = \Theta_\sigma$.

11. Index of a line field on a manifold. Let M be an oriented n-manifold (n even) and let σ be a line field defined in $\dot{U} = U - a$ where $a \in M$ and U is a neighbourhood of a. Choose a chart $(U_\alpha, u_\alpha, \mathbb{R}^n)$ such that $a \in U_\alpha$ and $U_\alpha \subset U$. Then σ defines a line field σ_α in $\dot{\mathbb{R}}^n$ given by $\sigma_\alpha = (u_\alpha)_* \sigma$.

Thus the index of σ_α at $x = 0$ is defined. It will be simply denoted by $j(\sigma_\alpha)$. We shall show that $j(\sigma_\alpha)$ is independent of the chart. Let g be a Riemannian metric in U with curvature zero and consider the (n-1)-form Θ_σ corresponding to σ. (cf. sec. 10).

Proposition III:

$$j(\sigma_\alpha) = \frac{1}{k_{n-1}} \int_{S_a^{n-1}} \Theta_\sigma .$$

Proof: Since g is flat, we can choose a Euclidean inner product in $\mathbb{R}^n$ such that the map u_α is an isometry. Hence formula (16) implies that $\Theta_\sigma = u_\alpha^* \Theta_{\sigma_\alpha}$ and so

$$\int_{S_a^{n-1}} \Theta_\sigma = \int_{S_a^{n-1}} u_\alpha^* \Theta_{\sigma_\alpha} = \int_{S_\alpha^{n-1}} \Theta_{\sigma_\alpha}$$

where S_α^{n-1} is the image of S_a^{n-1} under u_α. On the other hand, by definition, (cf. sec. 9)

$$j(\sigma_\alpha) = \frac{1}{k_{n-1}} \int_{S_\alpha^{n-1}} \Gamma_{\sigma_\alpha} .$$

Finally, Lemma II shows that $\Theta_{\sigma_\alpha} = \Gamma_{\sigma_\alpha}$ and so the proposition follows.

CorollaryI: The index $j(\sigma_\alpha)$ is independent of the chart.

Corollary II: The integral of Θ_σ over S_a^{n-1} is independent of the choice of the (flat) metric.

Thus it makes sense to define the index of σ at a by

$$j_a(\sigma) = j(\sigma_\alpha) \qquad \sigma_\alpha = (u_\alpha)_* \sigma$$

or equivalently by

$$j_a(\sigma) = \frac{1}{k_{n-1}} \int_{S_a^{n-1}} \Theta_\sigma .$$

It can be shown that every compact manifold admits line fields, even vector fields with only finitely many singularities (cf. [3], chapter VIII, §5). Now let M be a compact 4-manifold and let σ be a line field defined on $M - A$ where A is a finite set, $A = \{a_1, \dots, a_k\}$.

Then, the index sum of σ is defined by $j(\sigma) = \sum_{r=1}^{k} j_r(\sigma)$

where $j_r(\sigma)$ denotes the index of σ at a_r (r = i...k) .

12. Total curvature. Let M be a compact oriented Riemannian 4-manifold with curvature form Ω. Recall from sec. 5 that the Gaussian curvature K of M is defined by

$$< \Delta_M , \Omega \wedge \Omega > = K \cdot \Delta_M \quad .$$

Theorem II. The index sum of a line field σ with finitely many singularities is given by

$$j(\sigma) = \frac{1}{8\pi^2} \int_M K \cdot \Delta_M \quad .$$

Remark: This is a special case of the Gauss-Bonnet-Chern theorem for dimension 4. A similar theorem is true for all compact Riemannian manifolds of even dimension. A proof, which follows the same ideas, is given in [4], chapter X.

Proof of Theorem II: First observe that the integral on the right hand side is independent of the Riemannian metric (cf. Proposition I in sec. 6). Hence we may assume that the metric is flat in U_i (U_i a neighbourhood of a_i)

Now consider the 3-form Φ_σ defined in sec. 6. Then, by (12)

$$\delta \Phi_\sigma = < \Delta_M , \Omega \wedge \Omega > \quad .$$

Hence Stokes' theorem yields

$$(17) \qquad \int_{M-U} < \Delta_M , \Omega \wedge \Omega > = - \sum_r \int_{S_r} \Phi_\sigma \quad ,$$

where $U = \bigcup_r U_r$ and S_r is a 3-sphere in U_r around a_r with the orientation induced from M . Now fix r . The index of σ at a_r is given by

$$(18) \qquad j_r(\sigma) = \frac{1}{k_3} \int_{S_r} \Theta_\sigma = \frac{1}{2\pi^2} \int_{S_r} \Theta_\sigma$$

(cf. sec. 11). Since the metric is flat in U_r , formula (13) yields

$$\Phi_\sigma(X_1, X_2, X_3) = -4 \frac{1}{(X,X)^2} \Delta_M (X, D_{X_1}X, D_{X_2}X, D_{X_3}X)$$
$$= -4 \frac{1}{(X,X)^2} \Delta_M (X, \nabla_{X_1}X, \nabla_{X_2}X, \nabla_{X_3}X) = -4\, \Theta_\sigma(X_1, X_2, X_3) \quad .$$

Thus, $\Phi_\sigma = -4\,\Theta_\sigma$. Now (18) reads

$$(19) \qquad j_\tau(\sigma) = -\frac{1}{8\pi^2}\int_{S_\tau} \Phi_\sigma \quad .$$

Finally, since $\Omega_4 = 0$ in U , we have

$$(20) \qquad \int_{M-U} \langle \Delta_M, \Omega_4 \wedge \Omega_4 \rangle = \int_M \langle \Delta_M, \Omega_4 \wedge \Omega_4 \rangle \quad .$$

Combining equations (19), (17) and (20) we obtain

$$j(\sigma) = \int_M \langle \Delta_M, \Omega_4 \wedge \Omega_4 \rangle$$

and so, in view of (6)

$$j(\sigma) = \int_M K \cdot \Delta_M \quad .$$

13. Euler characteristic. Theorem II shows that the index sum of a line field on M is independent of the line field and hence an invariant of M . It is called the Euler characteristic of M and is denoted by $\chi(M)$,

$$\chi(M) = j(\sigma) \quad .$$

Next let X be a vector field without zeros on $M - A$, $A = \{a_1, \ldots, a_k\}$, and define its index sum by

$$j(X) = \sum_{\tau=1}^{k} j_k(X) \quad .$$

Thus, if σ_X is the corresponding line field, $j_\tau(X) = j_\tau(\sigma_X)$ and so

$$j(X) = j(\sigma) = \chi(M) \quad .$$

Theorem III: If a compact 4-manifold admits a line field, it also admits a vector field without zeros.

Proof: Since M admits a global line field, $\chi(M) = 0$. Let X be a vector field on M with a single singularity at a point a , then

$$j_a(X) = j(X) = \chi(M) = 0 \quad .$$

Hence X can be deformed into a vector field without zeros.

Remark: Observe that the theorem does not claim that a given line field can be lifted to a vector field. In fact, it will be shown in the next section that this is not always possible.

Theorem IV: Every compact Lorentz manifold has characteristic zero. Conversely, a compact 4-manifold M with $\chi(M) = 0$ admits a Lorentz metric.

Proof: By Proposition II, a Lorentz manifold admits a line field. Thus, by Theorem II, $\chi(M) = 0$. Conversely, if $\chi(M) = 0$, then M admits a vector field without zeros and hence a line field. Now apply proposition II.

14. Example: We shall now construct a line field on a compact orientable 4-manifold which can not be lifted to a vector field.

Let $M = S^1 \times SO(3)$. Then M is parallizable, and so we have bundle isomorphisms

$$T_M \xrightarrow{\cong} M \times \mathbb{R}^4 , \quad \dot{T}_M \xrightarrow{\cong} M \times \dot{\mathbb{R}}^4 , \quad P^M \xrightarrow{\cong} M \times \mathbb{R}P^3 .$$

Thus the vector fields (respectively line fields) on M can be identified with the maps $M \longrightarrow \mathbb{R}^4$ (respeptively the maps $M \longrightarrow \mathbb{R}P^3$).

Observe that $SO(3)$ is diffeomorphic to $\mathbb{R}P^3$. Hence a line field σ is given on M by

$$\sigma(x,y) = \varphi(y) \qquad x \in S^1 , \; y \in SO(3) ,$$

where $\varphi : SO(3) \longrightarrow \mathbb{R}P^3$ is a diffeomorphism. Let

$$\sigma_* : \pi_1(M) \longrightarrow \pi_1(\mathbb{R}P^3)$$

denote the induced homomorphism between the fundamental groups. Then, if $\alpha \in \pi_1(S^1)$ and $\beta \in \pi_1(SO(3))$ are the generators, we have

$$\sigma_*(\alpha,\beta) = \varphi_*(\beta) \neq 0 \tag{21}$$

Now assume, that σ lifts to a vector field X. Then the vector field X determines a map $\tau : M \longrightarrow \dot{\mathbb{R}}^4$ such that $\sigma = \rho \circ \tau$. It follows that

$$\sigma_* = \rho_* \circ \tau_* .$$

But $\tau_*(\alpha,\beta) \in \pi_1(\dot{\mathbb{R}}^4) = 0$ and so we obtain $\sigma_*(\alpha,\beta) = 0$, in contradiction with (21). Hence σ does not lift to a vector field.

15. Time-orientability. A Lorentz manifold M is called time-orientable if there exists a global time-like vector field on M. The Euler characteristic does not distinguish time orientable and not time orientable compact Lorentz manifolds since it is always zero. We shall however show that these two properties are independent by giving two examples.

Example I: Let $M = \mathbb{R}P^2 \times T^2$ (T^2 the 2-torus). Then M is non-orientable. On the other hand, since T^2 admits a vector field without zeros, so does M. Choose such a vector field X and use it to define a Lorentz metric on M such that X is time-like (cf. Proposition II, sec. 8). Thus M is time-orientable.

Example II: Let $M = S^1 \times SO(3)$ (cf. the example in sec. 14) and use the line field σ to make M into a Lorentz manifold such that σ is time-like (cf. Proposition II). We show that this Lorentz manifold is not time-orientable. In fact, assume that X is a time-like vector field. Fix $x \in M$ and write $X(x) = X_1(x) + X_2(x)$, where $X_1(x)$ is contained in the line $\sigma(x)$ and $X_2(x)$ is Lorentz orthogonal to $\sigma(x)$. Then $X_2(x)$ is a space-like vector or the zero-vector. Since

$$(X(x), X(x)) = (X_1(x), X_1(x)) + (X_2(x), X_2(x)) \quad ,$$

it follows that

$$(X_1(x), X_1(x)) < 0 \qquad , \quad x \in M \ ,$$

and so $X_1(x) \neq 0$, $x \in M$.

Thus X_1 is a lifting of σ. On the other hand, σ does not admit a lifting (cf. the example in sec. 14).

Remark: If S^1 is replaced by $\mathbb{R}$ in the example above, we obtain a non-compact Lorentz manifold which is orientable but not time-orientable.

Bibliography

[1] Avez, A., Formule de Gauss-Bonnet-Chern en métrigue de signature quelquonque, C.R. Acad, Sci., Paris 255 (1962)

[2] Chern, S.S., Pseudo-Riemannian geometry and Gauss-Bonnet formula, An. Acad. Brazil Ci. 35 (1963).

[3] Greub, W., Halperin, S. and Vanstone, R., Connections, Curvature and cohomology, volume I, Academic Press, New York, 1972.

[4] Greub, W., Halperin, S. and Vanstone, R., Connections, Curvature and Cohomology, volume II, Academic Press, New York, 1973.

[5] Marcus, L., Line element fields and Lorentz structures on differntiable manifolds, Ann. of Math. 8, volume 83, 1956.

[6] Samelson, H., A theorem on differentiable manifolds, Portigaliae Math., volume 10 (1951).

The manifold of embeddings of a closed manifold

by

E.Binz[+] and H.R.Fischer[++]

The present note is concerned with some basic properties of the set $E(M,N)$ of all smooth embeddings $M \longrightarrow N$ where M and N are (finite-dimensional) paracompact C^∞-manifolds. For technical reasons, we assume M to be compact without boundary, so that the currently available techniques of the theory of manifolds of smooth maps are applicable to our case. Since $E(M,N)$ is known to be open in the manifold of all smooth maps $M \rightarrow N$ (cf.e.g.[M]), it is a Fréchet manifold. The group Diff(M) of smooth diffeomorphisms of M operates in the obvious manner on this manifold and our main result is that under this operation, $E(M,N)$ is a principal Diff(M)-bundle over the quotient $U(M,N) = E(M,N)/\mathrm{Diff}(M)$. In particular $U(M,N)$ - which is the set of all submanifolds of type M in N - carries a natural differentiable structure and $E(M,N) \longrightarrow U(M,N)$ is a submersion. A differentiable structure was obtained on $U(M,N)$ for compact M with boundary some time ago (cf.e.g.[Ko]), so that now natural manifold structures are available on all $U(M,N)$ for compact manifolds M. The non-compact case so far remains quite intractable because of some fundamental difficulties arising already in connection with the choice of an appropriate topology on, say, $C(M,N)$ (=the smooth maps $M \rightarrow N$), of reasonable "candidates" for parameter spaces and with differential calculus.

The motivation for our investigation stems from geometrodynamics ([MW] , [W] and further references in section 6), the "3+1 approach" to relativity, especially in the formulation it obtained in the framework of "hyperspace" (Kuchar's terminology,- cf.[K] , [HKT] and their bibliographies). In fact, "hyperspace" essentially is a special case of manifold of the type of $E(M,N)$, arrived at as follows: M,N are (orientable) manifolds of dimension 3,4 resp. and one considers now the set of all space-like embeddings $M \longrightarrow N$ where N carries an Einstein structure and where "space-like" means that the metric induced on the image is Riemannian. Since the set $E_o(M,N)$ of such embeddings is open in $E(M,N)$, it is a Fréchet manifold and is saturated under the right action of Diff(M), hence again a principal bundle over the manifold $U_o(M,N)$ of all space-like "slices" of type M in N. "Motions" of such a slice $i(M)$ then correspond to (smooth) curves through i in $E_o(M,N)$ - which thus represent the kinematics of submanifolds of type M in N, as well as the dynamics in the presence of suitable constraints. The details of these matters, using the notations and results of this note, will be dealt with elsewhere and we limit our exposition here to a few brief remarks in section 6).

After introducing some notations and preliminary results in 1, we establish the topological bundle structure of $E(M,N)$ in sections 2 to 4 by proving the existence of "enough" local sections of $E(M,N) \longrightarrow U(M,N)$ and we add the necessary differentiability considerations in section 5, using a form of differential calculus developed in [F] and [G] (cf. also [O]) in order to obtain precise information on the differentiability of the various constructions of the preceding sections.

1. The finite-dimensional manifolds $M,N,Z,\ldots$ to be considered in the sequel are understood to be paracompact C^∞-manifolds, so that they always admit smooth partitions of unity. Where used, the term "differentiable" will mean "smooth", i.e. "of class C^∞". It is no loss of generality to assume the manifolds $M,N,\ldots$ connected.

Let now M be compact. For each finite $k, C^k(M,N)$ denotes the set of C^k-maps $M \longrightarrow N$ equipped with the usual C^k-topology (cf.e.g. [M]). Of importance to us is the well-known result that composition

$$C^k(N,Z) \times C^k(M,N) \xrightarrow{\beta} C^k(M,Z) ,$$

$\beta(f,g) = f \circ g$, is a continuous map. Moreover, if $\mathrm{Diff}^k(M)$ denotes the group of C^k-diffeomorphisms of M, inversion in $\mathrm{Diff}^k(M)$ also is continuous, so that $\mathrm{Diff}^k(M)$ is a topological group, cf.[M], and is an open subset of $C^k(M,M)$ for $k \geq 1$. Since the latter is known to be a Banach manifold, the same holds for $\mathrm{Diff}^k(M)$; however, it is not a differentiable group in this structure. For the sake of clarity, we shall give some details concerning the corresponding Fréchet-structure on spaces of smooth maps below.

$E^k(M,N)$ denotes the set of all C^k-embeddings $M \longrightarrow N$; again, for $k \geq 1$, this is open in $C^k(M,N)$ - in fact, C^1-open - and thus again is a Banach manifold: the proof that E^k is open is analogous to the one of prop.5 in [M] .

Similar remarks apply in the case of analogously defined structures on spaces of smooth maps - where we also simplify the notations by setting $C(M,N) = C^\infty(M,N)$, $\mathrm{Diff}(M) = \mathrm{Diff}^\infty(M)$, $E(M,N) = E^\infty(M,N)$, etc. The manifold structures now are Fréchet, not Banach, and are obtained as follows:

Choose a smooth spray $TN \longrightarrow T^2N$ on N and denote its expontential map by exp. It is known ([L]) that there exists an open neighbourhood O' of the zero-section $O_N \subset TN$ such that exp induces diffeomorphisms of the $O' \cap T_yN$ onto open neighbourhoods of the points $y \in N$ ("normal charts"). There is more: define the map Exp from TN to $N \times N$ by:

$$\mathrm{Exp}(v_y) = (y, \exp_y(v_y)) \ , \ v_y \in T_yN \ .$$

Then there is an open neighbourhood $\mathcal{O}$ of O_N such that Exp induces a (smooth) diffeomorphism of $\mathcal{O}$ onto an open neighbourhood of the diagonal $\Delta \subset N \times N$.

Given now $f \in C(M,N)$, we denote by $\mathcal{U}(f)$ the set of all smooth maps $g:M \longrightarrow N$ such that $(f \times g)(\Delta_M) \subset \mathrm{Exp}(\mathcal{O})$ or, which is the same thing, $(f \times id_N)(\langle g\rangle) \subset \mathrm{Exp}(\mathcal{O})$, $\langle g\rangle \subset M \times N$ the graph of g. The set $\mathcal{U}(f)$ is an open neighbourhood of f in $C(M,N)$ with respect to the usual C^∞-topology, M being compact.

Next, let

$$\Gamma_f(TN) = \{l \in C(M,TN) \mid \tau_N \circ l = f\} \ ,$$

$\tau_N : TN \longrightarrow N$ the projection, be the set of all "lifts" of f to TN. Clearly, this is a vector space under pointwise operations and it becomes a Fréchetspace under the C^∞-topology. The space $\Gamma_f(TN)$ is naturally isomorphic to $\Gamma(f^*TN)$, the space of smooth sections of the pull-back bundle f^*TN. Let $W(f) = \{l \in \Gamma_f(TN) \mid l(M) \subset \mathcal{O}\}$; this is an open neighbourhood of o and the exponential map induces a homeomorphism $W(f) \longrightarrow \mathcal{U}(f)$ by assigning to $l \in W(f)$ the map g defined by: $g(x) = \exp_{f(x)}(l(x))$, also written $g = \exp_f(l)$. We denote the inverse map by $\Phi_f : \mathcal{U}(f) \longrightarrow W(f)$. The pair $(\mathcal{U}(f),\Phi_f)$ is a <u>standard chart</u> at f. One shows that these charts define on $C(M,N)$ a Fréchet-manifold structure whose underlying topology is the C^∞-topology; for all details, we refer to [F] and [G], in particular, for the precise sense in which this structure is "differentiable" (ILB-differentiable in a strong sense!). We also note that the construction goes back to [E].

Since $E(M,N)$ is (C^1-)open in $C(M,N)$, it inherits an induced manifold structure; this will be referred to as the natural manifold structure of $E(M,N)$.

Similarly, Diff(M) is open in C(M,M) and therefore again a Fréchet manifold. As opposed to the C^k-case for finite k, however, the group operations in Diff(M) are (ILB-)differentiable: Diff(M) is a smooth Fréchet Liegroup. Again, the details may be found in [F] , [G] and in [E-M] , [O] . Note that the tangent space $\Gamma_e(TM)$ to Diff(M) at the identity e simply is the algebra $\Gamma(TM)$ of smooth vector fields on M. The construction of charts (at e, say) still requires the choice of a spray on M and the earlier arguments, since the classical "exponential map", assigning to a vector field its flow, will not, in general, be surjective onto any neighbourhood of e, cf.e.g.[O] .

2. Henceforth, M is assumed to be closed (compact without boundary). The group Diff(M) acts in a natural way on the right on E(M,N) by: $(f,g) \longrightarrow f\circ g$ and the action is continuous by what was mentioned earlier. By e.g. [G] , it even is (ILB-)differentiable in a sense which can be made precise quite explicitly. For reasons of convenience, we shall write $\beta(f,g)$ for $f\circ g$. The partial maps defined by β are

$$\beta_f : \mathrm{Diff}(M) \longrightarrow E(M,N) , \quad \text{for a fixed } f \in E(M,N);$$

$$\beta_g : E(M,N) \longrightarrow E(M,N) , \quad \text{for fixed } g \in \mathrm{Diff}(M) .$$

Clearly, β_g is a diffeomorphism of E(M,N) whose inverse is $\beta_g{-1}$. Moreover, the injectivity of the $f \in E(M,N)$ immediately implies that the right action of Diff(M) is free: $f\circ g_1 = f\circ g_2$ implies $g_1 = g_2$ by cancellation of f . Thus, β_f certainly is a bijection of Diff(M) onto the orbit of f. If we set $L = f(M) \subset N$, a closed submanifold of N, then the orbit of f consists precisely of all diffeomorphisms of M onto L: indeed, it is easy to see that $i,j \in E(M,N)$ are equivalent under Diff(M) if and only if im(i) = = im(j). In fact, the orbit of f may also be identified with E(M,L): one inclusion is clear and for the converse, if h is an embedding of M into f(M) = L, then h(M) is compact, hence closed; it also is open by the inverse function theorem since T_xh is an isomorphism of T_xM onto $T_{h(x)}L$ for each $x \in M$. Thus, M being connected, the assertion follows. We shall see below that the identification of the orbit of $i \in E(M,N)$ with E(M,i(M)) is topological as well, even differentiable.

Next, we note that the orbit $i\cdot\mathrm{Diff}(M) = \tilde{i}$ of $i \in E(M,N)$ is closed in E(M,N): let (g_α) be a net in Diff(M) such that $i\cdot g_\alpha$

converges in $E(M,N)$ to, say, f. Since $\beta_i(g_\alpha)(M) = i(M)$ for every α and since $\beta_i(g_\alpha)$ converges in particular for the C^0-topology we conclude $f(M) \subset i(M)$, thus $f(M) = i(M)$ as before. Further details will be added below (prop.3,cor.).

Lastly, we note that the quotient $U(M,N)$ of $E(M,N)$ mod. the action of Diff(M) simply is the set of all submanifolds "of type M" of N. This set carries the quotient topology. With this, we state the main result of this note:

Theorem: Let M be closed, $\dim(M) < \dim(N)$. Then $E(M,N)$ is a (topological) principal Diff(M)-bundle over $U(M,N)$.

Moreover, $U(M,N)$ carries a natural (ILB-)differentiable structure for which the fibration $E(M,N) \xrightarrow{u} U(M,N)$ is smooth.

The proof of the theorem is divided into several steps which will follow in the next sections, but is essentially classical: it consists of establishing the existence of "enough" local sections of the quotient map u and, eventually, the necessary differentiability considerations. The first part of the argument is topological in nature and will establish the first (main) part of the theorem.

3. We choose, once and for all, a quadratic structure G on N ("pseudo-Riemannian" or Riemannian structure). For $i \in E(M,N)$, we denote by ν_i the "geometric" normal bundle of i defined by means of G : the fibre of ν_i at $p = i(x) \in i(M)$ is the orthogonal in T_pN of $T_xi(T_xM)$. Clearly, then,

$$TN \mid i(M) = Ti(TM) \oplus \nu_i \,. \tag{1}$$

We now have the

Prop.1: For $i \in E(M,N)$, the map

$$\Omega_{Ti}\colon \Gamma(TM) \longrightarrow \Gamma_i(TN)$$

which maps a vector field X to $Ti \circ X$ is a linear and topological isomorphism of $\Gamma(TM)$ onto a topological direct summand, namely the subspace $\{l \in \Gamma_i(TN) \mid l(M) \subset Ti(TM)\}$.

For the proof one observes, firstly, that the indicated subspace is a topological summand - using the orthogonal projection onto $Ti(TM)$ in the standard Ω-lemma. Secondly, one easily verifies that Ω_{Ti} is a linear bijection onto this subspace which also is continuous. The closed graph theorem now implies the assertion.

In TN, we use the exponential maps exp and Exp defined by the metric G (so that the normal charts are geodesic charts for G). Given the open neighbourhood $\mathcal{O}$ of the zero-section O_N as used in 1., there now are open neighbourhoods $\mathcal{O}_M$ of the zero-section of TM and $\mathcal{O}_{\nu_i}$ of the zero-section of ν_i such that $Ti(\mathcal{O}_M) + \mathcal{O}_{\nu_i} \subset TN|i(M)$ still is open.
Then:

Prop.2: Given $i \in E(M,N)$, $\mathcal{O}_{\nu_i}$ and $\mathcal{O}_M$ may be chosen so as to satisfy the conditions

(1) $\exp|\mathcal{O}_{\nu_i} : \mathcal{O}_{\nu_i} \longrightarrow \exp(\mathcal{O}_{\nu_i})$ is a diffeomorphism,

(2) $\exp(\mathcal{O}_{\nu_i}) \subset N$ is open.

For the proof, let $pr_2 : N \times N \longrightarrow N$ be the second projection, so that $pr_2 \circ Exp = \exp$, and let $o_{\nu_i(p)} \in \nu_i(p)$ be the zero-element of the fibre. Exp yields a diffeomorphism of $\mathcal{O}$ onto some neighbourhood of the diagonal in $N \times N$ and, moreover, $T(pr_2 \cdot Exp|\mathcal{O} \cap \nu_i)$, mapping $T_{o_{\nu_i(p)}} \nu_i$ to T_pN , $p \in i(M)$, is an isomorphism, being a linear injection between vector spaces of the same finite dimension. The assertion now follows from Prop.7.3,p.69, of [G-G].

In addition, we assure that N possesses a Finsler structure $\| \ \| : TN \longrightarrow R$. Because of the compactness of M, we may assume that there is $\varepsilon > o$ such that

$$\mathcal{O}_{\nu_i} = \{v \in \nu_i | \ \|v\| < \varepsilon\} .$$

The open submanifold $\exp(\mathcal{O}_{\nu_i})$ then is the "ε-tube around i(M)" and will be denoted by S_i^ε . The image $\exp(\nu_i(p))$ is the fibre at $p \in i(M)$ in S_i^ε .

In the next step, we construct a map $P_i : S_i^\varepsilon \longrightarrow i(M)$ with $P_i|i(M) = id.$: S_i^ε is embedded into the fibre-product $i(M) \times_{i(M)} S_i^\varepsilon$ under the map ι_1 which maps $\exp_p(v) \in S_i^\varepsilon$, $v \in \nu_i(p) \cap \mathcal{O}_{\nu_i}$, to $(p, \exp_p(v)) = Exp(v)$. ι_2 denotes the inclusion of the fibre-product into $N \times N$. Set

$$P_i = \tau_N \circ Exp^{-1} \circ \iota_2 \circ \iota_1 ,$$

a smooth map $S_i^\varepsilon \longrightarrow i(M)$. Now $(Exp^{-1} \circ \iota_2 \circ \iota_1)(p) = o_{\nu_i(p)}$ for $p \in i(M)$, so that $P_i|i(M) = id.$: P_i is a retraction for the inclusion ι of $i(M)$ into S_i^ε , mapping an open neighbourhood of $i(M)$, namely S_i^ε , onto $i(M)$ along geodesics.

Clearly, $\Omega_{P_i} \bullet \Omega_\iota = \Omega_{P_i \circ \iota}$ = id. on C(M,i(M)). This implies that C(M,i(M)) is mapped homeomorphically onto its image in $C(M,S_i^\varepsilon)$, an open subset of C(M,N). Therefore, if we interpret C(M,i(M)) as a subspace of C(M,N), the induced topology will coincide with the C^∞-topology of C(M,I(M)). Thus:

<u>Prop.3:</u> The elements of E(M,i(M)) are precisely the maps $i \circ g$, $g \in \in$ Diff(M), i.e. E(M,i(M)), interpreted as a subset of E(M,N), is the orbit of i. Moreover, the C^∞-topology of E(M,i(M) coincides with the topology induced from E(M,N).

<u>Cor.:</u> With the earlier notations, β_i is a homeomorphism of Diff(M) onto the orbit of i in E(M,N) and this orbit is closed.

To obtain the corollary, note that $\beta_i^{-1} = \Omega_{i^{-1}}$ is continuous on E(M,i(M)) by the usual Ω-lemma. The remaining claims already were established (cf.also 2.).

Suppose now that the map $j:M \longrightarrow S_i^\varepsilon$ satisfies $P_i \circ j \in E(M,i(M))$. Then j is an <u>embedding</u> whose inverse is given by $(P_i \circ j)^{-1} \circ (P_i | j(M))$. In addition,

$$T_x j(T_x M) \oplus T_v \exp_{P_i(j(x))}(\nu_{P_i(j(x))}) = T_{j(x)} S_i^\varepsilon \text{ , where } j(x) = \exp_{P_i(j(x))}(v$$

The first part is easily verified directly; for the second assertion one notes that $TP_i(T_j(T_x M)) = T_{P_i(j(x))} S_i^\varepsilon = TP_i(T_{j(x)} S_i^\varepsilon$, from which the rest follows by a dimension argument.

In an obvious sense, the embeddings j with $P_i \circ j \in E(M,i(M))$ are transversal to the fibres of P_i; the set of these embeddings is denoted by $E^t(M,S_i^\varepsilon)$. E(M,i(M)) is open in C(M,i(M)) and Ω_{P_i} is continuous. Thus, we obtain

<u>Prop.4:</u> The complete inverse image of $E(M,i(M)) \subset C(M,i(M))$ under Ω_{P_i} is $E^t(M,S_i^\varepsilon)$ which therefore is open in $C(M,S_i^\varepsilon)$ as well as in E(M,N).

The composition

$$E(M,i(M)) \xrightarrow{\Omega_\iota} E^t(M,S_i^\varepsilon) \xrightarrow{\Omega_{P_i}} E(M,i(M))$$

is the identity and, in particular, Ω_{P_i} is a quotient map.

4. We now are going to investigate the quotient $E(M,N) \longrightarrow U(M,N)$ in more detail: we wish to show that u admits "enough" local sections and that, consequently, $U(M,N)$ possesses a (topological) manifold structure over a Fréchet space; the questions of differentiability will be discussed below.

As before, we let $(\mathcal{U}_i,\Phi_i)$ be a natural chart at $i \in E(M,N)$, $\Phi_i = \exp_i^{-1} | \mathcal{U}_i$, taking values in $\Gamma_i(TN)$. We may assume, moreover, that $\mathcal{U}_i$ is constructed using the neighbourhood $\mathcal{O} = Ti(\mathcal{O}_M) + \mathcal{O}_{\nu_\iota}$ of the zero-section of $TN|i(M)$.

Recall that $(i \circ g)(M) = i(M)$ for $g \in \mathrm{Diff}(M)$. In particular $\Omega_{P_i}^{-1} \circ \beta_i$ is well-defined. With this remark, we now obtain the

Prop.5: Given $i \in E(M,N)$ there exists a subset $\mathcal{V}_i \subset E(M,N)$ with the following properties:

(a) $i \in \mathcal{V}_i$ and $u|\mathcal{V}_i$ is injective;

(b) for every open neighbourhood W of the identity in $\mathrm{Diff}(M)$, the set $\mathcal{V}_i \cdot W$ is open.

In particular, $u(\mathcal{V}_i) \subset U(M,N)$ is an open neighbourhood of $\hat{i} = u(i)$ and $s_i = (u|\mathcal{V}_i)^{-1}$ is a continuous section of u such that $s_i(\hat{i}) = i$.

Proof: Recall that $E^t(M,S_i^\varepsilon)$ is open. If W is an open neighbourhood of the identity in $\mathrm{Diff}(M)$, then $U = \Omega_{P_i}^{-1}(\beta_i(W)) \subset E^t(M,S_i^\varepsilon)$ is open and contains i. We set

$$\mathcal{V}_i = \{j \in U | P_i \circ j = i\} = \Omega_{P_i}^{-1}(i) .$$

Clearly, $i \in \mathcal{V}_i$ and, furthermore, $u|\mathcal{V}_i$ is injective: if $j,j' \in \mathcal{V}_i$ and $u(j) = u(j')$, then $j' = j \circ g$ for some $g \in \mathrm{Diff}(M)$, therefore $i = P_i \circ j' = P_i \circ j \circ g = i \cdot g$, hence $g = \mathrm{id.}$ and $j' = j$.

Next we claim that $U = \mathcal{V}_i \cdot W$: Firstly, suppose that $P_i \circ j \in \beta_i(W)$, i.e. that $P_i \circ j = i \circ g$. Then $j \circ g^{-1} \in E^t(M,S_i^\varepsilon)$, $P_i \circ (j \circ g^{-1}) = i$, i.e. $j \circ g^{-1} \in \mathcal{V}_i$ and therefore $j \in \mathcal{V}_i \cdot W$. Conversely, let $j \in \mathcal{V}_i$, $g \in W$. Then $P_i \circ j = i$, i.e. $P_i \circ (j \circ g) = i \circ g \in \beta_i(W)$ and therefore $j \circ g \in U$.

It is clear that $U \cdot \mathrm{Diff}(M)$ is open in $E(M,N)$. The identity just established says that $\mathcal{V}_i \cdot \mathrm{Diff}(M) = \mathcal{V}_i \cdot W \cdot \mathrm{Diff}(M) = U \cdot \mathrm{Diff}(M)$. In particular, $u(\mathcal{V}_i)$ is an open neighbourhood of $\hat{i}$. The proposition now follows.

Let now O_i be the open neighbourhood $u(V_i)$ of $\hat{i}$, $s_i = (u|V_i)^{-1}$ the section just obtained. The fibres of u over the points of O_i meet V_i in precisely one point and this now means that the composition map $(j,g) \longrightarrow j\circ g$: $V_i \times \mathrm{Diff}(M) \longrightarrow u^{-1}(O_i)$ has an inverse Z. Composition is continuous and the continuity of Z may be derived from the assertion (b) of the proposition or, somewhat more directly, as follows: Let $j \in u^{-1}(O_i) = V_i \cdot \mathrm{Diff}(M)$. Then $Z(j) = (j_o, g)$ where j_o is the common point of V_i and the orbit of j. Since $P_i \circ j_o = i$, this implies that g is determined by the equation $P_i \circ j = i \circ g$, i.e. that $g = \beta_i^{-1} \circ \Omega_{P_i}(j)$ and this is continuous in j. Similarly $j_o = j \circ g^{-1}$ is continuous in j by the properties of composition and inversion. These remarks establish the

Prop.6: With the notations just introduced, the fibration $E(M,N \xrightarrow{u} U(M,N)$ is trivial over O_i, a trivilization being given as usual by the map $(y,g) \longrightarrow s_i(y)g$ of $O_i \times \mathrm{Diff}(M)$ onto $u^{-1}(O_i)$. Thus, $E(M,N) \xrightarrow{u} U(M,N)$ is a (topological) principal Diff(M)-bundle over U(M,N).

Evidently, this proves the topological assertions of the theorem of section 2.

Lastly, we determine the (topological) manifold structure of U(M,N): The charts will be defined by the sets O_i of prop.6 and the maps $\Phi_i \circ s_i$ of O_i into $\Gamma_i(TN)$. There remains the description of the image $\Phi_i(V_i)$ of O_i under $\Phi_i \circ s_i$: The condition $P_i\, j = i$ means that for each $x \in M$, $j(x)$ lies in the fibre of S_i^{ε} over $i(x)$. This in turn simply says that j is given by $\exp_i \circ\, l$ for some lift l of i which lies in ν_i, cf. 3. In other words: Φ_i maps V_i (homeomorphically) onto some open neighbourhood of $o \in \Gamma_i(\nu_i)$ which therefore is the parameter space of U(M,N) for the chart $(O_i, \Phi_i \circ s_i)$.

The sets O_i are metrizable, hence in particular completely regular. This may be used to show that U(M,N) is Hausdorff. We omit the elementary argument.

5. In this section, we briefly outline in what sense the various constructions of the preceding sections will be differentiable, retaining the earlier notations. Once again, we refer to [E-M], [F],[G] and [O] for all technical details (cf.also [B]). We already pointed out that the operation of Diff(M) on E(M,N) is Γ-differentiable, as also are composition and inversion in Diff(M); the respective manifold structures are "$HC^{\infty}_{\sigma_o}$" ([F],[G]), i.e. Γ-differentiable in a rather strong sense. The "Ω-lemma", known from the theory of Banach

manifolds of maps (e.g.[B] still holds: "Ω-constructions" do not lead to any loss of differentiability.

One observes that the sets $\mathcal{V}_i = s_i(O_i)$ constructed in 4. are submanifolds of $E(M,N)$ carrying a global chart Φ_i with values in $\Gamma_i(\nu_i)$, a topological summand of $T_iE(M,N) = \Gamma_i(TN)$.

The images $O_i = s_i(\mathcal{V}_i)$ are manifolds diffeomorphic to the $\mathcal{V}_i$ under s_i by construction. One also notes that the open neighbourhood W of the identity in $\mathrm{Diff}(M)$ used in prop.5 is arbitrary and may, in particular, be chosen to be $\mathrm{Diff}(M)$: the projection P_i makes sense in all of $u^{-1}(O_i)$. Thus, if t is any section of u over O_i, we can construct a map $g:O_i \longrightarrow \mathrm{Diff}(M)$ by solving the equation $P_i \circ t = i \cdot g$, i.e. by means of $g = \beta_i^{-1} \circ \Omega_{P_i}(t)$. By the Ω-lemma, this map will be Γ-differentiable in the same sense as t and, moreover, $t = s_i \cdot g$, as one readily sees.

Now set $L = i(M)$. The "ε-tube" S_i^{ε} introduced earlier clearly does not depend on i, only on L, so that it will also be denoted by S_L^{ε}. Similarly, ν_i only depends on L and is also denoted by ν_L : $Ti(TM)$simply is the bundle of those elements of TN which are tangent to (curves in) L. Lastly, the exponential map used in the construction of S_L^{ε} is given by the quadratic structure of N and again is independent of i, so that finally P_i depends only on L and will be denoted by P_L. The data $L, \nu_L, S_L^{\varepsilon}$ and P_L therefore are common to all embeddings j on the orbit of i. As a consequence, one sees that the "natural" transversal slices $\mathcal{V}_j$ satisfy $\mathcal{V}_{i \cdot g} = \mathcal{V}_i \cdot g$, i.e. are translates of each other and thus are Γ-diffeomorphic (even in the sense of $HC^{\infty}_{\sigma_o}$, as an adaption of the argument for the differentiability of right translations in $\mathrm{Diff}(M)$ will show). In particular, the differentiable structure of O_i may be obtained using any one of the natural sections s_j as long as j lies on the orbit of i.

Suppose now that $O_1 = O_{i_1}$ and $O_2 = O_{i_2}$ are given such that $O_1 \cap O_2 \neq \emptyset$; we assume that i_1 and i_2 do not lie on the same orbit. Set $L = i_1(M)$, $K = i_2(M)$. Finally, s_1 and s_2 are the natural sections obtained after choosing the representatives i_1, i_2 of the two orbits. Then $s_2(O_1 \cap O_2) = s_2(O_2) \cap u^{-1}(O_1 \cap O_2) =$ $= \mathcal{V}_2 \cap u^{-1}(O_1 \cap O_2)$. For $j \in \mathcal{V}_2$, $P_K \circ j = i_2$ and therefore $j = \exp_K \circ l_2$ for some $l_2 \in \Gamma_{i_2}(\nu_K)$. In addition, if $j \in s_2(O_1 \cap O_2)$, we also have that $P_L \circ j = i_1 \cdot g(\hat{j})$ for some $g(\hat{j}) \in \mathrm{Diff}(M)$ as above. Solving for g, we obtain

$$g(\hat{j}) = (\beta_{i_1}^{-1} (P_L \circ \exp_K \circ l_2);$$

by the Ω-lemma, this is Γ-differentiable in l_2, i.e. in $\hat{j}$. Once again, writing y for the points $\hat{j}$ of $O_1 \cap O_2$, we obtain

$$s_2(y) = s_1(y) \cdot g(y)$$

on $O_1 \cap O_2$. This implies that the parameter transformation $(\Phi_1 \circ s_1)(\Phi_2 \circ s_2)^{-1}$ is Γ-differentiable in the same sense as composition and inversion are. Thus, $U(M,N)$ carries a natural differentiable structure. In fact, with this structure, $U(M,N)$ is the quotient of $E(M,N)$ mod.u: Given the map $f: E(M,N) \longrightarrow H$ which factors through u, $f = h \circ u$, f will be smooth if and only if h is.

The map u is a submersion: u is Γ-differentiable, for if $i \in E(M,N)$ we may choose an open neighbourhood of i of the form $V_i \cdot W$ as before (so that $u(V_i \cdot W) = O_i$) and then $s_i \circ u$ is the "projection" of $V_i \cdot W$ onto V_i constructed earlier; this map is Γ-differentiable and $T(s_i \circ u)$ maps $\Gamma_i(TM)$ onto the tangent space at i to V_i, i.e. onto $\Gamma_i(\nu_i)$. Since u is a diffeomorphism of V_i onto O_i, $T_i u$ will be an isomorphism of $\Gamma_i(\nu_i)$ onto the tangent space at $\hat{i}$.

Since $u \circ \beta_i = \hat{i}$, we have $Tu \bullet T\beta_i = o$ and, in particular, $T_i u \bullet T_e \beta_i = o$, so that $\mathrm{im}(T_e \beta_i) = \Gamma_i(Ti(TM) \subset \ker(T_i u)$. Now $\Gamma_i(TN) = \Gamma_i(Ti(TM)) \oplus$ $\oplus\, \Gamma_i(\nu_i)$ and $T_i u | \Gamma_i(\nu_i)$ is an isomorphism. One concludes from this that $\ker(T_i u) = \Gamma_i(T_i(TM))$. With this, the vertical bundle $\ker(Tu)$ is determined. We shall return later to the problem of the existence of horizontal bundles, i.e. of principal connections on the bundle $E(M,N)$.

6. In this final section, we give brief outlines of some of the constructions which will be of importance in the applications mentioned in the introduction. The details will appear elsewhere.

I) So far, the tangent bundle of $E(M,N)$ has been tacitly identified with the space of all smooth maps $l: M \longrightarrow TN$ which satisfy $\tau_N \bullet l \in E(M,N)$. It can be verified directly that this mapping space is, in fact, a bundle over $E(M,N)$ with the expexted transition maps and local trivializations and is trivial over the natural charts introduced in 1. ; we refer to [EL] and [F] for some details on this point.

This construction agrees in an obvious manner with the "geometric" one which interprets a tangent vector at a point as an equivalence class of (locally defined) smooth curves through the point, etc. It also agrees with the classical construction of "contravariant vectors" at a point, cf.[E] , [L] . One observes that the assertions are not entirely trivial since, among other things, it has to be shown that the natural manifold structure of the set $\{l:M \longrightarrow TN \mid \tau_N \circ l \in E(M,N)\} = TE(M,N)$ agrees with the one obtained from the local trivializations mentioned. The natural manifold structure is the obvious one: $TE(M,N)$ is an open subset of $C(M,TN)$, $E(M,N)$ being open in $C(M,N)$. Note that the classical motivation for the use of $\Gamma_i(TN)$ as the tangent space $T_iE(M,N)$ is contained in the following observation: Let $\sigma: I \longrightarrow E(M,N)$, $I \subset R$ an open interval containing o, be a smooth curve such that $\sigma(o) = i$. Then σ corresponds to a smooth map $\sigma: I \times M \longrightarrow N$ under $\sigma(t,x) = \sigma(t)(x)$, etc. The tangent vector $\sigma(o)$ now corresponds to the map $x \longrightarrow \frac{\partial\sigma}{\partial t}(o,x)$ of M into TN which clearly is a lift of $i = \sigma(o)$.

The projection $TE \longrightarrow E$ now simply is Ω_{τ_N} , i.e. the map $l \longrightarrow \tau_N \circ l$ of TE onto E whose fibre at i is precisely $\Gamma_i(TN)$. In the same manner, $T^2E(M,N)$ is the manifold of smooth maps $f:M \longrightarrow T^2N$ with the property that $\tau_N^{(2)} \circ f \in TE(M,N)$, $\tau_N^{(2)}: T^2N \longrightarrow TN$ the usual projection.

The notion of a spray on $E(M,N)$ is introduced as it is on Banach manifolds (for which we refer to [L]): denoting multiplication on the fibres of $TE(M,N)$, $T^2E(M,N), \ldots$ by μ_λ $(\lambda \in R)$, the definition is the following:

A spray on $E(M,N)$ is a smooth section ξ of $T^2E(M,N)$ (a vector field on $TE(M,N)$!) which satisfies the additional conditions

(a) $T\tau_N \circ \xi = \text{id.}$ on TE (ξ is a second order differential equation on E!)

(b) $\xi \circ \mu_\lambda = \mu_\lambda \cdot T\mu_\lambda \circ \xi$.

Due to the absence of simple inverse function theorems in the general differential calculus for Fréchet spaces, there do not seem to be any reasonable, general existence theorems for flows of such sprays (or for their "base integral curves"). However, the following elementary considerations often are sufficient for practical purposes:

It is immediate from the definition that if $S:TN \longrightarrow T^2N$ is a spray on N, then Ω_S is a spray on E, called an "Ω-spray" on E for simplicity's sake. For these, one has a local existence theorem for a flow and, in particular, for base integral curves (geodesics) in E.

Let S be a spray on N. Then, given $i \in E(M,N)$ and $l \in \Gamma_i(TN)$, there is $\varepsilon > o$ such that there exists a unique geodesic $\sigma : (-\varepsilon,\varepsilon) \longrightarrow E(M,N)$ for Ω_S with the properties that $\sigma(o) = i$ and $\dot{\sigma}(o) = l$.

The argument uses the compactness of M, thus of $i(M) \subset N$, and is quite standard: for $p = i(x) \in i(M)$, there are $\varepsilon_x > o$, a neighbourhood $V_x = i(U_x)$ in $i(M)$ and a map s_x from $(-\varepsilon_x,\varepsilon_x) \times U_x$ into N such that $s_x(o,y) = i(y)$, that $t \longrightarrow s_x(t,y)$ is the unique S-geodesic through $i(y)$ with $\dot{s}_x(o,y) = l(y) \in T_{i(y)}N$, and that no two such geodesics meet. Finitely many U_x now cover M and one chooses ε equal to the smallest one of the corresponding numbers ε_x. One obtains $s:(-\varepsilon,\varepsilon) \times M \longrightarrow N$ satisfying the conditions mentioned on all of its domain. Lastly, one defines σ by: $\sigma(t)(x) = s(t,x)$ and obtains the desired geodesic through $i = \sigma(o)$ in $E(M,N)$.

II) Assume once more that N carries a quadratic structure G. For each $i \in E(M,N)$, let again ν_i be the normal bundle of i with respect to G, so that $TN|i(M) = Ti(TM) \oplus \nu_i$ and one has the induced splitting $T_iE(M,N) = \Gamma_i(Ti(TM)) \oplus \Gamma_i(\nu_i)$ which we know to be topological. Using e.g. the "product neighbourhood" $\mathcal{V}_i \cdot W$ introduced earlier one can show that both $\bigcup \Gamma_i(Ti(TM))$ and $\bigcup \Gamma_i(\nu_i)$ are subbundles of $TE(M,N)$, the former being the vertical bundle $VE(M,N)$ of the principal bundle $E(M,N) \longrightarrow U(M,N)$, the latter being a summand $HE(M,N)$ of VE. One remarks, by the way, that both subbundles are invariant under the right action of Diff(M) and that, as a consequence, $HE(M,N)$ defines a principal connection on $E(M,N)$. We shall return to this point elsewhere.

The splitting $TE = VE \oplus HE$ also has the following immediate consequence: Suppose that ξ is a spray on $E(M,N)$ which admits local solutions through i and let $\sigma: I \longrightarrow E(M,N)$ be a geodesic, $o \in I$ and $\sigma(o) = i$. Then, for each $t \in I$, the tangent vector $\dot{\sigma}(t) \in T_{\sigma(t)}E$ admits a decomposition into its vertical and horizontal components:

$$\dot{\sigma}(t) = v_1(t) + v_2(t)$$

with $v_1(t) \in \Gamma_{\sigma(t)}(T\sigma(t)(TM))$ and $v_2(t) \in \Gamma_{\sigma(t)}(\nu_{\sigma(t)})$. In this case, there exists a time-dependent vector field X_t on M such that

$$v_1(t) = T\sigma(t) \circ X_t ,$$

the "shift" (of M) corresponding to the motion of $L = i(M) = \sigma(o)(M)$

determined by σ .

In addition, we now assume that $n = \dim(N) = \dim(M) + 1 = m + 1$ and that both M and N are orientable. Then the normal bundle ν_i of $i \in E(M,N)$ is a trivial line bundle : that it is of rank 1 is clear and the orientability assumption means that $\Lambda^m TM$, $\Lambda^n TN$ are trivial line bundles. Moreover, $TN|i(M) = Ti(TM) \oplus \nu_i$ implies, as usual, that

$$\Lambda^n(TN|i(M)) = \Lambda^n TN|i(M) = \bigoplus_{r+s=n} (\Lambda^r Ti(TM) \oplus \Lambda^s \nu_i) = \Lambda^m Ti(TM) \oplus \nu_i;$$

the triviality of ν_i therefore is clear. In particular, ν_i admits non-vanishing global sections. This is of great importance in the following special case which yields a framework for the geometrodynamical view of general relativity:
Let $\dim(M) = 3$, $\dim(N) = 4$, G of index 1 (Lorentz structure) and suppose again that M and N are orientable. Denote by $E_o(M,N)$ the set of all space-like embeddings $M \longrightarrow N$:

$$i \in E_o(M,N) \text{ iff. } i^*G \text{ is positive-definite.}$$

Then $E_o(M,N)$ is an open submanifold of $E(M,N)$, saturated under $\mathrm{Diff}(M)$, and thus is a principal $\mathrm{Diff}(M)$-bundle over the open submanifold $U_o(M,N) \subset U(M,N)$ which consists of all space-like submanifolds of type M in N . Given $i \in E_o(M,N)$, ν_i now admits a global section T such that $G(T,T) = -1$ and any other section is of the form fT for some smooth real-valued function f on M . Therefore, given a geodesic σ through i as before, the decomposition of $\dot\sigma(t)$ now is of the form

$$\dot\sigma(t) = T\sigma(t)\circ X_t + N_t T .$$

In this formula, T stands for a section of TN , defined in some neighbourhood of $L = i(M)$, e.g. some "ε-tube" S_i^ε , such that T takes values in ν_j for $j \in E^t(M,S_i^\varepsilon)$ and that it has no zeroes. The time-dependent function N_t on M is called the lapse (function) in Geometrodynamics. For these notions, we refer to [D], [DW], [ADM], [HKT], [K], [FM] and the bibliographies of these papers. Note that the approach of [FM], directly following [D], [ADM] and [DW], is formulated essentially on the manifold $\mathcal{M}$ of Riemannian structures on M (or on "superspace" $\mathcal{M}/\mathrm{Diff}(M)$, which is not a manifold), while the model used in [K], [HKT], etc. is quite different: Since the term "hypersurface" in [HKT] means "smooth embedding of M into N", "hyperspace" of these papers simply is $E_o(M,N)$, described using local coordinates in M and in N . A more detailed description of "hyperspace" in our current frame-work will appear elsewhere, as will the connection with the "superspace approach": this part is based on the observation that there is a natural smooth map $\varphi : E_o(M,N) \longrightarrow \mathcal{M}$ given by $\varphi(i) = i^*G$. This allows us, e.g., to pull back the (DeWitt) Lagrangian from $\mathcal{M}$ and describe those geodesics of its spray on M which are "realizable" for given M and N , i.e. which correspond to curves in $E_o(M,N)$.

References

[ADM] R. Arnowitt-S. Deser-C. Misner, "The dynamics of general relativity" in "Gravitation: An introduction to current research". L.Witten ed., Wiley, New York (1962).

[B] N. Bourbaki, "variétés différentiables et analytiques: Fascicule des résultats". Herman, Paris (1969).

[D] P.A.M. Dirac, "Fixation of coordinates in the hamiltonian theory of gravitation". Phys.Rev. 114 (1959), 924.

[DW] B.S. DeWitt, "Quantum theory of gravitation, I. The canonical theory". Phys.Rev. 160, (1967), 1113.

[EM] D.G. Ebin-J. Marsden, "Groups of diffeomorphisms and the motion of an incompressible fluid".

[E] J. Eells, "On the geometry of function spaces". Symp.Top.Alg. Mexico City (1958).

[E] H.I. Eliasson, "Geometry of manifolds of maps". J.Diff. Geometry1 (1967).

[F] H.-R. Fischer, "On manifolds of mappings".

[FM] A.E. Fischer-J. Marsden, "The Einstein equations of evolution - A geometric approach. J.Math.Phys. 12 (1972).

[GG] M. Golubitski-V. Guillemin, "Stable mappings and their singularities". Graduate Text, Springer-Verlag, Berlin, Heidelberg, New York, (1973).

[G] J.Gutknecht, "Die C^{∞}_{Γ} - Struktur auf der Diffeomorphismengruppe einer kompakten Mannigfaltigkeit". Diss.ETH Zürich, 5879 (1977).

[HKT] S.A. Hojman-K. Kuchar-C. Teitelboim, "Geometrodynamics regained". Ann. Phys. 96 (1976), 88.

[Ko] J. Komorowski,"A geometrical formulation of the general free boundary value problems and the theorem of E. Noether connected with them". Rep.Math.Phys. 1 (1970).

[K] K. Kuchar, "Geometry of hyperspace", I.II.III., J.Math.Phys. 17 (1976), 777.

[L] S. Lang, "Introduction of differentiable manifolds", Addison-Wesley Publishing Company Inc. Reading, Mass. (1972)

[M] J. Mather, "Stability of C^{∞}-mappings II". Ann.Math. 89 (1969).

[M-W] C.W.Misner-J.A. Wheeler, "Classical physics as geometry". Ann.Phys. 2 (1957), 525.

[O] H. Omori, On the group diffeomorphisms on a compact manifold, Proc.Symp. Pure Math. XV, Providence (1970).

[W] J.A. Wheeler, "Geometrodynamics". Academic Press, New York (1962).

Appendix to the contribution by E. Binz and H.R. Fischer

THE MANIFOLD OF EMBEDDINGS OF A NON-COMPACT MANIFOLD

P. Michor

The following is a review of results generalizing the foregoing paper to the case of embeddings of a non-compact smooth manifold.

1. Let X,Y be finite dimensional manifolds, let $C^\infty(X,Y)$ denote the space of all smooth mappings from X to Y. The problem of making $C^\infty(X,Y)$ into a manifold is mainly the problem of choosing the right topology. Denote by W^∞ the Whitney C^∞-topology on $C^\infty(X,Y)$ (called the strong C^∞ topology in [2]). Now call $f,g \in C^\infty(X,Y)$ equivalent ($f \sim g$) iff f and g coincide off some compact subset of X, and refine the W^∞-topology in such a way that this becomes an open equivalence relation (so that equivalence classes are open too). Let us denote this topology by (FW^∞) (fine W^∞-topology) In [4] - [7] a finer topology is used, the $(F\mathcal{D})$-topology which comes from refining the $\mathcal{D}$-topology on $C^\infty(X,Y)$ (see [1]). These topologies have the following properties:

1.1. $W^\infty, \mathcal{D}$ are Baire spaces ($(FW^\infty),(F\mathcal{D})$ not!). Embeddings, immersions, submersions, surjective submersions, diffeomorphisms, proper mappings (i.e. f^{-1}(compact) is compact) are all open subsets of $C^\infty(X,Y)$ in W^∞ and all finer topologies.

1.2. $C^\infty(X,Y\times Z)=C^\infty(X,Y)\times C^\infty(X,Z)$ is a homeomorphism for all topologies.

1.3. If $f \in C^\infty(X,Z)$, then $f_*:C^\infty(X,Y)\to C^\infty(X,Z)$, given by $f_*(g)=f\circ g$, is continuous for all topologies.

1.4. Composition: $C^\infty_{prop}(X,Y)\times C^\infty(Y,Z)\to C^\infty(X,Y)$ is jointly continuous.

1.5. Inversion: $\mathrm{Diff}(X)\to\mathrm{Diff}(X)$ is continuous.

1.6. A sequence f_n converges to f in $C^\infty(X,Y)$ (for all topologies), iff f_n equals f off some compact set (up to finitely many n's) and $f_n \to f$ "uniformly in all derivatives" on this compact set.

2. Consider $C^\infty(X,Y)$ with the (FW^∞)-topology. The manifold structure on $C^\infty(X,Y)$ is given in the following way:

Let τ: $TY \to Y$ be a local addition (i.e. a mapping such that (π_Y,τ): $TY \to Y\times Y$ is a diffeomorphism onto a open subset, $\tau(0_y)=y$). This is like an exponential mapping, pulled over the whole tangent bundle for simplicity's sake. Fix $f \in C^\infty(X,Y)$, and consider the space $\mathcal{D}_f(X,TY)$ of all "vector fields along f with compact support" (i.e. $s \in \mathcal{D}_f(X,TY)$ iff $\pi_Y \circ s=f$, $s \sim 0_Y \circ f$), which is isomorphic to $\Gamma_c(f^*TY)$, the space of all notions with compact support of the pullback f^*TY of the bundle TY. Let $\mathcal{U}_f \subseteq C^\infty(X,Y)$ be the open subset of all g such that the image of (f,g) in $Y \times Y$ is contained in the (open) image of (π_Y,τ) and $g \sim f$.

Now let $\varphi_f: \mathcal{U}_f \to \Gamma_c(f^*TY)$ be defined by $\varphi_f(g)=(\pi_Y,\tau)^{-1}(f,g)$, $\Psi_f: \Gamma_c(f^*TY)=\mathcal{D}_f(X,TY) \to \mathcal{U}_f$, $\Psi_f(s)=\tau \circ s$. Then φ_f, Ψ_f are continuous by 1.2. and 1.3. and are easily seen to be inverse to each other. So $C^\infty(X,Y)$ is a topological manifold, modeled on topological vector spaces $\Gamma_c(f^*TY)$, which bear the usual inductive limit topology known from distribution theory if $C^\infty(X,Y)$ bears the $(F\mathcal{D})$-topology, and which are supposed to bear a topology similar to the topology on $C_c^\infty(\mathbb{R})$ whose dual space is the space of all distributions of finite type. ($\Gamma_c(f^*TY)=\varinjlim_K \Gamma_K(f^*TY)$, K compact in X, where $\Gamma_K(f^*TY)$ is the space of all sections with support in K, in case of $(F\mathcal{D})$; $\Gamma_c(f^*TY)=\varprojlim_r \Gamma_c^r(f^*TY)$, where $r < \infty$, and $\Gamma_c^r(f^*TY)$ bears the inductive limit topology, $\Gamma_c^r(f^*TY)=\varinjlim_K \Gamma_K^r(f^*TY)$, where r denotes the differentiability class.)

3. The chart change in 2 is easily computed to look like α_*, $\alpha_*(s) = \alpha \circ s$, where $\alpha: f^*TY \to g^*TY$ is a smooth fibre respecting mapping (defined on an open subset only). To make $C^\infty(X,Y)$ to a differentiable manifold one has to choose a differential calculus on locally convex spaces such the above mapping is of class C^∞. There are several solutions (which seem to be all the same in our case), the simplest being C_c^∞ of [3]:
$f: E \to F$ is C_c^1 if $Df(x)\cdot y = \lim_{t\to 0} \frac{1}{t}(f(x+ty) - f(x))$ exists for all x,y with $Df(x): E \to F$ a linear map such that $(x,y) \to Df(x)\cdot y$ is jointly continuous on $E \times E$. C_c^r, C_c^∞ is then defined by recursion. So $C^\infty(X,Y)$ is a C_c^∞-manifold. The tangent bundle $T\,C^\infty(X,Y)$ concides with the open subset $\mathcal{D}(X,TY) \subseteq C^\infty(X,TY)$ of all mappings $s: X \to TY$ "with compact support", i.e. $s^{-1}(TY\setminus\text{Zero section})$ is relatively compact in X; $\mathcal{D}(X,TY)$ is again a manifold (from $C^\infty(X,TY)$) and this structure coincides with the tangent bundle structure. See [7] for a detailed account.

4. It can be proved that Diff(X), the open subset of all diffeomorphisms of X in $C^\infty(X,X)$, is a C_c^∞-Lie-group; i.e. Composition and Inversion are differentiable, of class C_c^∞. The Lie-Algebra is $\Gamma_c(TX)$, the space of all smooth vector fields with compact support. The canonical exponential mapping is <u>not</u> surjective on any neighbourhood of the identity in general. See [5]. In [7] it is shown that this result is true, if X is a manifold with corners.

5. E(X,Y), the space of all embeddings from X into Y, is open in $C^\infty(X,Y)$, so it is a C_c^∞-manifold too. There is a right action ρ of Diff(X) onto E(X,Y) which is C_c^∞, and E(X,Y) turns out to be a C^∞ principal fibre bundle with structure

group Diff(X) in a similar way as in [1] .

REFERENCES

[1] E.Binz, H.R.Fischer: The manifold of embeddings of a closed manifold these Proceedings.

[2] M.W.Hirsch: Differential Topology, Springer GTM 33, (1976).

[3] H.H.Keller: Differential calculus in locally convex spaces, Springer Lecture Notes in Math. 417 (1974).

[4] P.Michor: Manifolds of smooth maps, Cahiers Top. Geom.Diff. XIX (1978), 47-78.

[5] P.Michor: Manifolds of smooth maps, II: The Lie-group of diffeomorphisms of a non-compact smooth manifold, Cahiers Top. Geom. Diff.XXI (1980).

[6] P.Michor: Manifolds of smooth maps, III: The principal bundle of embeddings of a non-compact smooth manifold, to appear in Cahiers Top. Geom. Diff.

[7] P.Michor: Manifolds of differentiable mappings I, 280 pages, to appear, Shiva Publ. Company.

+ Institut für Mathematik, Universität Mannheim, Mannheim, Federal Republic of Germany

++ Department of Mathematics, University of Massachusetts, Amherst, U.S.A.

A. Jaffe, J. Glimm

Quantum Physics

A Functional Integral Point of View

1981. 43 figures. Approx. 416 pages
ISBN 3-540-90551-0

Contents: Introduction. – Conventions and Formulas. – List of Symbols. – An Introduction to Modern Physics. – Function Space Integrals. – The Physics of Quantum Fields. – Bibliography. – Index.

A thorough investigation of the mathematical structures underlying statistical physics, quantum mechanics and quantum fields is provided in this work. Although an advanced and self-contained text, it presents material of highest priority in research both in mathematics proper and in mathematical physics. It consists of three parts, each of them written with a different scientific perspective. Part I is an introduction to modern physics. It is designed to make the treatment of physics self-contained for a mathematics audience. Part II presents quantum fields. This treatment is mathematically complete and self-contained. Part III concerns particle interaction, scattering, bound states, phase transitions and critical point theory.

Springer-Verlag
Berlin
Heidelberg
New York